Silver Nanoparticles

Edited by
Dr. David Pozo Perez

Silver Nanoparticles
Edited by Dr. David Pozo Perez

Republished by InTech
Janeza Trdine 9, 51000 Rijeka, Croatia

Technical Editor Maja Jakobovic

Cover Designer Dino Smrekar

Additional hard copies can be obtained from orders@intechopen.com

Silver Nanoparticles,
Edited by Dr. David Pozo Perez
p. cm.
ISBN 978-953-307-028-5

For EU product safety concerns :
IN TECH d . o . o . , Prolaz Marije Krucifikse Kozulić 3 , 51000 Rijeka , Croatia ,
info@intechopen . com or visit our website at intechopen . com .

Meet the editor

Professor of Biochemistry and Molecular Biology, University of Seville Medical School, Seville, Spain. Principal Investigator at the Center for Molecular Biology and Regenerative Medicine (CABIMER), Seville, Spain. Dr. Pozo Perez received his Ph.D (1998) from the University of Seville. He subsequently joined the Division of Immunology at Cambridge University (UK) as an EMBO postdoctoral fellow and Marie Curie Fellow from 2000-2003. He did a second post-doctoral stint at the Department of Immunology of Weizmann Institute of Science (2004) in Israel and short stages as visiting scientist at the Center for Neurologic Diseases, Brigham & Women Hospital, Harvard Medical School (2013, 2015). His research interests in immunology are in the area of neurodegenerative diseases. He served as President of the Beit Rambam Jewish Community of Andalusia (2014).

Preface

Herzl stated "If you will it, it is no dream", and, one of the long-lasting wills of our common history as humankind has been the manipulation of Nature's inner components. Remember the old art of Alchemy, for example.

In our days, by using nanoscale technologies to create nanoparticles and nanomaterials, we are now able to dramatically transform existing materials and to design new ones in an unprecedented way.

Nanotechnology will be soon required in most engineering and science curricula. It can not be questioned that cutting-edge applications based in nanoscience are having a considerable impact in nearly all the fields of research, from basic to more problem-solved scientific enterprises. In this sense, books like "Silver Nanoparticles" aim to fill the gaps for comprehensive information to help both newcomers and experts, in a particular, fast-growing area of research. Besides, one of the key features of this book is that it could serve both academia and industry.

Silver nanoparticles have unique optical, electrical and biological properties that have attracted significant attention due to their potential use in many applications, such as catalysis, biosensing, drug delivery and nanodevice fabrication. This book highlights scientific advancements and recent applications of silver nanoparticles nanotechnology with a key focus on nanocomposite coatings and advanced characterization techniques.

"Silver nanoparticles" is a collection of seventeen chapters written by experts in their respective fields. These reviews are representative of the current research areas in silver nanoparticle nanoscience and nanotechnology.

The synthesis of dispersed nanoparticles is critical for further applications because of their novel properties greatly different from the equivalent bulk material. Chapters 1-9 deal with different aspect related to the synthesis and properties' characterisation of silver nanoparticles involving the use of physical and chemical methods. From there, we can exploit the linear and non-linear properties of silver nanoparticles, its thermodynamic features as well as alternative methods as low energy energy implementation and laser assistance. The ultimate goal is to obtain non-agglomerated nanoparticles that able us to exploit the unique properties of isolated nanoparticles.

Chapters 10-12 reviewed the analytical and bioanalytical applications of silver nanoparticles associated to enhanced optical spectroscopies.

Chapter 13 shifts the book contents to more biological oriented questions. In this chapter, Sadowski explores biosynthetic methods as alternative methods to both chemical and physical procedures reviewed in chapters 1-9. Koga and Kitaoka in chapter 14 deal with the well-known antibacterial applications of silver nanoparticles, but they also show a versatile technique for the immobilization of bioactive silver nanoparticles onto a "paper" matrix. Besides the more classic antibacterial effects of silver nanoparticles, Chapter 15 by Lee and collaborators expands the application of silver nanoparticle to the field of antifungal agents.

The biological effects of silver nanoparticles involve not only their direct-mediated effects as a microbicide agent, but also those related to alterations of key immune responses as it is discussed by Klippstein and collaborators in chapter 16. Chapter 17 by Fondevila grounded the case for using silver nanoparticles as a prebiotics in the animal industry.

Finally, it is the hope of the editor and the authors that this book provides support in developing an understanding of silver nanoparticle nanotechnology, and that it will become a useful resource for engineers and scientists in mastering this topic area.

Brussels, December 2009.

Dr. David Pozo Perez
CABIMER
Andalusian Center for Molecular Biology and Regenerative Medicine
CSIC-University of Seville-UPO-Junta de Andalucia
Seville, Spain

Contents

1

Thermodynamic properties of nano-silver and alloy particles

Wangyu Hu, Shifang Xiao, Huiqiu Deng,
Wenhua Luo and Lei Deng
Department of Applied Physics, Hunan University, Changsha 410082, PR China

In this chapter, the analytical embedded atom method and calculating Gibbs free energy method are introduced briefly. Combining these methods with molecular dynamic and Monte Carlo techniques, thermodynamics of nano-silver and alloy particles have been studied systematically.

For silver nanoparticles, calculations for melting temperature, molar heat of fusion, molar entropy of fusion, and temperature dependences of entropy and specific heat capacity indicate that these thermodynamic properties can be divided into two parts: bulk quantity and surface quantity, and surface atoms are dominant for the size effect on the thermodynamic properties of nanoparticles.

Isothermal grain growth behaviors of nanocrystalline Ag shows that the small grain size and high temperature accelerate the grain growth. The grain growth processes of nanocrystalline Ag are well characterized by a power-law growth curve, followed by a linear relaxation stage. Beside grain boundary migration and grain rotation mechanisms, the dislocations serve as the intermediate role in the grain growth process. The isothermal melting in nanocrystalline Ag and crystallization from supercooled liquid indicate that melting at a fixed temperature in nanocrystalline materials is a continuous process, which originates from the grain boundary network. The crystallization from supercooled liquid is characterized by three characteristic stages: nucleation, rapid growth of nucleus, and slow structural relaxation. The homogeneous nucleation occurs at a larger supercooling temperature, which has an important effect on the process of crystallization and the subsequent crystalline texture. The kinetics of transition from liquid to solid is well described by the Johnson-Mehl-Avrami equation.

By extrapolating the mean grain size of nanocrystal to an infinitesimal value, we have obtained amorphous model from Voronoi construction. From nanocrystal to amorphous state, the curve of melting temperature exhibits three characteristic regions. As mean grain size above about 3.8 nm for Ag, the melting temperatures decrease linearly with the reciprocal of grain size. With further decreasing grain size, the melting temperatures almost keep a constant. This is because the dominant factor on melting temperature of nanocrystal shifts from grain phase to grain boundary one. As a result of fundamental difference in structure, the amorphous has a much lower solid-to-liquid transformation temperature than that of nanocrystal.

The surface and size effects on the alloying ability and phase stability of Ag alloy nanoparticles indicated that, besides the similar compositional dependence of heat of formation as in bulk alloys, the heat of formation of alloy nanoparticles exhibits notable size-dependence, and there exists a competition between size effect and compositional effect on the heat of formation of alloy system. Contrary to the positive heat of formation for bulk immiscible alloys, a negative heat of formation may be obtained for the alloy nanoparticles with a small size or dilute solute component, which implies a promotion of the alloying ability and phase stability of immiscible system on a nanoscale. The surface segregation results in an extension of the size range of particles with a negative heat of formation.

1. Thermodynamic properties of silver nanoparticles

Nanoparticle systems currently attract considerable interest from both academia and industry because of their interesting and diverse properties, which deviate from those of the bulk. Owing to the change of the properties, the fabrication of nanostructural materials and devices with unique properties in atomic scale has become an emerging interdisciplinary field involving solid-state physics, chemistry, biology, and materials science. Understanding and predicting the thermodynamics of nanoparticles is desired for fabricating the materials for practical applications.[1] The most striking example of the deviation of the corresponding conventional bulk thermodynamic behavior is probably the depression of the melting point of small particles of metallic species. A relation between the radius of nanoparticles and melting temperature was first established by Pawlow,[2] and the first experimental investigation of melting-temperature dependence on particle size was conducted more than 50 years ago.[3] Further studies were performed by a great number of researchers.[4-12] The results reveal that isolated nanoparticles and substrate-supported nanoparticles with relatively free surfaces usually exhibit a significant decrease in melting temperature as compared with the corresponding conventional bulk materials. The physical origin for this phenomenon is that the ratio of the number of surface-to-volume atoms is enormous, and the liquid/vapor interface energy is generally lower than the average solid/vapor interface energy.[9] Therefore, as the particle size decreases, its surface-to-volume atom ratio increases and the melting temperature decreases as a consequence of the improved free energy at the particle surface.

A lot of thermodynamic models of nanoparticles melting assume spherical particles with homogeneous surfaces and yield a linear or almost linear decreasing melting point with increasing the inverse of the cluster diameter.[2,6,10-12] However, the determination of some parameters in these models is difficult or arbitrary. Actually, the melting-phase transition is one of the most fundamental physical processes. The crystal and liquid phases of a substance can coexist in equilibrium at a certain temperature, at which the Gibbs free energies of these two phases become the same. The crystal phase has lower free energy at a temperature below the melting point and is the stable phase. As the temperature goes above the melting point, the free energy of the crystal phase becomes higher than that of the liquid phase and phase transition will take place. The same holds true for nanoparticles. We have calculated the Gibbs free energies of solid and liquid phases for silver bulk material and its surface free energy using molecular dynamics with the modified analytic embedded-atom method (MAEAM). By representing the total Gibbs free energies of solid and liquid clusters as the sum of the central bulk and surface free energy,[5,13,14] we can attain the free energies for the liquid and solid phase in spherical particles as a function of temperature. The melting temperature of nanoparticles is obtained from the intersection of these free-energy curves. This permits us to characterize the thermodynamic effect of the surface atoms on size-dependent melting of nanoparticles and go beyond the usual phenomenological modeling of the thermodynamics of melting processes in nanometer-sized systems. In addition, we further calculate the molar heat of fusion, molar entropy of fusion, entropy, and specific heat capacity of silver nanoparticles based on free energy calculation.

In order to explore the size effect on the thermodynamic properties of silver nanoparticles, we first write the total Gibbs free energy G^{total} of a nanoparticle as the sum of the volume free energy G^{bulk} and the surface free energy $G^{surface}$

$$G^{total} = G^{bulk} + G^{surface} = Ng(T) + \gamma(T)A_s \tag{1}$$

The detailed description on calculation of G^{bulk} and $G^{surface}$ has been given in Ref. 15-17. Assuming a spherical particle leads to a specific surface area of[5,10,18]

$$A_s = \frac{6}{D} N v_{at}(T) \tag{2}$$

where N is the total number of atoms in the particle, D is the radius of the particle, and $v_{at}(T)$ is the volume per atom. Second-order polynomials are adjusted to the simulation results of the internal energy for the solid and liquid phase shown in Fig. 1. The Gibbs free energies per atom for the solid and liquid phase are written as

$$\begin{aligned} g(T) = g(T_0)(T/T_0) - T[a_2(T-T_0) + a_1 \ln(T/T_0) \\ - a_0(1/T - 1/T_0)] \end{aligned} \tag{3}$$

where a_i are the polynomial coefficients, resulting from molecular dynamics (MD) simulations.[17]

The surface free energy of a solid spherical particle may be determined by the average surface free energy of the crystallite facets and the Gibbs–Wulff relation[19]

$$\sum_i A_i \gamma_{si} = \text{mininum} \tag{4}$$

The equilibrium crystal form develops so that the crystal is bound by low surface energy faces in order to minimize the total surface free energy.[20] For two surfaces i and j at equilibrium, $A_i\gamma_i = A_j\gamma_j = \mu$, where μ is the excess chemical potential of surface atoms relative to interior atoms. A surface with higher surface free energy (γ_i) consequently has a smaller surface area (A_i), which is inversely proportional to the surface free energy. Accordingly, the average surface free energy of the crystal, weighted by the surface area, is

$$\gamma_S = \frac{\sum_{i=1}^{n} A_i \gamma_{si}}{\sum_{i=1}^{n} A_i} = \frac{\sum_{i=1}^{n} \frac{\mu}{\gamma_i} \gamma_{si}}{\sum_{i=1}^{n} \frac{\mu}{\gamma_{si}}} = \frac{n}{\sum_{i=1}^{n} \frac{1}{\gamma_{si}}} \tag{5}$$

where n is the number of facets under consideration. Each crystal has its own surface energy, and a crystal can be bound by an infinite number of surface types. Thus, we only consider three low index surfaces, (111), (100), and (110), because of their low surface energies, and the surface free energy γ_i of the facet i is calculated as follows

$$\gamma_i(T) = \gamma_i(T_0)\frac{T}{T_0} - T[b_{2i}(T-T_0) + b_{1i}\ln(\frac{T}{T_0}) - \frac{b_{0i}}{T} + \frac{b_{0i}}{T_0}] \tag{6}$$

where b_{ki} (k = 0,1,2) are the coefficients for the surface free-energy calculation for facet i, and

$\gamma_i(T_0)$ is surface free energy at the reference temperature T_0.[17] On the basis of the expression for the Gibbs free energy, general trends for thermodynamic properties may be deduced. For example, the melting temperature T_m for nanoparticles of diameter D can be obtained by equating the Gibbs free energy of solid and liquid spherical particles with the assumption of constant pressure conditions, and temperature and particle size dependence of the entropy per atom for solid nanoparticles can then be defined using the following expression

$$s^s(T,D) = -(\frac{\partial g^s}{\partial T})_p$$
$$= 2a_2{}^sT + a_1{}^s\ln(\frac{T}{T_0}) - \frac{g^s(T_0)}{T_0} - a_2{}^sT_0 + a_1{}^s + \frac{a_0{}^1}{T_0} - \frac{2v_{at}(T)\,(\gamma^s)^2}{D}\sum_{i=1}^{3}\frac{\gamma_i'}{\gamma_i^2} \tag{7}$$

where the primes denote derivatives with regard to temperature. The contribution from the derivative of atomic volume is trivial; it is reasonable to neglect. Using the relation between the specific heat capacity at constant pressure and the entropy, we can write the expression for the specific heat capacity per mole as

$$C_p^s(T,D) = N_0T(\frac{\partial s^s}{\partial T})_p$$
$$= 2N_0a_2^sT + N_0a_1^s - \frac{6N_0Tv_{at}(T)}{D}\left\{\frac{2}{\gamma^s}[(\gamma^s)']^2 + \frac{(\gamma^s)^2}{3}\sum_{i=1}^{3}\left[\frac{\gamma_i''}{\gamma_i^2} - \frac{2(\gamma_i')^2}{\gamma_i^3}\right]\right\} \tag{8}$$

where N_0 is Avogadro's number. The internal energy per atom for nanoparticles can be written as[5,10]

$$h_{v,D}(T,D) = h_v(T) + \frac{6v_{at}(T)\,\gamma(T)}{D} \tag{9}$$

where h_v represents the internal energy per atom of bulk material. The molar heat of fusion and molar entropy of fusion for nanoparticles can be derived from the internal energy difference of solid and liquid nanoparticles easily.

$$\Delta H_m = N_0\left[h_{v,D}^l(T_m,D) - h_{v,D}^s(T_m,D)\right]$$
$$= \Delta H_{mb}\left[1 - \frac{6v_{at}^s(T_m)}{LD}(\gamma^s(T_m) - \gamma^l(T_m)\frac{v_{at}^l(T_m)}{v_{at}^s(T_m)})\right] \tag{10}$$
$$= \Delta H_{mb}\left[1 - \frac{\beta_1(T_m)}{D}\right]$$

where ΔH_{mb} is the molar heat of fusion for bulk, and L is the latent heat of melting per atom. The superscript "*s*" and "*l*" represent solid phase and liquid phase, respectively.

Figure 1 shows the behavior of the solid and liquid internal enthalpies as a function of temperature, and an abrupt jump in the internal energy during heating can be observed, but this step does not reflect the thermodynamic melting because periodic boundary condition calculations provide no heterogeneous nucleation site, such as free surface or the solid-liquid interface, for bulk material leading to an abrupt homogeneous melting transition at about 1500 K (experimental melting point 1234 K), as it is revealed that the confined lattice without free surfaces can be significantly superheated.[21] The latent heat of fusion is 0.115 eV/atom, in good agreement with the experimental value of 0.124 eV/atom.[22]

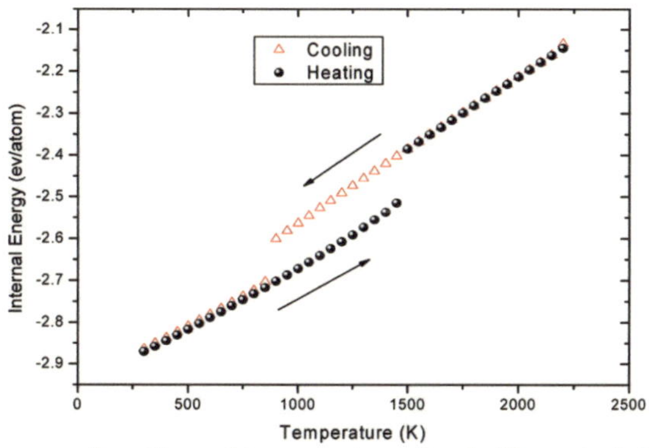

Fig. 1. Internal energy as a function of temperature for bulk material. Heating and cooling runs are indicated by the arrows and symbols. (Picture redrawn from Ref. 17)

The free-energy functions for the solid and liquid phases have been plotted in Fig. 2. The melting temperature T_{mb} is obtained from the intersection of these curves. From Fig. 2, two curves cross at T_{mb}=1243 K, which is in good agreement with the experimental melting point T^{exp}=1234 K. The good agreement in melting point is consistent with accurate prediction of the Gibbs free energies.

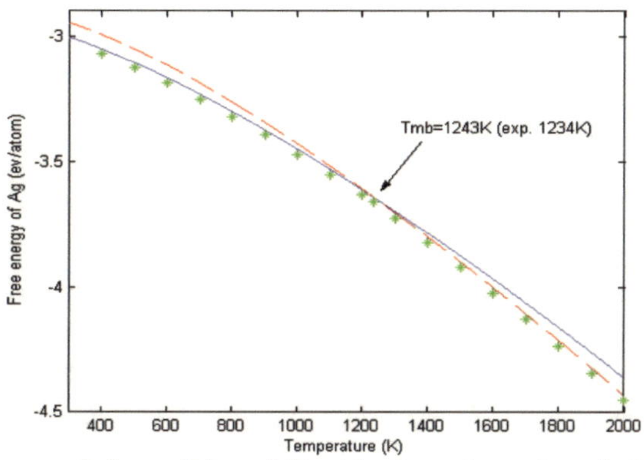

Fig. 2. Gibbs free energy of the solid and liquid phase in units of eV/atom. The asterisks denote the experimental values [22]. The solid curve is the MAEAM solid free energy, and the dashed curve is the MAEAM liquid free energy. The temperature at which Gibbs free energy of the solid and liquid phase is identical is identified as the melting point. (Picture redrawn from Ref. 17)

The calculation results of solid surface free energy for the (111), (100), and (110) surfaces with thermodynamic integration approach (TI)[23] is depicted in Fig. 3. It can be seen that the free energies of the surfaces at low temperatures are ordered precisely as expected from packing of the atoms in the layers. The close-packed (111) surface has the lowest free energy, and loosely packed (110) the largest. As temperature increases, the anisotropy of the surface free energy becomes lower and lower because the crystal slowly disorders. For comparison, we also utilize Grochola et al's "simple lambda" and "blanket lambda" path (BLP)[24,25] to calculate the solid surface free energy for the three low-index faces. The results are in good agreement with the TI calculation for temperature from 300 to 750 K. As an example, the simulation results for the integrand $<\partial E(\lambda)/\partial\lambda>\lambda$ + $<\partial\varphi_{rep}{}^{AB}/\partial\lambda>\lambda$ for the (110) face at 750 K is shown in Fig. 4. It is obvious that the results are very smooth and completely reversible. In order to create the slab, we also show the expansion process using *z*-density plots for $(L_z-L_z^0)/L_z^0$ = 0, 0.045, and 0.08, in Fig. 5. At $(L_z-L_z^0)/L_z^0$=0.08, the adatoms appearing between A and B sides can be seen. According to Grochola et al.,[25] it indicates that the BLP samples the rare events more efficiently than the cleaving lambda method [26] because the two surfaces interact via the adatoms when separated, as seen in Fig. 5. These adatoms would tend to have greater fluctuations in the z direction interacting with each other than if they were interacting with a static cleaving potential. They should therefore be more likely to move onto other adatoms sites or displace atoms underneath them, which should result in better statistics. The work obtained from the system in this expansion is roughly 5% of the work put into the system in the first part. For comparison, ab initio calculation results at T=0 K performed by L.Vitos et al. adopting the FCD method[27] is shown in Fig. 3.

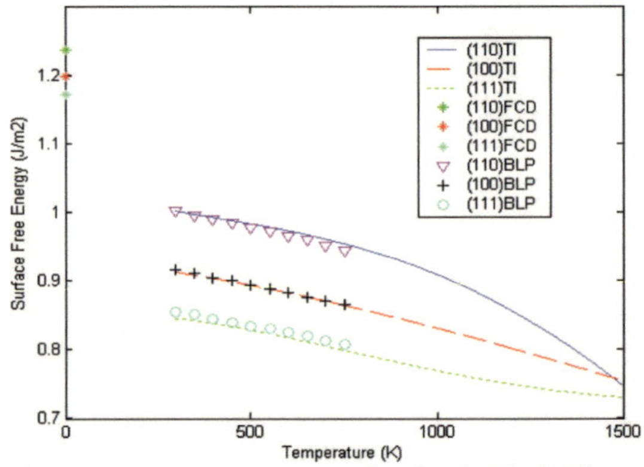

Fig. 3. Solid surface free energies vs temperature for the (111), (100), and (110) faces obtained using the thermodynamic integration technique and the lambda integration method. Also shown are L. Vitos et al.'s FCD results at 0 K. (Picture redrawn from Ref. 17)

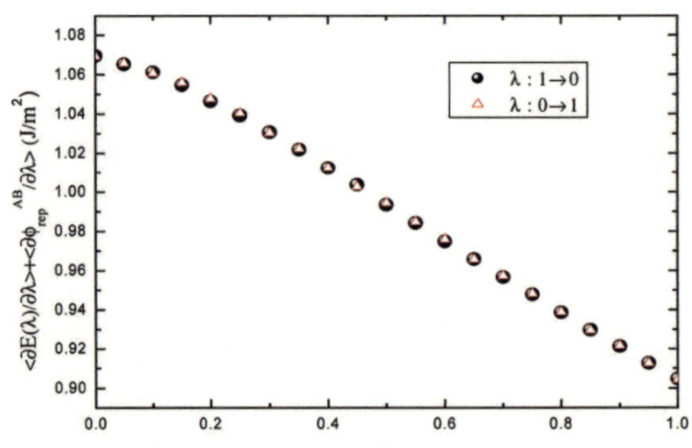

Fig. 4. Simulation results for the integrand $<\partial E(\lambda)/\partial\lambda>\lambda$ + $<\partial\varphi_{rep}{}^{AB}/\partial\lambda>\lambda$ for the (110) face at 750 K. (Picture redrawn from Ref. 17)

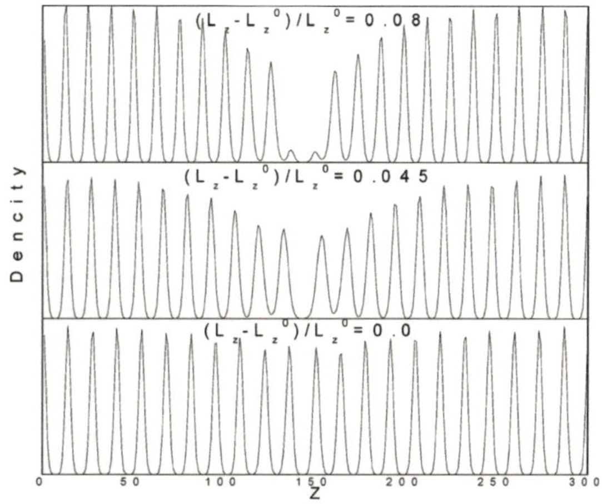

Fig. 5. *z* density plots for the expansion part of Grochola et al.'s "blanket lambda" path at $(L_z-L_z^0)/L_z^0 = 0$, 0.045, and 0.08 applied to the (110) face at a temperature of 750 K. (Picture redrawn from Ref. 17)

The calculated average solid surface free energy is shown in Fig. 6. Also shown are the liquid surface free energies and their linear fitting values, $\gamma_L(T) = 0.5773-2.3051\times10^{-4}(T-1243)$. At melting point, we acquire the solid surface free energy and the liquid surface free-energy values of 0.793 J/m^2 and 0.577 J/m^2, respectively. The semi-theoretical estimates of Tyson and Miller[28] for the solid surface energy at T_{mb} are 1.086 (J/m^2), and the experimental value[29] for the surface energy of the solid and the liquid states at T_{mb} are 1.205 and 0.903 (J/m^2), respectively. It should be emphasized that surface free energies of crystalline metals are notoriously difficult to measure and the spread in experimental values for well-defined low-index orientations is substantial, as Bonzel et al.[30] pointed out.

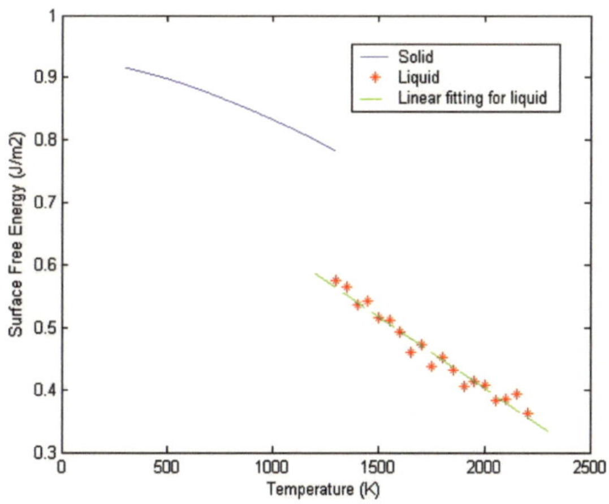

Fig. 6. Surface free energy of the solid and liquid phase in units of J/m^2 as a function of temperature. The data for liquid surface free energy is fitted to a linear function of temperature. (Picture redrawn from Ref. 17)

It is obvious that because MAEAM is developed using only bulk experimental data, it underestimates surface free energy in both the solid and the liquid states as many EAM models do.[31,32] Though there is the difference between the present results and experimental estimates, we note that the surface free-energy difference between the solid and liquid phase is 0.216 (J/m^2) and is between Tyson and Miller's result of 0.183 (J/m^2) and the experimental value of 0.3 (J/m^2). Furthermore, the average temperature coefficient of the solid and liquid phase surface free energy is 1.32×10^{-4} (J/m^2K) and 2.3×10^{-4} (J/m^2K), respectively. Such values compare reasonably well with Tyson and Miller's estimate of 1.3×10^{-4} (J/m^2K)[28] for the solid and the experimental results of 1.6×10^{-4} (J/m^2K)[33] for the liquid. Therefore, we expect the model to be able to predict the melting points of nanoparticles by means of determining the intersection of free-energy curves. Because the liquid surface free energy is lower than the solid surface free energy, the solid and liquid free-energy curves of nanoparticles change differently when the size of the nanoparticle decreases so that the melting points of nanoparticles decrease with decreasing particle size, as is depicted by Fig. 7. This indicates actually that the surface free-energy difference between the solid and liquid phase is a decisive factor for the size-dependent melting of nanostructural materials.

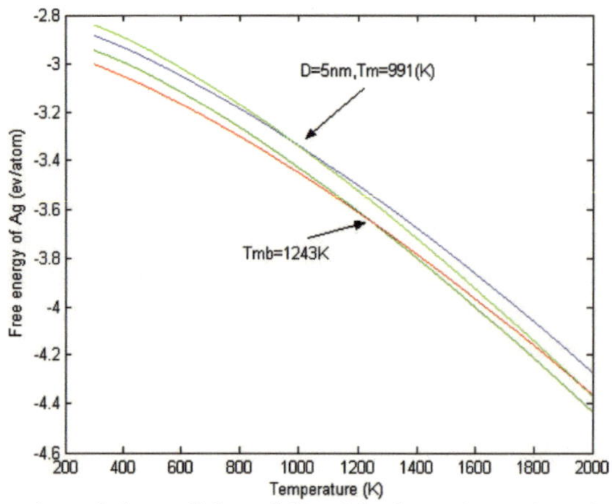

Fig. 7. Gibbs free energies of the solid and liquid phase in units of eV/atom for the bulk material and 5 nm nanoparticle. (Picture redrawn from Ref. 17)

In order to test our model, we plotted the results for the melting temperature versus inverse of the particle diameter in Fig. 8. Because there is no experimental data available for the melting of Ag nanoparticles, the predictions of Nanda et al.[10] and Yang et al.'s[34] theoretical model are shown in Fig. 8 for comparison. It can be seen that agreement between our model and Nanda et al.'s[10] theoretical predictions for Ag nanoparticles is excellent. The nonlinear character of the calculated melting curve results from the temperature dependence of the surface free-energy difference between the solid and liquid phase, which is neglected in Nanda et al.'s[10] model. Alternatively, Yang et al.'s[34] theoretical predictions may overestimate the melting point depression of Ag nanoparticles.

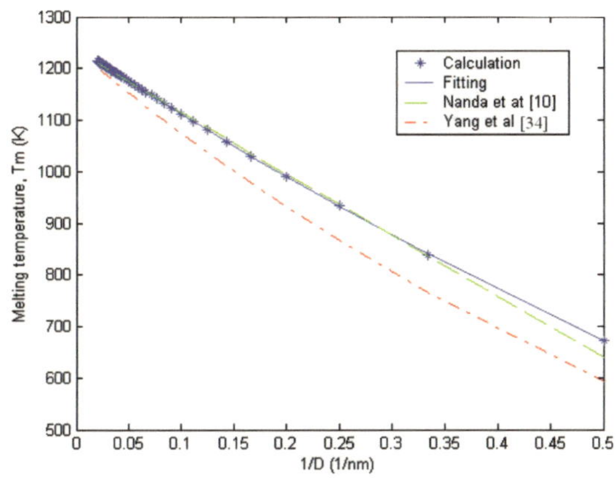

Fig. 8. Melting point vs the reciprocal of nanopartical diameter. The solid line is the fitting result. The dashed line is the result calculated from the thermodynamic model $T_m = T_{mb}(1 - \beta/d)$,[10] ($\beta = 0.96564$). (Picture redrawn from Ref. 17)

It is believed that understanding and predicting the melting temperature of nanocrystals is important. This is not only because their thermal stability against melting is increasingly becoming one of the major concerns in the upcoming technologies[1,34,35] but also because many physical and chemical properties of nanocrystals follow the exact same dependence on the particle sizes as the melting temperature of nanocrystals does. For example, the size-dependent volume thermal expansion coefficient, the Debye temperature, the diffusion activation energy, the vacancy formation energy, and the critical ferromagnetic, ferroelectric, and superconductive transition temperature of nanocrystals can be modeled in a fashion similar to the size-dependent melting temperature.[34,36,37] However, Lai et al.[38] pointed out that in order to understand the thermodynamics of nanosized systems comprehensively an accurate experimental investigation of "the details of heat exchange during the melting process, in particular the latent heat of fusion" is required. Allen and co-workers developed a suitable experimental technique to study the calorimetry of the melting process in nanoparticles and found that both the melting temperature and the latent heat of fusion depend on the particle size.[38-40] Here we calculate the molar heat of fusion and molar entropy of fusion for Ag nanoparticles, and the results are shown in Fig. 9. It can be seen that both the molar heat and entropy of fusion undergo a nonlinear decrease as the particle diameter D decreases. In analogy with the melting point, Figure 9 shows that the system of smallest size possesses the lowest latent heat of fusion and entropy of fusion. In a particle with a diameter of 2.5 nm or smaller, all of the atoms should indeed suffer surface effects, and the latent heat of fusion and the entropy of fusion are correspondingly expected to vanish. It is also observed that the size effect on the thermodynamic properties of Ag nanoparticles is not really significant until the particle is less than about 20 nm.

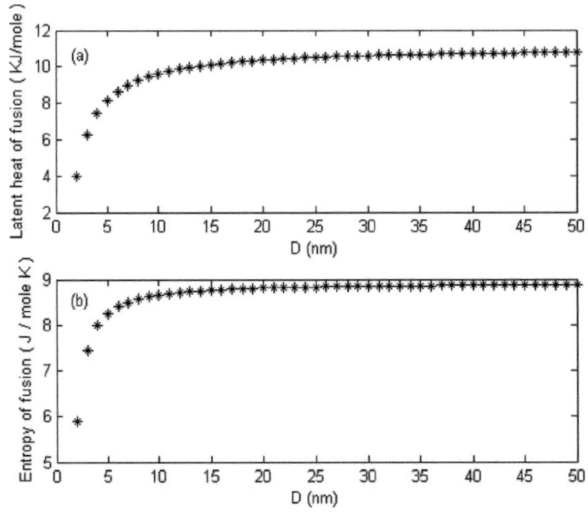

Fig. 9. (a) Molar latent heats of fusion ΔH_m and (b) molar entropy of fusion ΔS_m of Ag nanoparticals as a function of particle diameter D. (Picture redrawn from Ref. 17)

Figure 10 plots the molar heat capacities as a function of temperature for bulk material and nanoparticles. One can see that the molar heat capacity of nanoparticles increases with increasing temperature, as the bulk sample does. The temperature dependence of molar heat capacity qualitatively coincides with that observed experimentally. Figures 10 and 11 show that the molar heat capacity of bulk sample is lower compared to the molar heat capacity of the nanoparticles, and this difference increases with the decrease of particle size. The discrepancy in heat capacities of the nanoparticles and bulk samples is explained in terms of the surface free energy. The molar heat capacity of a nanoparticle consists of the contribution from the bulk and surface region, and the reduced heat capacity C/C_b (C_b denotes bulk heat capacity) varies inversely with the particle diameter D. Likhachev et al.[41] point out that the major contribution to the heat capacity above ambient temperature is determined by the vibrational degrees of freedom, and it is the peculiarities of surface phonon spectra of nanoparticles that are responsible for the anomalous behavior of heat capacity. This is in accordance with our calculation. Recently, Li and Huang[42] calculated the heat capacity of an Fe nanoparticle with a diameter around 2 nm by using MD simulation and obtained a value of 28J/mol·K, which is higher than the value of 25.1J/mol·K[22] for the bulk solid. It might be a beneficial reference data for understanding the surface effect on the heat capacity of nanoparticles. The ratio C/C_b=1.1 they obtained for 2 nm Fe nanoparticles is comparative to our value of 1.08 for 2 nm Ag nanoparticles. Because we set up a spherical face by three special low-index surfaces, the molar heat capacity of nanoparticles necessarily depends on the shape of the particle.

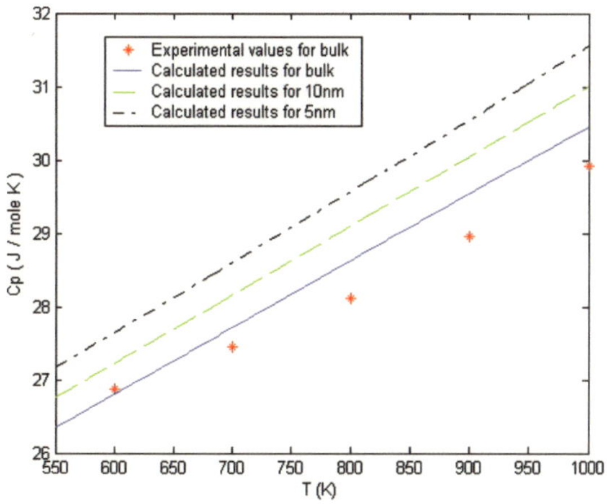

Fig. 10. Molar heat capacity as a function of temperature for Ag nanoparticles and bulk sample. (Picture redrawn from Ref. 17)

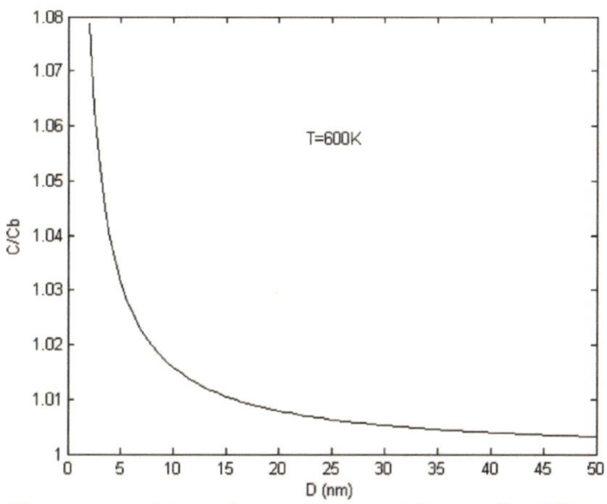

Fig. 11. Dependence of heat capacities of Ag nanoparticles with different sizes relative to the bulk sample. C_b is the heat capacity of the bulk sample. (Picture redrawn from Ref. 17)

The molar entropy as a function of temperature is shown in Fig. 12. It can be seen that the calculated molar entropies are in good agreement with experimental values.[22] The molar entropy of nanoparticles is higher than that of the bulk sample, and this difference increases with the decrease of the particle size and increasing temperature. According to Eq. 7, the reduced molar entropy S/S_b (S_b denotes bulk entropy) also varies inversely with the particle diameter D, just as the heat capacity of a nanoparticle does. Because entropy is only related to the first derivatives of Gibbs free energy with regard to temperature, and we have obtained the average temperature coefficient of solid surface energy agreeing with the value in literature,[28] it may be believed that Fig. 12 rightly reveals the molar entropy of nanoparticles.

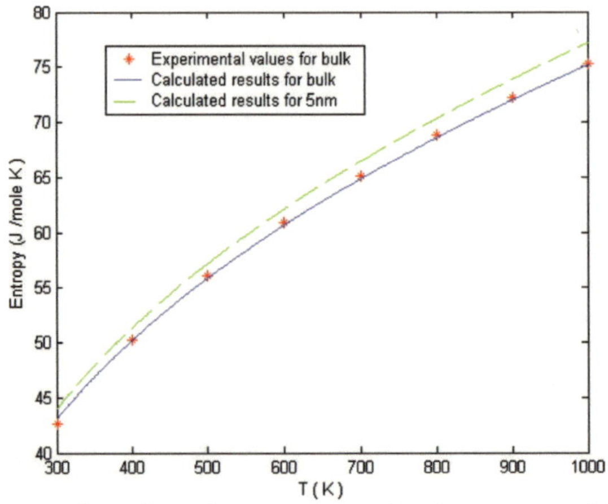

Fig. 12. Molar entropy as a function of temperature for Ag nanoparticles and bulk material. (Picture redrawn from Ref. 17)

2. Grain growth of nanocrystalline silver

Nanocrystalline materials are polycrystalline materials with mean grain size ranging from 1 to 100 nm. Affected by its unique structural characteristics, nano-sized grains and high fraction of grain boundary, nanocrystalline materials possess a series of outstanding physical and chemical properties, especially outstanding mechanical properties, such as increased strength/hardness and superplastic [43]. However, just because of the high ratio of grain boundary, the nanocrystalline materials usually show low structural stability, the grain growth behavior directly challenges the processing and application of nanocrystalline materials. How to improve the thermal stability of nanocrystalline materials became a challenging study.

Since the network of grain boundary (GB) in a polycrystalline material is a source of excess energy relative to the single-crystalline state, there is a thermodynamic driving force for reduction of the total GB area or, equivalently, for an increase in the average grain size [44]. Especially as grain size decreases to several nanometers, a significant fraction of high excess energy, disordered GB regions in the nanostructured materials provide a strong driving force for grain growth according to the classic growth theory [45]. In contrast to the microcrystal, recent theoretical and simulation studies indicate that grain boundary motion is coupled to the translation and rotation of the adjacent grains [46]. How Bernstein found that the geometry of the system can strongly modify this coupling [47]. We simulate the grain growth in the fully 3D nanocrystalline Ag. It is found that during the process of grain growth in the nanocrystalline materials, there simultaneously exist GB migrations and grain rotation movements [46,48,49]. The grain growth of nanocrystalline Ag exhibits a Power law growth, followed by a linear relaxation process, and interestingly the dislocations (or stacking faults) play an important intermediary role in the grain growth of nanocrystalline Ag.

For conventional polycrystalline materials, the mechanism of grain growth is GB curvature-driven migration [44]. Recently, the grain rotation mechanism has been found both in the experiments and simulations [46,48,49]. These two mechanisms are also found in our simulations as illustrated in Fig. 13 and Fig. 14, respectively. Fig. 13 shows the GB migration in a section perpendicular to Z-axis in the course of grain growth for the 6.06 nm sample at 1000 K. It is clear that the grain 1, as a core, expands through GB migrating outwards until the whole nanocrystal closes to a perfect crystal. Fig. 14 shows the atomic vector movement in the same section as in Fig.13 from 200ps to 320ps, the grain 1 and grain 2 reveal obvious rotation, although these two grains don't coalesce fully by their rotations.

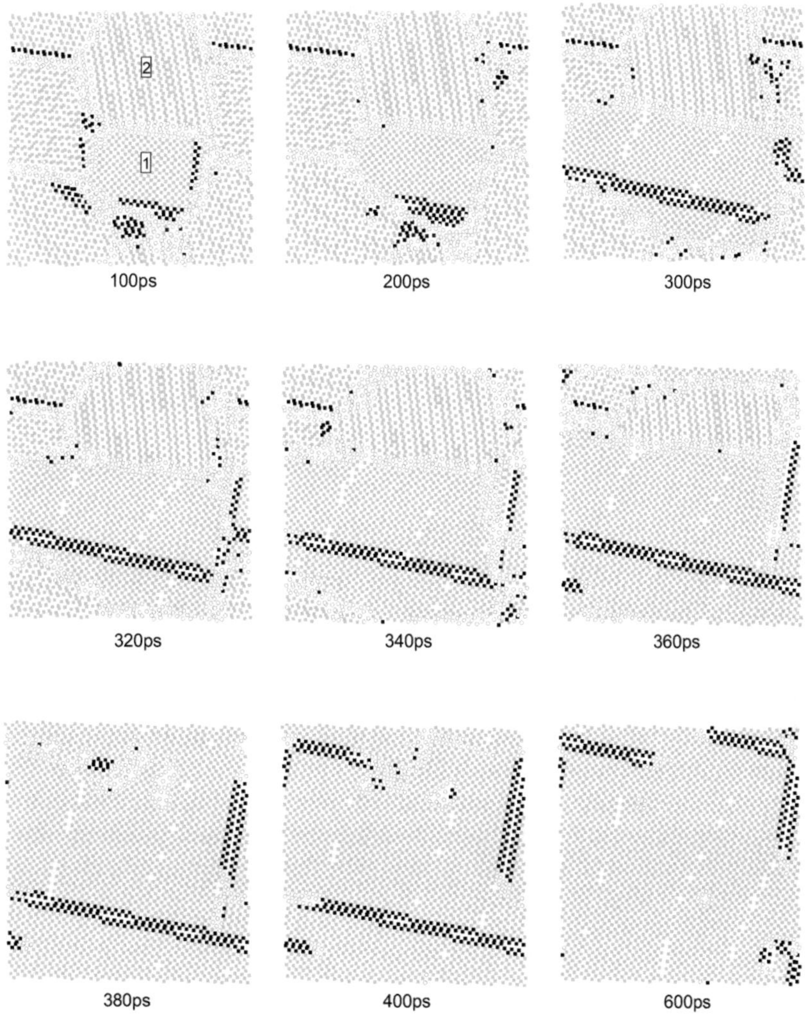

Fig. 13. The typical structural evolution of a section perpendicular to Z-axis during grain growth by GB migration for the 6.06nm specimen at 1000K, the grey squares represent FCC atoms, the black squares for HCP atoms and the circles for the other type atoms, respectively. (Picture redrawn from Ref. 48)

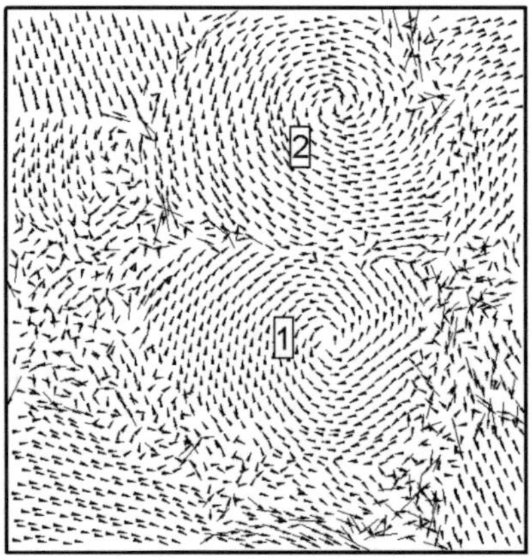

Fig. 14. Atomic vector movement in the same section as in Fig.3 from 200ps to 320ps, the grains 1 and 2 exhibit obvious rotations. (Picture redrawn from Ref. 48)

Figure 15 shows the quantitative evolution of FCC atoms for the 6.06 nm sample at 900 K and 1000 K, respectively. Affected by thermally activated defective atoms at the beginning of imposing thermostat, the number of atoms with FCC structure decreases with the relaxation time (as triangle symbols shown in Fig. 15). Subsequently the grains begin to grow up. It is evident that the process of grain growth can be described as a Power law growth, followed by a linear relaxation stage. In the Power law growth stage, the growth curves are fitted as follows:

$$C = C_0 + Kt^n \tag{11}$$

where C is the proportion of FCC atoms, K is a coefficient and n is termed the grain growth exponent, and the fitted values of n are 3.58 and 3.19 respectively for annealing temperature 1000 K and 900 K. The values of n, which indicate their growth speeds, increase with increasing the annealing temperature. In the Power law growth stage, the grain growth is mainly dominated by the GB migration. In the succedent linear relaxation process, the fraction of FCC atoms increases linearly with time, and this increment mostly comes from the conversion of the fault clusters and dislocations (or stacking faults) left by GB migration (as shown in Fig. 13). In addition, it is noted that the change from Power law growth to linear relaxation is overly abrupt, this is because the sampled points in the structural analysis are very limited and there exists a bit of fluctuation in the structural evolution.

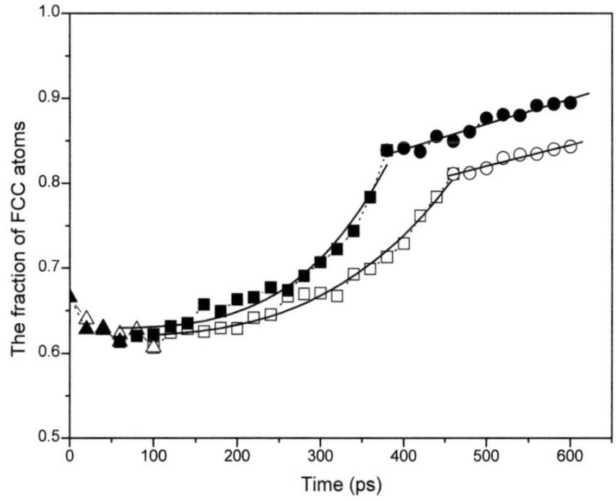

Fig. 15. The isothermal evolution of the fraction of FCC atoms with time for the 6.06nm sample, the solid and the open symbols represent the fraction at 1000K and at 900K, respectively. The triangles represent the thermally disordered stage, the squares for the Power law growth stage, and the circles for the linear relaxation process. (Picture redrawn from Ref. 48)

Besides GB migration and grain rotation, the dislocations (or stacking faults) may play an important role in grain growth of FCC nanocrystals. From the structural analysis it is found that along with the grain growth, the fraction of HCP atoms undergoes a transformation from increasing to decreasing. Comparing the evolution of FCC and HCP atoms with time, the critical transformation time from HCP atoms increasing to decreasing with annealing time is in accordance with the turning point from the Power law growth stage to the linear relaxation stage. At the Power law growth stage, dislocations (or stacking default) are induced after the migration of GB (as shown in Fig. 13), and the fraction of HCP atoms increases. Their configuration evolves from the dispersive atoms and their clusters on GB to aggregative dislocations (or stacking faults) as shown in Fig. 16. Turning into the linear relaxation stage, the fraction of HCP atoms decreases with the annealing time gradually, and some dislocations (or stacking defaults) disappear (as shown in Fig. 13 and Fig. 16). Comparing with the interface energy (about 587.1 mJ/m^2 and 471.5 mJ/m^2 for the 6.06 nm and 3.03 nm samples, respectively, if supposing grains as spheres and neglecting the triple-junction as well as high-junction GBs), although stacking default energy is very small for Ag (14.1 mJ/m^2), the evolutive characteristic of HCP atoms during grain growth is probably correlative with the stacking fault energy, which lowers the activation energy for atoms on GB converting into stacking faults than directly into a portion of grains, so the dislocations (or stacking faults) may act as the intermediary for the atom transforming from GB to grains.

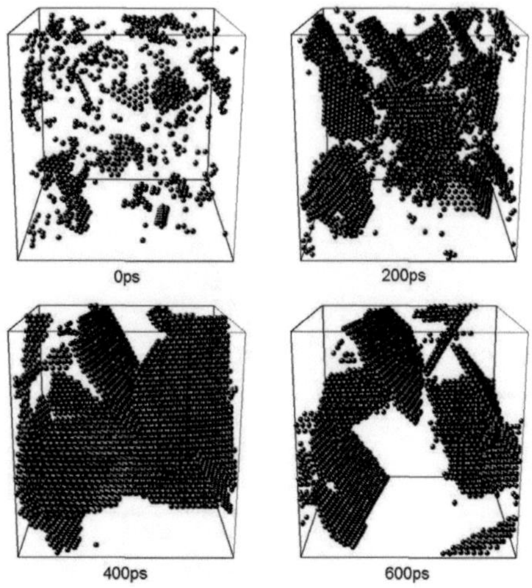

Fig. 16. HCP atoms configuration evolution during grain growth process for the 6.06nm specimen at 1000K. (Picture redrawn from Ref. 48)

3. Melting behaviour of Nanocrystalline silver

Melting temperature (T_m) is a basic physical parameter, which has a significant impact on thermodynamic properties. The modern systematic studies have provided a relatively clear understanding of melting behaviors, such as surface premelting [50,51], defect-nucleated microscopic melting mechanisms [52], and the size-dependent T_m of the low dimension materials [10]. In recent years, the unusual melting behaviors of nanostructures have attracted much attention. On a nanometer scale, as a result of elevated surface-to-volume ratio, usually the melting temperatures of metallic particles with a free surface decrease with decreasing their particle sizes [7,35], and their melting process can be described as two stages, firstly the stepwise premelting on the surface layer with a thickness of 2-3 perfect lattice constant, and then the abrupt overall melting of the whole cluster.[53] For the embedded nanoparticles, their melting temperatures may be lower or higher than their corresponding bulk melting temperatures for different matrices and the epitaxy between the nanoparticles and the embedding matrices [54]. Nanocrystalline (NC) materials, as an aggregation of nano-grains, have a structural characteristic of a very high proportion of grain boundaries (GBs) in contrast to their corresponding conventional microcrystals. As the mean grain size decreases to several nanometers, the atoms in GBs even exceed those in grains, thus, the NC materials can be regarded as composites composed by grains and GBs with a high excess energy. If further decreasing the grain size to an infinitesimal value, at this time, the grain and GB is possibly indistinguishable. What about its structural feature and melting behavior? We have reported on the investigation of the melting behavior for "model" NC Ag at a limited grain-size and amorphous state by means of MD simulation, and give an analysis of thermodynamic and structural difference between GB and amorphous state.[55,56]

Figure 17 shows the variation of T_m of NC Ag and the solid-to-liquid transformation temperature of amorphous state Ag. It can be seen that, from grain-size-varying nanocrystal to the amorphous, the curve of T_m exhibits three characteristic regions named I, II and III as illustrated in Fig. 17. In addition, considering the nanocrystal being an aggregation of nanoparticles, the T_m of nanoparticles with FCC crystalline structure is appended in Fig. 17.

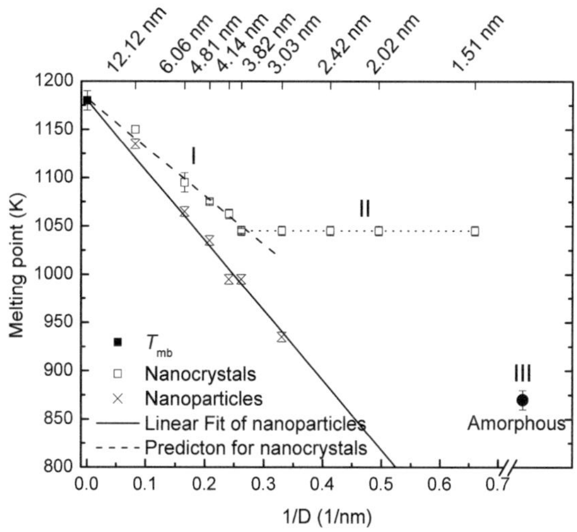

Fig. 17. Melting temperature as a function of the mean grain size for nanocrystalline Ag and of the particle size for the isolated spherical Ag nanoparticles with FCC structure, and the solid-to-liquid transformation temperature of amorphous state. (Picture redrawn from Ref. 56)

In region I, in comparison to T_m (bulk) = 1180±10 K simulated from the solid-liquid coexistence method with same modified analytical embedded atom method (MAEAM) potential, the melting temperatures of the NC Ag are slightly below T_m (bulk) and decrease with the reduction of the mean grain size. This behavior can be interpreted as the effects of GBs on T_m of a polycrystal. MD simulations on a bicrystal model have shown that an interfacial melting transition occurs at a temperature distinctly lower than T_m (bulk) and the width of interfacial region behaving like a melt grows significantly with temperature [57]. This will induce the grains in a polycrystal melted at a temperature lower than T_m (bulk) when the mean grain size decreases to some extent and results in the depression of T_m for the NC materials.

Comparing with the corresponding nanoparticle with the same size, the NC material has a higher T_m. This is to be expected since the atoms on GB are of larger coordination number than those on a free surface, and the interfacial energy (γ_{GB}) is less than the surface energy (γ_{Sur}). It is well known that the main difference between the free particles and the embedded particles or grains in a polycrystal is the interfacial atomic structure. A thermodynamic prediction of T_m from a liquid-drop model [10] for free particles was proposed as follows

$$T_m = T_{mb}(1-(\beta / d)) \qquad (12)$$

where T_{mb} is the melting temperature for conventional crystal, β is a parameter relative to materials, d is the mean diameter of the grain. Through the linear fitting of the data from the MD simulations in terms of Eq.(12), the value of β for Ag nanoparticles is determined as 0.614, which is lower than that from thermodynamic prediction, and this tendency is the same for several other elements [58]. Considering the difference between the interfacial energy and the surface energy, an expression describing the mean grain size-dependent T_m for the NC materials is proposed as follows:

$$T_m = T_{mb}(1-(\beta / d)(1-\gamma_{GB} / \gamma_{Sur})) \tag{13}$$

where γ_{GB} =375.0 mJ/m^2 [59] and γ_{Sur} =1240.0 mJ/m^2 [28] for Ag. Based on the fitted value β=0.614, the prediction of T_m of the NC Ag from Eq.(13) gives a good agreement with the values from the MD simulations as shown in Fig. 17. Thus, the melting temperatures of the NC materials can be qualitatively estimated from the melting temperatures of the corresponding nanoparticles.

Contrasting Eq.(13) with Eq.(12), the difference of T_m of nanocrystal from that of corresponding size spherical nanoparticle is induced by the distinction of properties between grain boundary and free surface. This further indicates the grain size dominated T_m of nanocrystalline Ag in region I.

Whereas in region II, the value of T_m almost keeps a constant, that is to say, the T_m of nanocrystalline Ag in this size region is independent on grain size. This tendency of T_m for nanocrystals is not difficult to understand from its remarkable size-dependent structure. According to the structural characteristic of small size grains and high ratio of grain boundary network, the nanocrystal can be viewed as a composite of a grain boundary phase and an embedded grain phase. As mean grain size is large enough, the grain phase is the dominant one, and the T_m of nanocrystals is correlated with their grain size. With mean grain size decreasing to a certain degree, the grain boundary phase becomes dominant, and the T_m of nanocrystal is possibly dominated by grain boundary phase. So the melting temperature of limited nanocrystalline materials can be considered as the T_m of grain boundary phase. Actually, these similar size-dependent physical properties of nanocrystals have been found recently in the mechanical strength of nanocrystalline Cu.[60] In region III (i.e. from nanocrystal to amorphous state), there exists a sharp decrease in T_m. This remarkable difference in T_m is a manifestation of fundamental difference in structure between nanocrystal/GB and amorphous state.

To further the investigation of size effect on the T_m of nanocrystal, the local atomic structure is analyzed with CNA. According to the local atomic configurations from the CNA, the atoms are classified into three classes: FCC, HCP and the others. Comparing the structural characteristic before and after annealing process (as shown in Fig. 18), the difference increases with decreasing the grain size because of the enhanced GB relaxation and instability in smaller size samples. Especially for the sample with a grain size of 1.51 nm, it shows minor grain growth. For nanocrystalline Ag, as the mean grain size below about 4 nm, the fraction of GB exceeds that of grain. This size exactly corresponds to the critical transformation size from the size-dependent T_m region to size-independent one. Within the small grain-size range, although the fraction of GB and mean atomic configurational energy keep on increasing with grain size decreasing, the T_m of nanocrystal is almost invariable, which provides the evidence of GB dominated T_m in this size range.

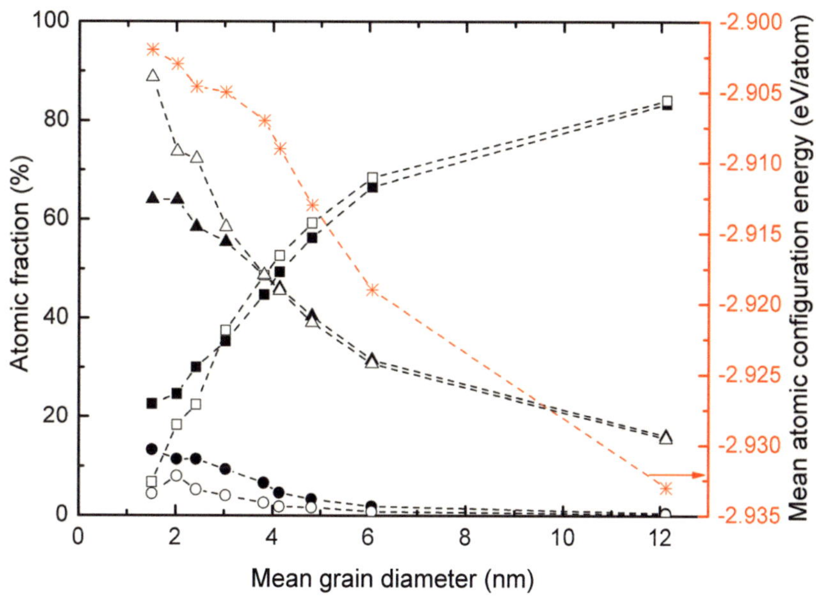

Fig. 18. The grain size-dependent structure and mean atomic configuration energy (Stars) of nanocrystalline Ag. The Squares, Circles and Triangles represent the atoms with local FCC, HCP and other type structure, and the open symbols and solid ones denote the case before and after annealing process, respectively. (Picture redrawn from Ref. 56)

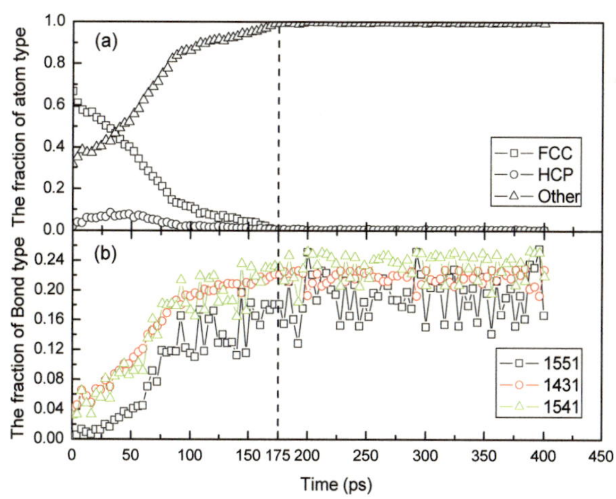

Fig. 19. The typical bond-type existing in the liquid phase and atom-type (from CNA analysis) evolution with MD relaxation time for specimen with a mean grain size of 6.06 nm (T_m: 1095±5K) at 1100K. (Picture redrawn from Ref. 57)

According to the local atomic configurations from the CNA, we observed the evolution of grains and GBs during the melting process. Figure 19 shows the relative number of three structural classes of atoms (a) and that of typical bonded pairs existing in metallic liquid (b) as a function of heating time during melting. With the melting developing, the fraction of atoms with a local FCC structure drops rapidly from an initial value of 66.6% to 0 at 175ps, as the NC material turns into a liquid phase completely. On the contrary, the relative numbers of the three typical bonded-pairs (1551), (1431) and (1541), which indicates the liquid (or like liquid) structural characteristics, increase rapidly from 0.4%, 4.1% and 3.2% to the average values of 19.2%, 21.7% and 23.3%, respectively, i.e. the fraction of the liquid phase (or like liquid) increases. It reveals that the melting in the polycrystals is a gradual process with heating time. Corresponding to a quantitative description on structural evolution (shown in Fig. 19), Figure 20 illustrates the 3-dimentional snapshots of grains in the NC material during melting. It is found that the melting in the polycrystals stars from GB. Along with melting, the interfacial regions (liquid or like liquid) between grains widen and the grains diminish till absolutely vanish.

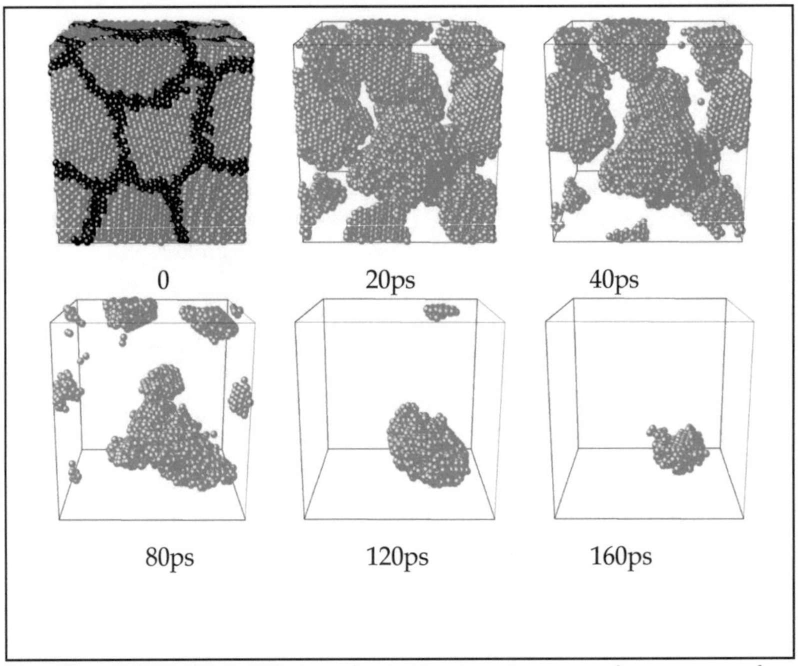

Fig. 20. Three-dimensional snapshots of atomic positions in the course of melting for specimen with a mean grain size of 6.06 nm at 1100K, the grey spheres denote atoms with a local FCC structure and the black spheres denote the other type of atoms on GB. For clarity, only FCC atoms are sketched during melting. (Picture redrawn from Ref. 57)

As well known, the crystallization from amorphous materials is an effective technique in the fabrication of nanocrystals.[61] On the contrary, it is found that the so-called nanocrystal, with the grain size extrapolating to an infinitesimal value during the Voronoi construction, has a similar structure as that of the amorphous from rapid quenching of liquid. They have similar RDF with the characteristic of the splitting of the second peak for an amorphous state. The relative numbers of the three typical bond pairs (1551), (1431) and (1541) in two amorphous samples from different processes (as listed in TABLE 1), are very close respectively. The slight difference on the percentage of corresponding type of bond pair is because the annealing temperature and quenching speed have a little influence on the amorphous structure. In addition, the two amorphous samples have the same solid-to-liquid temperature of 870±10 K. In other words, the Voronoi construction is an effective method in the construction of amorphous model.

Size (nm)	1551 (%)	1431 (%)	1541 (%)
Amorphous [a]	10.26	20.49	21.02
Amorphous [b]	14.56	21.47	23.73
1.51	2.25	16.75	18.69
2.02	2.51	17.38	20.23
2.42	2.03	16.65	18.24
3.03	1.78	15.75	15.85
3.82	1.78	15.27	14.26
4.14	2.05	15.49	14.03
4.81	1.90	15.18	13.56
6.06	1.39	13.14	10.01
12.12	1.36	12.73	9.83

[a] The amorphous is obtained from the rapid quenching of liquid.
[b] The amorphous is obtained from the annealing of the Voronoi construction.
Table 1. Fraction of three typical bond pairs (1551), (1431) and (1541) existing in non-crystalline structure from common neighbor analysis.

On the structural difference between GB and amorphous state, there is a long standing argument. Presently, the prevalent viewpoint is that the structure of GB is different from that of amorphous state[62,63], but the intrinsic difference is still not fully understood yet. Here if excluding the grain phase in nanocrystal, from TABLE 1, one can see the fraction of bond pairs (1431) and (1541) shows approximately consecutive increase with decreasing the mean grain size of nanocrystal gradually to an amorphous state, which indicates the increase of disordered degree in GB. Whereas the fraction of bond (1551) indicating fivefold symmetry have no evident change with grain size in nanocrystal. Once turning into amorphous state, there is a sharp increase in the proportion of bond pair (1551). So we concluded that the main difference between grain boundary and amorphous state lies in the higher fivefold symmetry in the latter.

4. Solidification of liquid silver and silver nano-drop

The crystallization from liquid is one of the important issues in crystal-growth technology. Just from the transformation of state of matter, crystallization from liquid is a reverse process of the melting, but there is much difference in thermodynamics and kinetics of phase transition between them. For instance, the spontaneous crystallization can only occur via homogeneous nucleation, a thermally activated process involving the formation of a growing solid nucleus. That is why the supercooling phenomenon is more ubiquitous than the superheating one experimentally.[64] So far, with the molecular dynamics simulation, a lot of work about crystallization has been made on the size of critical nucleus,[65,66] the structural feature of nucleus,[67,68] and the effect of cooling rate on the crystalline texture.[69] Beside the widely studied metallic materials, recently there is some work on the crystallization process of covalent and molecular crystals.[70-72] Here we mainly focus on the isothermal structural evolution in the course of crystallization. By means of tracking the evolution of local atomic structure, the processes of crystallization at a certain temperature are well observed.

As mentioned above, the melting temperature of conventional Ag from simulation is 1180 K. However, once the sample turned into liquid completely, it didn't crystallize till temperature cooling down to 850 K. This is because the homogeneous nucleation from liquid needs a certain driving force, which is closely correlated with supercooling temperature according to the classic theory of nucleation. Here we have considered three supercooling degree (from 850K to 800K, 750K and 700K respectively) to investigate its effect on crystallization process. The temperature-dependent mean atomic energy is shown in Fig. 21. One can find that there occurs abrupt drop of energy at the temperature of 800 K, 750 K, and 700 K against that at liquid state, which is a signature of phase transition taking place at these temperature points. This also has been confirmed by the structural transformation from the analysis of CNA.

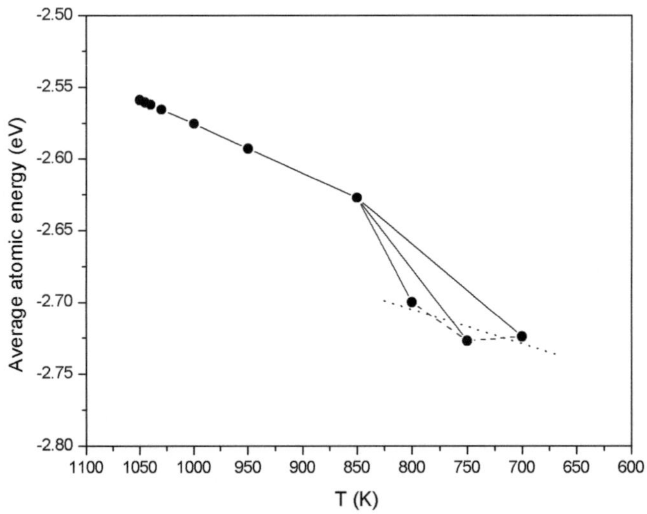

Fig. 21. The temperature-dependent mean atomic energy curves during cooling process. (Picture redrawn from Ref. 74)

Figure 22 shows the mean atomic configurational energy and mean atomic volume as a function of time during crystallization at 800 K, 750 K and 700 K. The energy and volume have similar changing trend for they each are correlative with the arrangement of atoms. From the evolution of energy and volume, the curves can be separated into three characteristic regions, which correspond to the three physical stages of the isothermal crystallization from liquid: nucleation, rapid growth of nuclei and slow structural relaxation. Contrasting the three curves at indicated temperatures, as temperature decreases, i.e. the enhancement of supercooling degree, the consumed time in the process of nucleation reduces obviously. This is because, according to the classic theory of nucleation, the critical size of nuclei decreases and the nucleation rate increases with supercooling degree increasing. As temperature decreasing to 700 K, the transition from nucleation to rapid growth of nuclei becomes ambiguous. Interestingly, after a relatively long period of slow structural relaxation, the mean atomic configurational energy at 700 K even is larger than that at 800 K. This is because the high rate of nucleation generally results in a large number of grains in unit volume, and thus a high proportion of grain boundary network. In addition, the much higher supercooling degree inhibits the atomic motion, and the defects within the grains increase. This high supercooling-degree technique has a potential application in the fabrication of nanocrystals by crystallization from liquid.

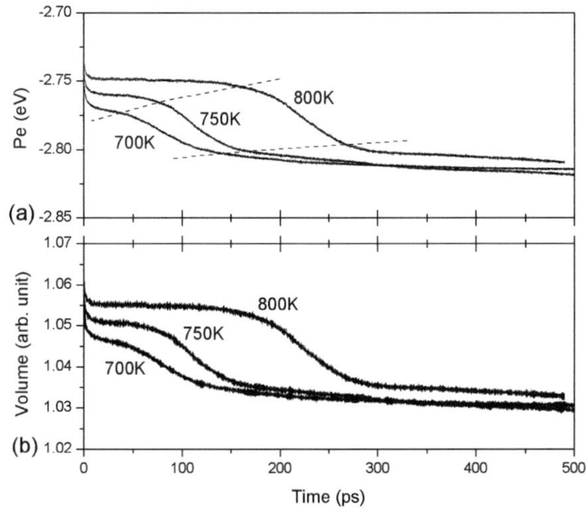

Fig. 22. The mean atomic configuration energy (a) and mean atomic volume (b) as a function of time during crystallization at 800 K. (Picture redrawn from Ref. 74)

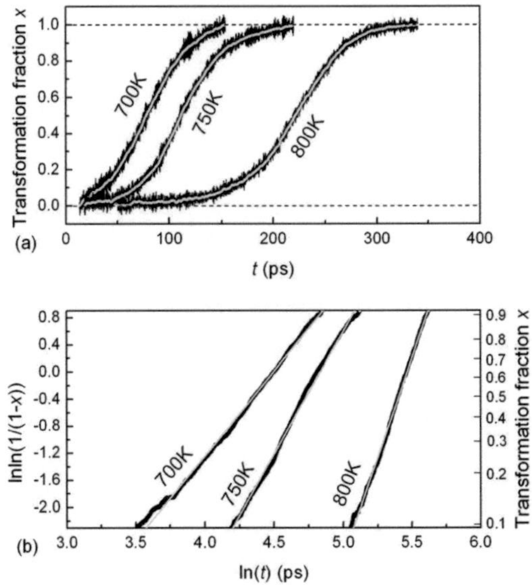

Fig. 23. The variation of volume transformation fraction of solid with time at indicated temperatures. (Picture redrawn from Ref. 74)

Supposing the systematic volume of a liquid-solid mixture is a linear superposition of its constituents, the volume transformation fraction can be determined from the following expressions:

$$V(t) = (1 - x(t))\, V_l + x(t)\, V_s \tag{14}$$

where V_l and V_s are the volume of unit liquid and solid at a certain temperature respectively, and $V(t)$ and $x(t)$ represent the variation of systematic volume and volume fraction of solid with time respectively. The values of V_l and V_s can be defined from the systematic volume before nucleation and after the completion of rapid growth of nucleus respectively. Based on the volume evolution in the stages of nucleation and rapid growth of nucleus as in Fig. 22(b), the typical "S" curves of transformation fraction of solid from liquid are shown in Fig. 23(a). For kinetic analysis of phase transition from liquid to solid, the Johnson-Mehl-Avrami (JMA) equation was used: [73,74]

$$x(t) = 1 - \exp(-K_T t^n) \tag{15}$$

where K_T is the constant of reaction rate and n is the Avrami exponent. The rate constant and the Avrami exponent are obtained from a plot of ln(-ln((1-$x(t)$))) vs. ln(t) as shown in Fig. 23(b) and the fitted parameters are listed in Table 2.

T (K)	n	K_T
800	5.84	1.33×10^{-14}
750	3.62	2.46×10^{-8}
700	2.50	1.29×10^{-5}

Table 2. The fitted kinetic parameters at different temperatures of crystallization by JMA equation.

The Avrami exponent decreases with increasing supercooling degree, which indicates the supercooling degree has a remarkable effect on nucleation rate. At 800 K, the Avrami exponent n has a value of 5.84(>4), which indicates an increasing nucleation rate with cooling time. As temperature decreasing to 750 K, the value of n =3.62 (<4) denotes a decreasing nucleation rate with cooling time. Especially at 700 K, the initial saturated nuclei and much higher supercooling degree block the succedent nucleation rate and result in a high ratio of grain boundary network, which is consistent with above analysis of atomic configuration energy.

Corresponding to the description of systematic ordering, the time evolution of local atomic structure and some characteristic bonded pairs from CNA are shown in Fig. 24. The three-staged process of crystallization is well reproduced in the evolution of local atomic structure. The enhancement of crystalline atoms mainly focuses on the stage of rapid growth. In addition, with the development of crystallization, the HCP-type atoms with laminar arrangement occupy a considerable proportion. This is because the FCC and HCP atoms have similar close-packed structure with tiny difference of atomic configurational energy. In the variation of bonded pairs, the change in the stage of nucleation is imperceptible. After the rapid growth of nuclei, the bonded pairs (421) become predominant, with its fraction achieving more than 60%. On the contrary, the relative numbers of the typical bonded pairs (433) and (544) only contribute to 5% of the total number of bonded pairs respectively, and the (555) almost disappear. Once a stable nucleus formed in supercooled liquid, the sample exhibited rapid growth of nuclei. This is the so-called instability for supercooled liquid relative to solid state.

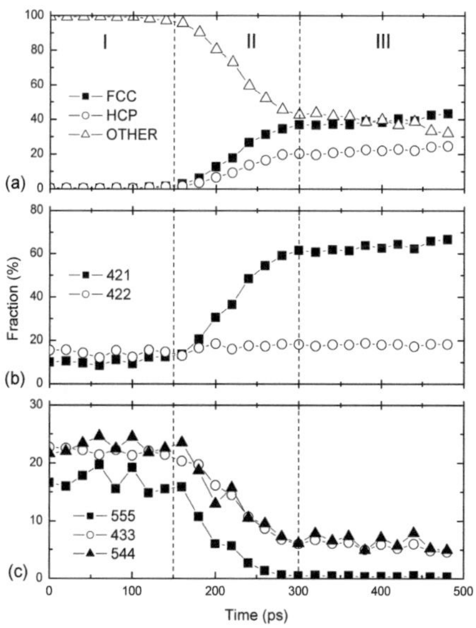

Fig. 24. The time evolution of local atomic structure and some characteristic bonded pairs from CNA in the course of crystallization at 800 K. (Picture redrawn from Ref. 74)

Different from conventional liquid, the nano-sized droplet exhibits particular freezing behavior. The investigation of gold nanoclusters revealed that ordered nanosurfaces with a fivefold symmetry were formed with interior atoms remaining in the disordered state and the crystallization of the interior atoms that proceeded from the surface towards the core region induced an icosahedral structure.[75] While Bartell et al. found that when molten particles in a given series were frozen, several different final structures were obtained even though conditions had been identical.[76] Due to the structural diversity for small sized nanoparticles, the freezing behavior of nanodroplets present size dependence. Using MD simulation and local atomic structure analysis technique, we have investigated the freezing behavior of Ag nano-droplets with diameters ranging from 2 nm to 14 nm.

Figure 25 shows the freezing temperature with droplet size. In order to discuss the crystallization kinetics, the supercooling temperature relative to melting temperature of same sized nanoparitcles is shown in Fig. 25. Same as size-dependent melting temperature of nanoparticles, it is found that for Ag nano-droplets, their freezing temperature decreases with droplet size. But for the three small sized samples, their freezing temperature have the same value of 790±10 K. While the droplet size increases to a certain value, the freezing temperature of droplet even higher than that of conventional liquid despite the melting temperature of nanoparticle being lower than conventional materials. These are because on the nanoscale, the freezing temperature of droplet is affected not only by melting temperature, but also by the different freezing mechanism resulting from surface effect. According the classic nucleation theory, the homogeneous nucleation in conventional liquid originates from the interior and the required supercooling temperature generally is 0.2 T_m. However, we found that in our simulation, the nucleation for all sized samples is on the surface of nano-droplet. This results the deviation of effective supercooling temperature of nano-droplets from classic rule.

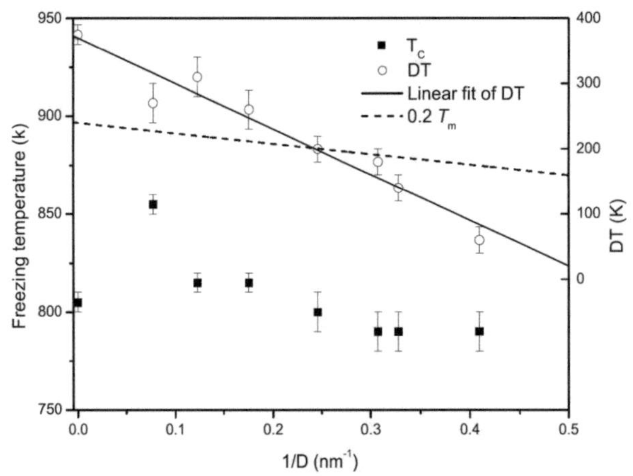

Fig. 25. The freezing temperature of Ag nano-droplt varying with droplet size.

According to the analysis of local atomic structure, Fig. 26 shows the freezing evolution for the sample of 2123 atoms. We can see that the crystalline nucleus first occurs on the surface layers of nano-droplet and with the preferential surface growth, the nano-droplet frozen into a particle with icosahedral structure.

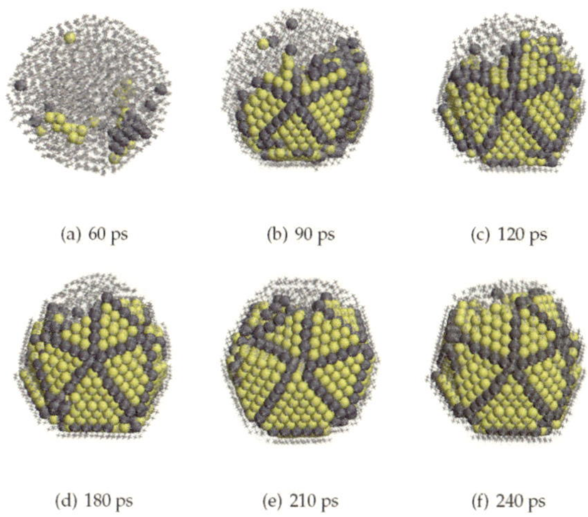

Fig. 26. The structural evolution in the course of freezing of the Ag nano-droplet with 2123 atoms, the yellow and black spheres represent the atoms with local FCC and HCP structure, respectively, the other atoms are denoted by star symbol.

5. Thermodynamic properties of alloy nanoparticles

In contrast to homogeneous nanoparticles composed of only one type of atom, the alloy nanoparticles exhibit more complicated structure and some special physical or chemical properties as a result of alloying effect. For instance, the Ag-Ni and Ag-Cu nanoparticles are of the core-shell structure with an inner Ni or Cu core and an Ag external shell,[77] and the Cu-Au nanoalloy clusters show an evident compositional dependence of structural characteristic.[78] Aguado et al. found that Li and Cs-doped sodium clusters have lower T_m than those of pure sodium ones for introducing a chemical defect.[79] However, a single Ni or Cu impurity in Ag icosahedral clusters considerably increases the T_m even for sizes of more than a hundred atoms.[80] Recently, the enhanced and bifunctional catalytic properties of bimetallic nanoparticles have made them attractive in the field of chemical catalysis.[81]

Driven by the high surface-to-volume ratio and surface free energy, the nanoparticles have a strong tendency of coalescence as they are put together. The molecular dynamics simulations has shown that the coalescence of iron nanoclusters occurs at the temperatures lower than the cluster melting point, and the difference between coalescence and melting temperatures increases with decreasing cluster size.[82] This feature are early expected to be applied in the alloying of components which are immiscible in the solid and/or molten state such as metals and ionics or metals and polymers.[83] As known from the Au-Pt alloy phase diagram, there exists a miscible gap for Au-Pt bulk alloy.[22] However, the Au-Pt alloy nanoparticles with several nanometers can be synthesized chemically almost in the entire composition range,[84] which demonstrates that the alloying mechanism and phase properties of nanoscale materials are evidently different from those of bulk crystalline state. For instance, Shibata et al. interpreted the size-dependent spontaneous alloying of Au-Ag nanoparticles under the framework of defect enhanced diffusion.[85] By calculating the formation heat of Au-Pt nanoparticles from their monometallic ones using a thermodynamic model and analytic embedded atom method, we have analyzed size effect on the alloying ability and phase stability of immiscible binary alloy on a nanometer scale. It is of importance for the study of alloying thermodynamics of nanoparticles and the fabrication of immiscible alloys.[86]

According to the definition of formation heat being the energy change associated with the formation of alloy from its constituent substances, the formation heat of alloy nanoparticle from the pure nanoparticles of their constituents can be expressed as

$$E_f^{pA-B} = \frac{NE_c^{pA-B} - N(1-x)E_c^{pA} - NxE_c^{pB}}{N} \tag{16}$$

where the superscripts A-B, A and B denote alloy and its constituent elements A and B, respectively. N is the total number of atoms and x is the chemical concentration of element B in alloy nanoparticles. E_c^p is the mean atomic cohesive energy of nanoparticles. The size-dependent cohesive energy of nanoparticles has the following expression:[87]

$$E_c^p = E_c^b\left(1 - \frac{d}{D}\right) \tag{17}$$

where E_c^b is the cohesive energy of the corresponding bulk material. d and D represent the diameters of a single atom and nanoparticle respectively. For alloy nanoparticles, the d denotes the mean atomic diameter derived from Vegard's law. If neglecting the difference of atomic volume for atoms resided in the interior of and on the surface layer of nanoparticles, there exists a relation among d, D and the number of atoms (n) in a nanoparticle as follows.

$$\frac{d}{D} = \sqrt[3]{\frac{1}{n}} \tag{18}$$

Substituting Eq. 17 and 18 into Eq. 16 yields

$$E_f^{pA-B} = \frac{1}{N}\left[NE_c^{bA-B}\left(1 - \sqrt[3]{\frac{1}{N}}\right) - N(1-x)E_c^{bA}\left(1 - \sqrt[3]{\frac{1}{N(1-x)}}\right) - NxE_c^{bB}\left(1 - \sqrt[3]{\frac{1}{Nx}}\right)\right] \tag{19}$$

One can find that, to obtain the formation heat of alloy nanoparticle from the pure nanoparticles of its constituents, it is only needed to calculate the cohesive energy of the corresponding bulk alloy.

In order to calculate the composition-dependent cohesive energy of binary alloy, we adopted the MAEAM to describe the interatomic interactions.[88] For the interaction between alloy elements, we take the formula in Ref. 89. Thus, the cohesive energy of disordered solid solution can be written as

$$E_c^{bA-B} = \left(\frac{1}{2}\varphi^A + F^A(\rho) + M^A(P)\right)(1-x) + \left(\frac{1}{2}\varphi^B + F^B(\rho) + M^B(P)\right)x \tag{20}$$

The pair potential between two different species of atom A and B is included in the terms of φ^A and φ^B. All the model parameters, determined from fitting physical attributes such as lattice parameter, cohesive energy, vacancy formation energy and elastic constants, for Au, Pt and Au-Pt intermetallic compound.[86] Figure 27(a) shows the formation enthalpy of Au-Pt disordered solid solution from the present model together with other calculated [90-92] and experimental values [93]. The results have a good agreement with experiment and other calculations, which indicates that the present AEAM model is reliable.

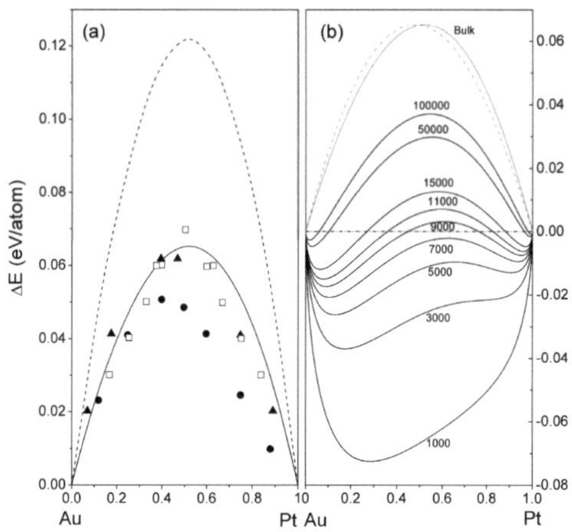

Fig. 27. (a) Formation heat for Au-Pt disordered solid solution as a function of Pt concentration. The solid line is the corresponding result from the present calculation; dash line and full circles present the results based on old EAM[90] and LMTO[91] respectively; open squares denote the calculation from Miedema theory[92]; full triangles denote the experimental data[93]. (b) The variation of the formation heat for Au-Pt nanoparticles of disordered structure along with Pt concentration at several indicated number of atoms in alloy nanoparticles. The dash line denotes mirror-image curve of that for bulk formation heat about the axis of $x = 0.5$, which give a clear comparison between the Au-Pt formation heat along with Au and Pt concentration. (Picture redrawn from Ref. 88)

Figure 27(b) shows the variation of formation heat of Au-Pt alloy nanoparticles with Pt atomic concentration for several samples with indicated total number of atoms (i.e. particle size). Naturally, as a result of alloying effect, the formation heat of alloy nanoparticles shows similar compositional dependence as in bulk materials. Comparing with bulk alloys, the most prominent characteristic on the formation heat of nanoparticles is its size-dependence. At a fixed Pt atomic concentration, the formation heat increases with the alloy particle size increasing, and its value even turns from negative to positive. This differs from the size-dependent formation enthalpy calculated by Liang et al.[94], where they only considered surface effect relative to the corresponding bulk materials. As the total number of atoms in Au-Pt alloy nanoparticles not exceeding 7 000 (about 6 nm in diameter of spherical particle), the formation heat within full concentration region is negative as a result of surface effect, which indicates that the alloying of Au and Pt nanoparticles becomes easy from the thermodynamic point of view, at the same time, indicates that the Au-Pt alloy nanoparticles within this size range having a better thermodynamic stability. In addition, the formation heat of bulk alloy has a great influence on that of nanoparticles. As shown in Fig. 27(b), the formation heat in Au-rich range for Au-Pt bulk alloy is lower than that in Pt-rich one. This difference is magnified in nanoparticles. Thus, in the Au-rich range, the Au-Pt nanoparticles show negative formation heat in a broad concentration range and a large particle size range, that is to say, the easy alloying region is extended. In Fig. 28, the contour of formation heat of disordered Au-Pt nanoparticles is shown as a function of alloy nanoparticle size and the chemical concentration of Pt atom. For the nanoparticles with a dilute solute of Pt in Au or Au in Pt, there exists negative formation heat in a large particle size range. This can be looked as the instability of a small-size particle relative to a large one.

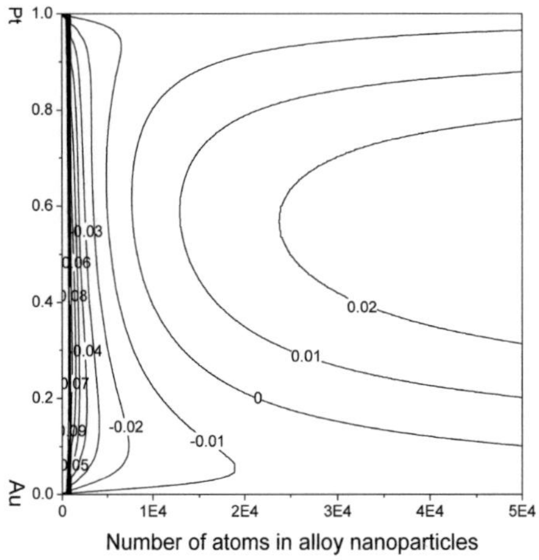

Fig. 28. The contour of formation heat for disordered Au-Pt nanoparticles as a function of alloy nanoparticle size and the chemical concentration of Pt atom. (Picture redrawn from Ref. 88)

As discussed above, the main difference between bulk material and nanoparticles is the surface effect of the latter. Figure 29 shows the changing of systematic surface area before and after alloying process under ideal condition (spherical nanoparticles without surface relaxation). Naturally, when the size of an alloy particle is fixed, there is a maximal difference of surface area for the alloying of two equal-volume monoatomic nanoparticles. Comparing Fig. 28 and 29, one can find that there is a maximal reduction in surface area after the alloying as Pt concentration is about 50%, but the formation heat is largest. This is because there exists a competition between surface effect and alloying effect on formation heat during alloying process for the immiscible nanoparticles.

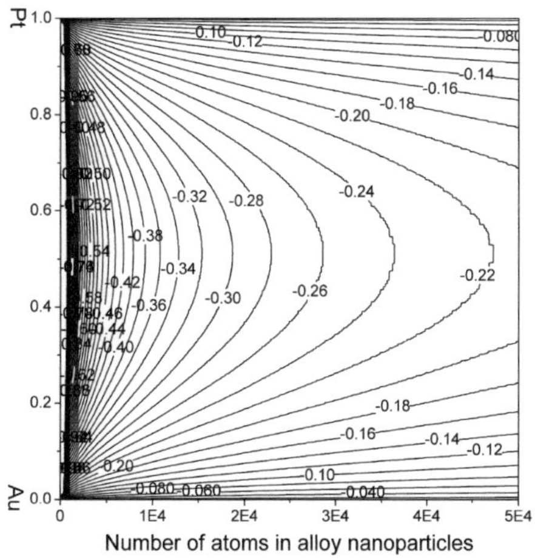

Fig. 29. The surface area changing contour of as a function of alloy nanoparticle size and the chemical concentration of Pt atom under the hypothesis of the disorered alloy obeying Vegard's law and the atoms in the interior of and on the surface of nanoparticle having the same volume. (Picture redrawn from Ref. 88)

In addition, since the Au has low surface energy (1.50 J/m^2) than that of Pt (2.48 J/m^2),[92] there is a thermodynamic driving force for Au atoms segregating to surface.[95] The segregation behavior in alloy nanoparticles generally induces the core-shell structure. Here we ignore the difference of structural details resulted by surface segregation. According to the effect of surface segregation being decreasing the systematic free energy, simply, a segregation factor f_{seg} is introduced to describe the changing of cohesive energy by surface segregation, i.e.

$$E_c^{bA-B}(Segregation) = f_{seg} \cdot E_c^{bA-B}(Ideal) \tag{21}$$

Figure 30 shows the variation of formation heat for Au-Pt alloy nanoparticles with different f_{seg}. Comparing with the formation heat of ideal alloy nanoparticles as shown in Fig. 28, the effect of surface segregation is extending the size range of alloy nanoparticles with negative formation heat. As the segregation factor f_{seg} increases from 1.001 to 1.008, the size of alloy nanoparticle, with negative formation heat in an entire composition range, increases from about 7 nm to 14 nm (number of atoms from 10^4 to 10^5).

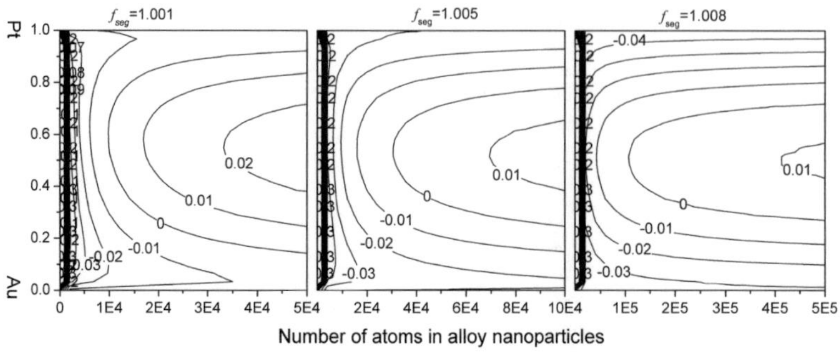

Fig. 30. The effect of surface segregation on the formation heat of alloy nanoparticles. (Picture redrawn from Ref. 88)

6. References

(1) H. Gleiter, Acta Mater. 2000, 48, 1.
(2) P. Pawlow, Z. Phys. Chem. 1909, 65, 1.
(3) M. Takagi, J. Phys. Soc. Jpn. 1954, 9, 359.
(4) F. Ding, K. Bolton, A. Rosen, J. Vac. Sci. Technol., 2004, A22, 1471.
(5) Y. Qi, T. Cagin,; W. L. Johnson, W. A. Goddard, J. Chem. Phys. 2001, 115, 385.
(6) P. Buffat, J. P. Borel, Phys Rev. 1976, A13, 2287.
(7) K. F. Peters, J. B. Cohen, Y. W. Chung, Phys. Rev. 1998, B57, 13430.
(8) C. L. Cleveland, W. D. Luedtke, U. Landman, Phys. Rev. Lett. 1998, 81, 2036.
(9) O. Gulseren, F. Ercolessi, E. Tosatti, Phys Rev. 1995, B51, 7377.
(10) K. K. Nanda, S. N. Sahu, S. N. Benera, Phys. Rev. 2002, A66, 013208.
(11) A. Safaei, M. A. Shandiz, S. Sanjabi, Z. H. Barber, J. Phys.: Condens. Matter 2007, 19, 216216.
(12) H. Reiss, P. Mirabel, R. L. Wetten, J. Phys. Chem. 1988, 92, 7241.
(13) J. G. Lee, H. Mori, Philos. Mag. 2004, 84, 2675.
(14) T. Tanaka, S. Hara, Z. Metallkd. 2001, 92, 467.
(15) J. B. Sturgeon, B. B. Laird, Phys Rev. 2000, B62, 14720.
(16) E. O. Arregui, M. Caro, A. Caro, Phys Rev. 2002, B66, 054201.
(17) W. H. Luo, W. Y. Hu, S. F. Xiao, J. Phys. Chem. 2008, C112, 2359
(18) H. Zhang, J. F. Banfield, J. Mater. Chem. 1998, 8, 2073.
(19) (a) J. W. Gibbs, In Collected Works; W. R. Longley, R. G. Van Name, Eds.; Longmans and Green: New York, 1931. (b) G. Wulff, Z. Kristallogr. 1901, 34, 449.
(20) H. Zhang, F. Huang, B. Gilbert, J. F. Banfield, J. Phys. Chem. 2003, B107, 13051.
(21) Z. H. Jin, K. Lu, Nanostruct. Mater. 1999, 12, 369.
(22) R. Hultgren, P. D. Desai, D. T. Hawkins, M. Gleiser, K. K. Kelly, D. D. Wagman, Selected Values of the Thermodynamic Properties of the Elements; American Society for Metals: Metals Park, OH, 1973.
(23) U. Hansen, P. Vogl, V. Fiorentini, Phys. Rev. 1999, B60, 5055.
(24) G. Grochola, S. P. Russo, I. K. Snook, I. Yarovsky, J. Chem. Phys. 2002, 116, 8547.
(25) G. Grochola, S. P. Russo, I. K. Snook, I. Yarovsky, J. Chem. Phys. 2002, 117, 7676.
(26) G. Grochola, S. P. Russo, I. K. Snook, I. Yarovsky, J. Chem. Phys. 2002, 117, 7685.

(27) L. Vitos, A. V. Ruban, H. L. Skriver, J. Kollar, Surf. Sci. 1998, 411, 186.
(28) W. R. W. A. Miller, Surf. Sci. 1977, 62, 267.
(29) V. K. Kumikov, K. B. Khokonov, J. Appl. Phys. 1983, 54, 1346.
(30) H. P. Bonzel, M. Nowicki, Phys Rev. 2004, B70, 245430.
(31) M. S. Daw, S. M. Foils, M. I. Baskes, Mater. Sci. Rep. 1993, 9, 251.
(32) (a) L. M. Holzman, J. B. Adams, S. M. Foils, W. N. G. Hitchon, J. Mater. Res. 1991, 6, 298. (b) B. Sadigh, G. Grimvall, Phys. Rev. 1996, B54, 15742. (c) M. Asta, D. Morgan, J. J. Hoyt, B. Sadigh, J. D. Althoff, D. de Fontaine, S. M. Foiles, Phys Rev. 1999, B59, 14271.
(33) L. E. Murr, Interfacial Phenomena in Metals and Alloys; Addison-Wesley: London, 1975.
(34) C. C. Yang, S. Li, Phys Rev. 2007, B75, 165413.
(35) F. Ercolessi, W. Andreoni, E. Tosatti, Phys. Rev. Lett. 1991, 66, 911.
(36) C. C. Yang, M. X. Xiao, W. Li, Q. Jiang, Solid State Commun. 2006, 139, 148.
(37) C. C. Yang, Q. Jiang, Acta Mater. 2005, 53, 3305.
(38) S. L. Lai, J. Y. Guo, V. Petrova, G. Ramanath, L. H. Allen, Phys. Rev. Lett. 1996, 77, 99.
(39) M. Y. Efremov, F. Schiettekatte, M. Zhang, E. A. Olson, A. T. Kwan, L. S. Berry, L. H. Allen, Phys. Rev. Lett. 2000, 85, 3560.
(40) E. A. Olson, M. Y. Efremov, M. Zhang, Z. Zhang, L. H. Allen, J. Appl. Phys. 2005, 97, 034304.
(41) V. N. Likhachev, G. A. Vinogradov, M. I. Alymov, Phys. Lett. A 2006, 357, 236.
(42) X. Li, J. J. Huang, Solid State Chem. 2003, 176, 234.
(43) H. Gleiter, Nanostruct. Mater. 1995, 6, 3.
(44) H. V. Atkinson, Acta Metall. 1988, 36, 469.
(45) P. G. Shewmon, Transformation in metals, McGraw-Hill, New York, 1969, pp. 300-306.
(46) A. J. Haslam, S. R. Phillpot, D. Wolf, D. Moldovan, H. Gleiter, Mater. Sci. Eng. 2001, A318, 293.
(47) N. Bernstein, Acta Mater. 2008, 56 1106.
(48) S. F. Xiao, W. Y. Hu, J. Crystal Growth 2006, 286,512
(49) K. E. Harris, V. V. Singh, A. H. King, Acta Mater. 1998, 46, 2623.
(50) D. Moldovan, V. Yamakov, D. Wolf, S. R. Phillpot, Phys. Rev. Lett. 2002, 89, 206101.
(51) P. Stoltze, J. Chem. Phys. 1990, 92, 6306.
(52) K. F. Peters, Y. W. Chung, J. B. Cohen, Appl. Phys. Lett. 1997, 71, 2391.
(53) J. F. Lutsko, D. Wolf, S. R. Phillpot, S. Yip, Phys. Rev. 1989, B40, 2841.
(54) W. Y. Hu, S. F. Xiao, J. Y. Yang, Z. Zhang, Eur. Phy. J. 2005, B45, 547
(55) H. W. Sheng, K. Lu, E. Ma, Nanostr. Mater. 1998, 10, 865.
(56) S. F. Xiao, W. Y. Hu, J. Y. Yang, J. Phys. Chem. 2005, B109, 20339
(57) S. F. Xiao, W. Y. Hu, J. Y. Yang, J. Chem. Phys. 2006, 125, 184504
(58) T. Nguyen, P. S. Ho, T. Kwok, C. Nitta, S.Yip, Phys. Rev. Lett. 1986, 57, 1919.
(59) W. H. Qi, M. P. Wang, Mater. Chem. Phys. 2004, 88, 280.
(60) L. E. Murr, Interfacial Phenomena in Metals and Alloys; Addison Wesley: New York, 1975.
(61)J. Schiotz, K. W. Jacobsen, Science 2003, 301, 1357.
(62)K. Lu, Mater. Sci. Eng. Rep. 1996, 16, 161.
(63)Y. H. Zhao, K. Lu, T. Liu, Phys. Rev. 1999, B59, 11117.
(64)P. M. Derlet, H. van Swygenhoven, Phys. Rev. 2003, B67, 014202.
(65) U. Tartaglino, T. Zykova-Timan, F. Ercolessi, and E. Tosatti, Phys. Rep. 2005, 411, 291.

(66) C. S. Hsu, A. Rahman, J. Chem. Phys. 1979, 79, 5234.
(67) D. Moroni, P. R. ten Wolde, P. G. Bolhuis, Phys. Rev. Lett. 2005, 94, 235703.
(68) B. O'Malley, I. Snook, Phys. Rev. Lett. 2003, 90, 085702.
(69) J. M. Leyssale, J. Delhommelle, C. Millot, J. Chem. Phys. 2005, 122, 184518.
(70) C. S. Liu, J. Xia, Z. G. Zhu, D. Y. Sun, J. Chem. Phys. 2001, 114, 7506.
(71) M. Matsumodo, S. Saito, I. Ohmine, Nature 2002, 416, 409.
(72) P. Beaucage, N. Mousseau, Phys. Rev. 2005, B71, 094102.
(73) J. M. Leyssale, J. Delhommelle, C. Millot, J. Chem. Phys. 2005, 122, 104510.
(74) S. F. Xiao, W. Y. Hu, J. Chem. Phys. 2006, 125, 014503
(75)W. A. Johnson, R. F. Mehl, Trans. Am. Inst. Min. Metall. Pet. Eng. 1939, 135, 416.
(76)M. Avrami, J. Chem. Phys. 1939, 7, 1103; 1940, 8, 212.
(77) H. S. Nam, Nong M. Hwang, B. D. Yu, J. K. Yoon, Phys. Rev. Lett. 2002, 89, 275502.
(78) Y. G. Chushak, L. S. Bartell, J. Phys. Chem. 2001, B105, 11605.
(79) G. Rossi, A. Rapallo, C. Mottet, A. Fortunelli, F. Baletto, R. Ferrando, Phys. Rev. Lett. 2004, 93, 105503.
(80) S. Darby, T. V. Mortimer-Jones, R. L. Johnston, C. Roberts, J. Chem. Phys. 2002, 116, 1536.
(81) A. Aguado, L. E. Gonzalez, J. M. Lopez, J. Phys. Chem. 2004, B108, 11722.
(82) C. Mottet, G. Rossi, F. Baletto, R. Ferrando, Phys. Rev. Lett. 2005, 95, 035501.
(83) S. Koutsopoulos, K. M. Eriksen, R. Fehrmann, J. Catal. 2006, 238, 270.
(84) F. Ding, A. Rosen, K. Bolton, Phys. Rev. 2004, B70, 075416.
(85) R. Birringer, Mater. Sci. Eng. 1989, A117, 33.
(86) J. Luo, M. M. Maye, V. Petkov, N. N. Kariuki, L. Wang, P. Njoki, D. Mott, Y. Lin, C. J. Zhong, Chem. Mater. 2005, 17, 3086.
(87)T. Shibata, B. A. Bunker, Z. Zhang, D. Meisel, C. F. Vardeman II, ; J. D. Gezelter, J. Am. Chem. Soc. 2002, 124, 11989.
(88) S. F. Xiao, W. Y. Hu, W. H. Luo, Y. R. Wu, X. F. Li, H. Q. Deng, Eur. Phys. J. 2006, B54, 479.
(89)W. H. Qi, M. P. Wang, J. Mater. Sci. Lett. 2002, 21, 1743.
(90)J. Y. Yang, W. Y. Hu, H. Q. Deng, D. L. Zhao, Surf. Sci. 2004, 572, 439.
(91)Y. R.Wu, W. Y. Hu, S. C. Han, J. Alloy Comp. 2006, 420, 83.
(92)R. A. Johnson, Phys. Rev. 1990, B41, 9717.
(93)I. A. Abrikosov, H. L. Skriver, Phys. Rev. 1993, B47, 16532.
(94)F. R. De Boer, R. Boom, W. C. M. Mattens, A. R. Miedema, A. N. Niessen, Cohesion in Metals; North-Holland: Amsterdam, 1988.
(95)H. J. Schaller, Z. Metallkde 1979, 70, 354.
(96)L. H. Liang, G. W. Yang, B. Li, J. Phys. Chem. 2005, B109, 16081.
(97)T. T. Tsong, Y. S. Ng, S. B. McLane, J. Chem. Phys. 1980, 73, 1464.

Linear and nonlinear optical properties of aligned elongated silver nanoparticles embedded in silica

Raul Rangel-Rojo[1], J.A. Reyes-Esqueda[2], C. Torres-Torres[3], A. Oliver[2], L. Rodríguez-Fernandez[2], A. Crespo-Sosa[2], J.C. Cheang-Wong[2], J. McCarthy[4], H.T. Bookey[4] and A.K. Kar[4]
[1]*Departamento de Optica, Centro de Investigación Científica y de Educación Superior de Ensenada, Apartado Postal 2732, Ensenada BC 22860, México.*
[2]*Instituto de Física, Universidad Nacional Autónoma de México, Circuito de la Investigación Científica S/N, Ciudad Universitaria, Distrito Federal, México.*
[3]*Sección de Estudios de Posgrado e Investigación ESIME-Zacatenco, Instituto Politécnico Nacional, México, D.F. 07738, Mexico*
[4]*School of Engineering and Physical Sciences, David Brewster Building, Heriot-Watt University, Edinburgh, EH14 4AS, Scotland UK.*

1. Introduction

Nanoestructured materials have attracted a considerable amount of attention in the past few years for their potential use in different applications, especially in the field of optics. Because of this, the optical properties of semiconductor (Yildirim and Bulutay, 2008), organic (Kasai *et al*, 1992), and metallic nanoparticles embedded in different media, have been thoroughly studied. For metallic nanoparticles in particular, this has been driven by the possibility of the use of their nonlinear optical properties in information processing applications, and more recently for the potential implementation of plasmonic circuitry (Barnes *et al*, 2003; Maier *et al*, 2003). Regarding their nonlinear optical properties, metallic nanoparticles embedded in dielectric matrices have shown considerably large nonlinearities with response times in the picosecond regime (Inouye *et al*, 2000). One of the most important features of this class of materials is the possibility of tuning their optical properties by manipulation of the particle size, shape, and the appropriate choice of the host matrix, which has been explored to a certain extent. There are however few reports of the generation of elongated nanoparticles, and even fewer reports of the study of their linear and nonlinear optical properties (Kyoung & Lee, 1999; Elim *et al*, 2006; Ruda & Shirk, 2007; Lamarre *et al*, 2008).
Among the many techniques employed to generate nanoparticles, metal ion implantation in glass substrates has proven to be a reliable technique for producing samples with well controlled characteristics. Recently, further high energy ion irradiation with different ions, has been shown to produce highly elongated metallic nanoparticles, with a prolate spheroidal shape (Oliver *et al*, 2006). Although the position of the nanoparticles produced

through this technique is random, the resulting spheroids are aligned in the direction of incidence of the second set of ions. For such a composite with aligned particles, a strong birrefringence of the linear and nonlinear optical properties can be expected.

In this work we present a novel technique for producing elongated silver nanoparticles embedded in a silica matrix, that are aligned in a preferential direction. This is achieved through a two-step ion implantation technique, and we present high resolution electron microscopy studies of the morphology of the resulting nanoparticles. We also present a characterization of the linear and nonlinear optical properties of these materials. The nanoparticles do not form an ordered array, but because they are elongated in a preferential direction, the composite is highly anisotropic. A complete study of the anisotropy is presented, and absorption spectra results show that the usual surface plasmon resonance found in these composite materials is split into two distinct bands, due to the elongation of the nanoparticles. A correlation between the observed position of the bands and the morphology of the particles is presented.

Regarding the nonlinear response of the material, the resonant nonlinear response to femtosecond and picosecond pulses is probed, employing the z-scan technique in the former case, and a vectorial four-wave mixing technique, in the latter. The observed response consists mainly of saturable absorption, which is shown to be also highly anisotropic. An analysis of the nonlinear response obtained as function of the polarization of the light employed is made, and the results are explained based on a simple model in the case of femtosecond pulses. For the picoseconds case, we discuss some insights about the influence of the hot-electrons on the nonlinear response.

2. Sample preparation and structural characterization

Elongated silver nanoparticles were synthesized through a double ion implantation process. In the first step 2-MeV Ag^{2+} ions are implanted into host matrices consisting of high -purity silica glass plates (20×20×1 mm^3) from NSG ED-C (Nippon silica glass), at room temperature. The impurity content of the silica plates was not greater than 1 ppm, with less than 1 ppm of OH, and a total impurity content of less than 200 ppm. The samples were then thermally annealed at 600° C in a 50%N_2+ 50% H_2 reducing atmosphere. The Ag-ion fluence and projected range, measured by Rutherford backscattering spectrometry (RBS) were 5 × 10^{16} Ag/cm^2 and 0.9 μm, respectively. Both the ion implantation process and the RBS analysis were performed at the 3 MV Tandem accelerator (NEC 9SDH-2 Pelletron) at UNAM. Optical absorption spectra of these samples showed that the result of this process was a film containing spherical-like silver nanoparticles, of around 6 nm in diameter.

Afterwards, the samples were cut into several pieces, and in order to induce deformation of the particles, each piece was irradiated at room temperature with 8 MeV Si ions at an angle of 45° with respect to the sample surface, with Si ion fluence values in the 0.1-2.3 × 10^{16} Si/cm^2 range. According to previous results (Cheang-Wong *et al* 2006) concerning the ion beam-induced deformation of silica particles, 8 MeV Si ions were chosen because their electronic stopping power for SiO_2 is 200 times larger than the nuclear one. On the other hand, the ion projected range for this energy is 4.3 μm in SiO_2, i.e. far beyond the location of the Ag nanoparticles. The result of the second ion implantation process was elongated nanoparticles aligned in the direction of the second ion implantation.

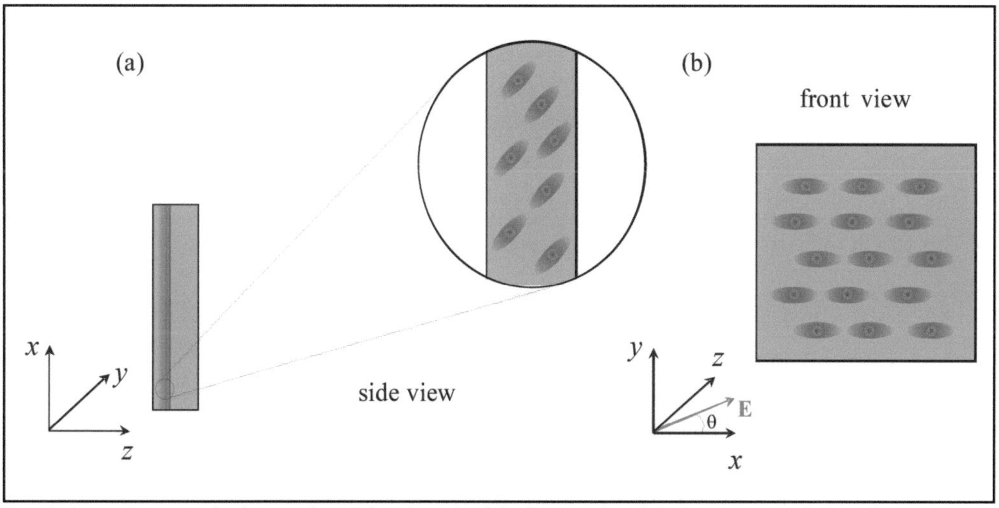

Fig. 1. Sample morphology, the side view in (a) shows the thin layer containing the elongated nanoparticles, which are aligned at 45° with respect to the normal to the surface, and (b) the front view showing the projection of the nanoparticles and the geometry chosen for the polarization of the incident light.

The resulting sample has a 0.5 μm thick layer containing the elongated nanoparticles, at a 1 μm depth inside the silica matrix, as shown schematically in fig. 1. The particles are aligned at 45° with respect to the substrate normal, as is also shown in fig. 1, and when viewed from the front, the projection of the nanoparticles long axes point in the direction we label as x. Evidence of the shape of the individual nanoparticles is obtained through HRTEM images of the sample, obtained with a 200 KV JEOL-2010FEG microscope in contrast-Z mode. Fig. 2 shows images taken for a sample with a 2.3×10^{16} Si/cm^2 fluence, where it can be seen that the randomly placed nanoparticles are aligned in the same direction, and that they are shaped as prolate spheroids, with an average minor axis diameter of 5 nm, and an aspect ratio of 1.7. The size distribution of the implanted sample was obtained from a digitized amplified micrograph by measuring the diameter of each nanoparticle. The size distribution obtained from the statistics over 290 measurements shows a diameter distribution centred at 5.9 nm with a standard deviation of 1.1 nm.

3. Linear optical properties. Anisotropic absorption and birrefringence

In order to characterize the linear optical properties of the samples produced by this double implantation technique, studies of the polarization dependence of the absorption spectrum were conducted, together with studies of the possible birefringence exhibited by the material.

3.1 Polarization dependent absorption spectra

An Ocean Optics Dual Channel SD2000 UV-visible spectrophotometer was used to collect the optical absorption spectra, using linearly polarized light with different polarization orientations. At the beginning, the absorption spectra had a constant 'baseline' which was attributed to the presence of a superficial layer of carbon that deposited on top of the sample in the implantation process. This layer was removed by further baking the samples at 400° C for 5 hours. Fig. 3 shows the absorption spectra of the sample after this process, recorded at normal incidence and using two mutually orthogonal polarizations, at *0º* and *90º* which are roughly parallel (labeled $E_{\parallel}$), and perpendicular (labeled $E_{\perp}$) to the long axis of the nanoparticles respectively. Two different absorption bands are clearly discerned from the spectra, one centered at *365 nm* for the spectrum taken with the $E_{\perp}$ polarization, and a broader one at 517 nm for the $E_{\parallel}$ polarization.

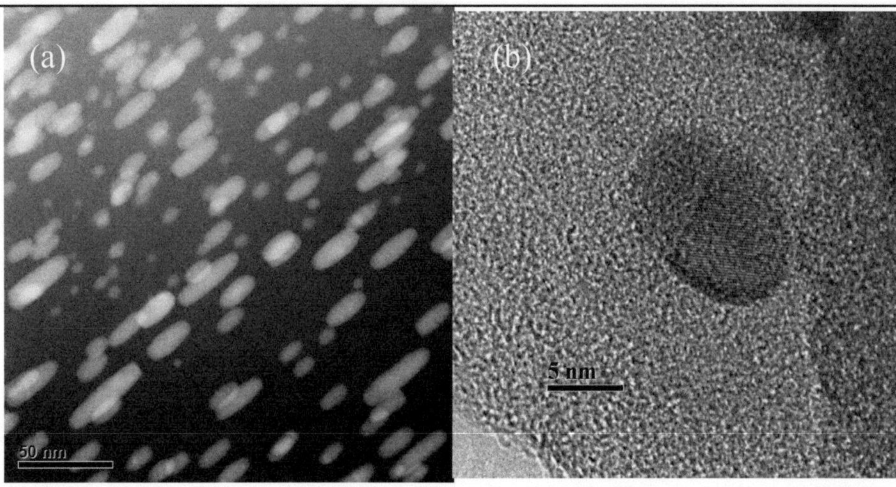

Fig. 2. High resolution TEM micrographs of the composite film: a) Z contrast (HAADF) image showing the Ag nanoparticles deformed by Si ion irradiation, obtained with a TEM, at 200 KV, with a point to point resolution of 0.19 nm. b) HRTEM micrograph of a deformed nanoparticle.

These peaks correspond to the different surface plasmon resonances for each polarization, as has been probed by studying the linear birrefringence of the sample (Reyes-Esqueda *et al*, 2008). From Fig. 3 it is easily seen that for the $E_{\parallel}$ polarization there is a remnant of the *365 nm* peak, this can be due to a residual misalignment with respect to the direction of elongation of the particles, or to a fraction of the nanoparticles remaining spherical after the second ion-implantation process.

Absorption spectra were also taken at intermediate polarization angles. Fig. 4 shows the values of the linear absorption coefficient extracted from the optical density (*OD*) measured at *527 nm* as function of the polarization angle θ (α=*OD* $\ln 10/L$ with L the thickness of the layer containing the particles, which was taken to be *0.5 μm*). This is the wavelength of the laser source employed in the z-scan measurements. For a collection of perfectly aligned anisotropic particles, the linear absorption coefficient $\alpha(\theta)$ can be written as (Boyd, 1992):

$$\alpha(\theta) = (\alpha_0 - \alpha_{\pi/2})\cos^2\theta + \alpha_{\pi/2}, \tag{1}$$

where α_0 is the linear absorption coefficient for $\theta=0^o$ ($E_{\parallel}$ polarization), and $\alpha_{\pi/2}$ is the one corresponding for $\theta=90^o$ ($E_{\perp}$ polarization). Fig. 4 shows a fit to the experimental values of OD using expression (1) $\alpha_0 = 2.76 \times 10^4\ cm^{-1}$, and $\alpha_{\pi/2} = 5066\ cm^{-1}$. In this way, we can see that for a given wavelength, the linear absorption observed is strongly dependent on the polarization of the light. This anisotropic linear absorption is important for the characterization of the linear birefringence of the material, and for their nonlinear optical properties as well.

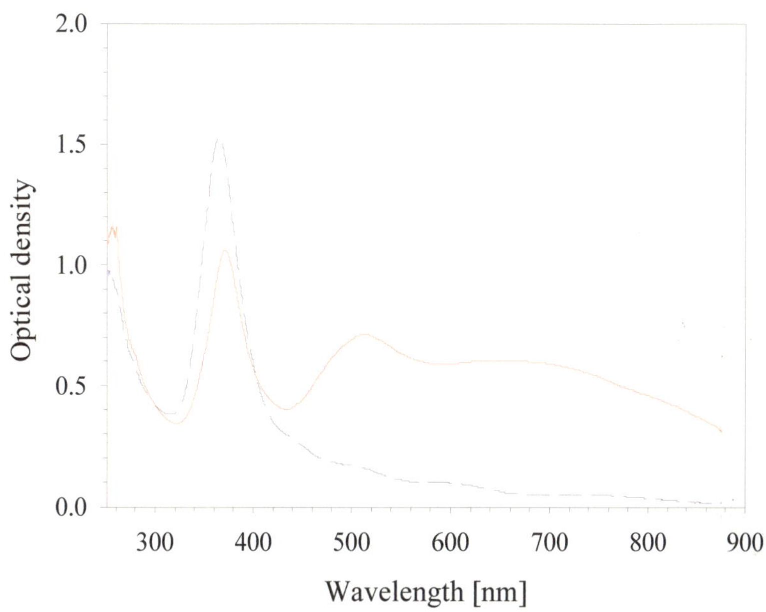

Fig. 3. Absorption spectra of the prolate spheroidal silver nanoparticle sample taken for mutually orthogonal linear polarizations, the continuous line corresponds to a polarization angle $\theta=0^o$ ($E_{\parallel}$), and the dotted line to $\theta=90^o$ ($E_{\perp}$).

3.2 Birrefringence measurements

In order to analyze the optical anisotropy (or birefringence) of the sample, we have used an ellipsometric technique where we measure the light transmission through our samples when placed and rotated between crossed and parallel polarizers. The experimental setup shown in Fig. 5(a) was used to measure the birefringence, $\Delta n^{\alpha}=n_p(\psi)-n_s$, experienced by the probe beam when traversing the sample. The subscripts s and p in the expression refer to the linear optical eigenpolarization components for the probe beam, when traveling in the direction **k** (normal to the surface of the sample, $\hat{N}_{sample}$) at an internal angle of propagation ψ with respect to the normal of the deformed nanoparticle, $\hat{N}_{NP}$, as shown in Fig. 5(b). The corresponding refraction indices are n_s and n_p, and the superscript α refers to the angle between the incident electric field and the *x* axis. The linearly polarized incident light may contain both linear eigenpolarizations, while the analyzer may be oriented parallel or crossed to transmit the linear polarization state parallel or orthogonal to the incident light, respectively. The complete mathematical details of the analysis for these ellipsometric measurements are given in (Gonzalez *et al,* 2008), here we only present some of the most important expressions necessary to analyze the experimental data.

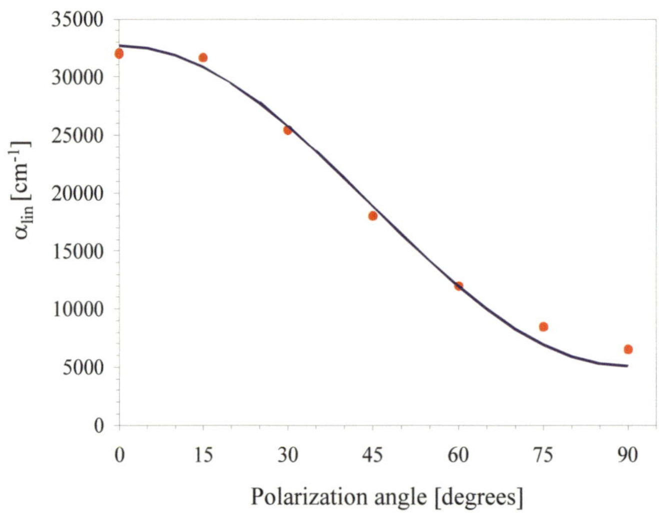

Fig. 4. Dependence of the linear absorption coefficient α_{lin} of the sample measured as a function of polarization angle θ. The dots represent the experimental points, while the line represents a fit made using expression (1).

In the coordinate system shown in Fig. 1(b), the normal to the nanoparticle can be represented by the vector

$$\hat{N}_{NP} = \begin{pmatrix} \sin\psi \\ 0 \\ \cos\psi \end{pmatrix}. \tag{2}$$

In consequence, the associated normalized eigenpolarization vectors will be [18]:

$$\hat{s} = \begin{pmatrix} 0 \\ 1 \\ 0 \end{pmatrix}, \hat{p} = \begin{pmatrix} 1 \\ 0 \\ 0 \end{pmatrix}, \tag{3}$$

and, since the probe beam propagates parallel to the normal of the macroscopic sample, $k \parallel \hat{N}_{sample}$, which makes an angle ψ_1=49° with the normal to the nanoparticle, one can suppose that light is refracted just until it arrives to the surface of the nanoparticle according to Snell's law $\sin\psi = n_{host} \sin\psi_1 / n_{NP}^{eff}$. This angle ψ_1 between both normals is the same angle between the major axis of the nanoparticle and the surface of the sample, and it is a consequence of the Si irradiation as explained further below.

On the other hand, the orientation of the incident electric field $\boldsymbol{E}_{in}$, will be determined by the polarizer orientation, which allows resolving it into the two eigenpolarization components of the optically anisotropic nanocomposite, $E_s = \hat{E}_{in} \cdot \hat{s}$, and $E_p = \hat{E}_{in} \cdot \hat{p}$, where $\hat{E}_{in} = \begin{pmatrix} \cos\alpha \\ \sin\alpha \\ 0 \end{pmatrix}$ is the unitary incident electric field and α is the angle of the polarizer (measured from x). Therefore, the birefringence for such a light path will be $\Delta n^{\alpha} = n_p(\psi) - n_s$, where the p polarization will experience a refraction index n_p, while the s polarization will experience n_s. In order to be able to relate it to the measured intensity, we define complex transmission factors for these two eigenpolarizations, which correspond to the transmitted electric field, as

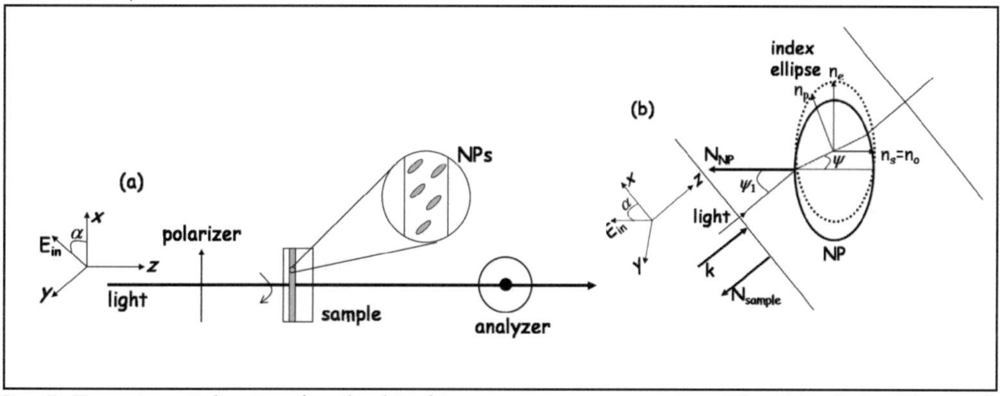

Fig. 5. Experimental setup for the birefringence measurements with white light; $\boldsymbol{E}_{in}$ stands for the incident electric field.

$$E'_s(\alpha) = A_s \exp\left(\frac{-i\pi L \Delta n^{\alpha}}{\lambda}\right) E_s \hat{s}, \text{ and } E'_p(\alpha) = A_p \exp\left(\frac{i\pi L \Delta n^{\alpha}}{\lambda}\right) E_p \hat{p}, \quad (4)$$

where A_s and A_p are the measured amplitude transmission factors for each eigenpolarization, L is the interaction length, *i.e.* the thickness of the NPs layer, and λ is the free-space incident wavelength. For the transmitted light, when the axes of the polarizer-analyzer system are aligned, it can be shown (Gonzalez *et al*, 2008) that the detected intensity at a given in-plane polarization α is given by

$$I(\alpha,0) = A_s^{\ 2} \sin^4 \alpha + A_p^{\ 2} \sin^4 \alpha + \frac{1}{2} A_s A_p \sin^2 2\alpha \cos \frac{2\pi L \Delta n^{\alpha}}{\lambda}, \tag{5}$$

while when they are crossed, the detected intensity is given by

$$I(\alpha,\frac{\pi}{2}) = \frac{1}{4}\sin^2 2\alpha \left[A_s^{\ 2} + A_p^{\ 2} - 2 A_s A_p \cos \frac{2\pi L \Delta n^{\alpha}}{\lambda} \right]. \tag{6}$$

The experiment consists then of measuring these two intensities, and extracting the values of A_p^2, A_s^2, and of the birefringence of the nanocomposite from the experimental results.
We have performed the birefringence measurements at 532 and 355 nm, which are close to the surface plasmon resonances associated with the major (470 nm) and the minor (375 nm) axes of the oriented prolate-spheroid nanoparticles, as discussed above. Fig. 6 shows a typical measurement at 532 nm. For α=0 and $\alpha = \pi/2$, we obtain A_p^2, A_s^2 respectively, from Eq. (5). On the other hand, from Eq. (6) with $\alpha = \pi/4$, we get the maximum measured birefringence as

$$\Delta n^{\max} = \frac{\lambda}{2\pi L} \cos^{-1} \left[\frac{A_s^{\ 2} + A_p^{\ 2}}{2 A_s A_p} - \frac{2 I_{meas}^{\ \max}}{A_s A_p} \right]. \tag{7}$$

Now, given the nanoparticles geometry and their orientation, the uniaxial symmetry of the nanocomposite allows describing its refractive-index anisotropy by the ellipse shown in Fig. 5(b) with principal axes n_e and n_o, *i.e.* the extraordinary and ordinary indices, respectively. For a refracted light path at an angle ψ relative to the normal of the nanoparticles, we have $[1/n_p^2(\psi)]=\sin^2\psi/n_o^2+\cos^2\psi/n_e^2$ for the *p*-polarized light (Saleh & Teich, 1991) , whereas the *s*-polarized light sees an ordinary index n_o. Finally, the refractive-index anisotropy of the nanocomposite will be related in a first approximation to the measured birefringence by

$$n_e - n_o = \frac{\Delta n^{\max}}{\cos^2 \psi} = \frac{\Delta n^{\max}}{\cos^2 \left[\sin^{-1} \left(\frac{n_{host}}{n_{NP}^{eff}} \sin \psi_1 \right) \right]}. \tag{8}$$

This previous analysis has been made under the supposition that the wavelength used to perform the measurements, 532 nm, is close to the surface plasmon resonance associated with the major axis of the oriented prolate nanoparticles (at 470 nm). In such a case, it is rather clear that the index ellipse superposes with the nanoparticle as shown in Fig. 5(b). However, when exciting its minor axis, the index ellipsoid is perpendicular to the major axis of the particle. Nevertheless, the analysis is exactly the same, although the respective coordinate system (x',y',z') is rotated by $\pi/2$ around the z axis with respect to the previous one shown in Fig. 5(b). This means that the new angle α' for this last case is related to the angle α with respect to the former (x,y,z) system by $\alpha = \pi/2 - \alpha'$, and since we used the same setup for both wavelengths, in order to model our results properly, we have to substitute α' for α. This substitution only affects Eq. (5), transforming it into

$$I(\alpha',0) = I(\alpha,0) = A_s^2 \sin^4 \alpha + A_p^2 \sin^4 \alpha + \frac{1}{2} A_s A_p \sin^2 2\alpha \cos \frac{2\pi L \Delta n^{\alpha'}}{\lambda}, \tag{9}$$

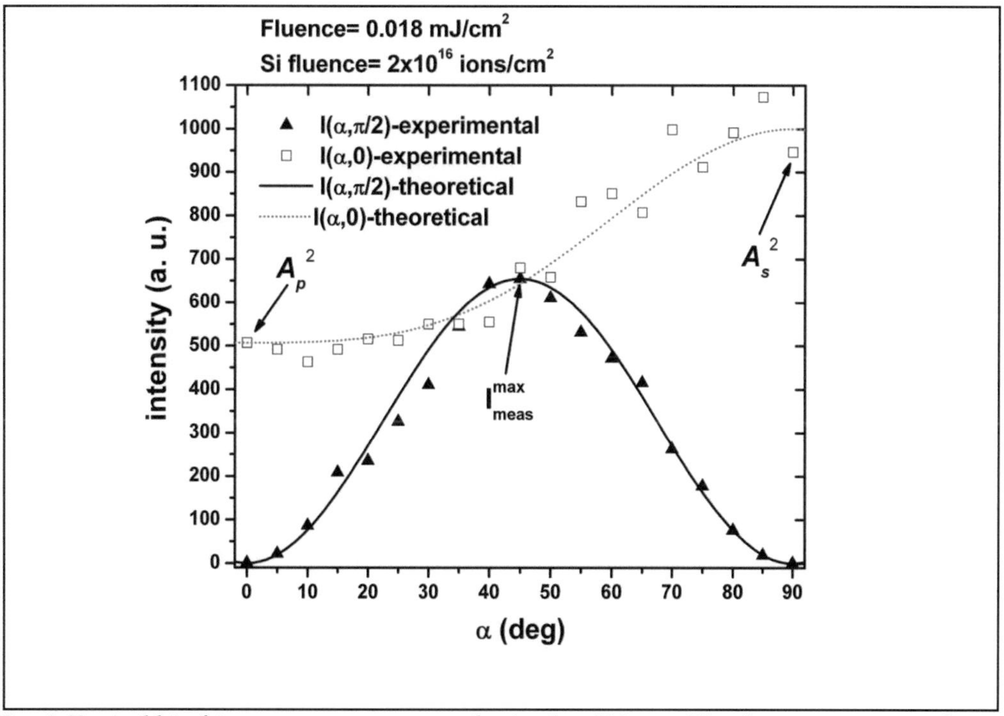

Fig. 6. Typical birefringence measurement obtained at 532 nm. The data were averaged over three consecutive measurements, taken within a given window of stability for the laser system by using a reference beam. Solid curves are the theoretical calculations given for Eqs. (5) and (6) by taking the birefringence calculated with Eq. (7).

since sin α'=cos α, cos α'=sin α, and sin 2 α'=sin 2 α. In consequence, for 355 nm, the typical measurement is a reflection of the one shown in Fig. 6 with respect to a vertical axis located at $\pi/4$. The other difference is that, now, for α=0, Eq. (9) gives A_s^2, while for $\alpha=\pi/2$, it gives A_p^2, while Δn^{max} is obtained again from Eq. (7).

Aside from the strongly polarization dependent absorption shown in section **3.1**, when placed between polarizers, there is considerable dichroism, that is, a selective absorption of one of the orthogonal components of the linear polarization of the incident beam (Saleh & Teich, 1991). By looking at Fig. 6, we can see that the perpendicular intensity is zero for α=0 and $\pi/2$, and nonzero in the middle, with a maximum at $\pi/4$. The performance of the nanocomposite is quite impressive since it is totally opaque when its optic axis is aligned with the polarizer or the analyzer, and highly transparent when is oriented at 45° with respect to them. This optic axis corresponds to the major axis of the nanoparticle for 532 nm and to its minor axis for 355 nm. This behavior can be qualitatively understood for 532 nm as follows: when α=0 and $\alpha=\pi/2$, the transmitted electric field is crossed to the analyzer and then totally filtered. On the contrary, for α differing from these two values, only the component of the incident electric field parallel to the major axis is absorbed, and the resulting transmitted field is not orthogonal with respect to the analyzer, obtaining a nonzero intensity measurement, which is maximum for $\alpha = \pi/4$. The same happens for 355 nm. We can quantify first this performance by looking at the extinction ratio defined as

$$\text{Extinction ratio} = 10\log_{10}\left(T_{\perp} / T_{||} \right) = 10\{abs(A_{\perp} - A_{||})\} \ [dB], \tag{10}$$

which is a typical figure of merit for telecommunications. The extinction ratio for the sample with Ag nanoparticles deformed with a dose of 0.5×10^{16} *Si/cm*^{2} is around *15 dB* for the resonance associated with the major axis of the nanoparticles and *10 dB* for the resonance associated with the minor axis, while similar values are obtained for the other samples. These values are close to 20 dB, which is a typical value at *1.55 μm,* indicating a very good performance in the visible region. For the anisotropic systems described in (Künzner *et al,* 2001; Genereux *et al,* 2001; Muskens *et al,* 2006), the corresponding extinction ratios were estimated to be around *17, 9* and *40 dB*, respectively, with corresponding thicknesses of around *100, 235* and *15 μm,* respectively.

On the other hand, by using our previous analysis, we have measured the birefringence of the anisotropic nanocomposites, where we have taken the interaction length as the FWHM of the Gaussian distribution of the implanted and deformed nanoparticles, which have a value of L=454 nm and an error of around 8%, as measured by RBS. The results are $\psi_1=49\pm0.5$ *deg*, $n_{host}=n_{SiO2}$=1.47, n_{NP}^{eff}(532nm)=|N_{NP} *(532nm)*|=3.32, and n_{NP}^{eff}(355nm)=|N_{NP} *(335nm)*| =1.42, these two last values were taken from (Johnson & Christy,1972). As we have already made clear in (Oliver *et al,* 2006), the nanoparticles absorb energy through the excitation of their resonances only, and, given the size and scattered distribution of the nanoparticles into the SiO_2 matrix, radiation effects and interaction between nanoparticles (or coupling between surface plasmon resonances) are negligible.

In Table 1 we present the measured birefringence for our samples as a function of the Si fluence in the second ion implantation process, for 355 and 532 nm, as well as their refractive-index anisotropy. The aspect ratio of our deformed nanoparticles is around 1.6, which indicates a rather small shape anisotropy. However, the macroscopic birefringence of these samples is very large, as a matter of fact, practically comparable to that measured for other nanostructured semiconductor materials reported in (Künzner *et al,* 2001; Genereux *et al,* 2001; Muskens *et al,* 2006). Furthermore, when increasing the Si fluence, the nanoparticles aspect ratio increases from 1.58 to 1.69. Within the context of this paper, this means that there is an increment of their shape anisotropy, and therefore of their optical birefringence, as it is corroborated from Table 1, principally for 532 nm. This is rather difficult to see at 355 nm, since it is known that the resonance corresponding to the minor axis is less sensitive to the changes of the aspect ratio of the deformation than the resonance of the major axis. In fact, the position of this last is proportional to the aspect ratio, while the resonance of the minor axis is inversely proportional to it (Noguez, 2007). Therefore, we could expect very similar values of the birefringence measured at 355 nm for the minor axis for different values of the Si fluence. This is more evident for the highest fluence, due to the large uncertainty measured in such a case.

Si fluence [ions/cm²]	aspect ratio	Δn^{max} (532 nm)	Δn^{max} (335 nm)	n_e-n_o (532 nm)	n_e-n_o (335 nm)
0.0	1.00	0.000	0.000	0.000	0.000
0.1	1.58	0.085±0.011	0.018±0.005	0.096±0.011	0.045±0.006
0.5	1.62	0.112±0.013	0.025±0.006	0.126±0.013	0.063±0.008
1.5	1.65	0.111±0.014	0.036±0.008	0.125±0.014	0.092±0.010
2.0	1.69	0.135±0.024	0.017±0.014	0.152±0.024	0.043±0.015

Table 1. Measured birefringence of the anisotropic silver nanoparticles with the corresponding measurement propagated uncertainties.

On the contrary, if we perform the birefringence measurements on a pure SiO_2 matrix or one with embedded spherical-like nanoparticles, no birefringence is detected. Similarly, for the deformed nanoparticles, when the wavelength used is 1064 nm, a null birefringence is detected again. These results allow us to conclude that the observed birefringence is only due to the silica embedded layer of deformed and aligned Ag nanoparticles.

Although the reported birefringence reported in (Reyes-Esqueda *et al,* 2008) is already larger than that measured in naturally anisotropic crystals (around ten times that of quartz), we believe that it can be further enhanced by controlling more precisely the morphology of the nanoparticles. Similarly, by using deformed Au or Cu nanoparticles instead of Ag, the nanocomposite would have the similar surface plasmon resonances but placed at different wavelengths, which gives other choices where to observe dichroism (Hoa *et al,* 2007). The totality of these results offers a new means of engineering highly birefringent materials on a nanoscale. These birefringent nanocomposites could be used to create a broad array of photonic integrated nanodevices, including waveplates, polarization rotators and beamsplitters; since they are very thin (0.5 μm) when compared to similar ones like those in (Künzner *et al,* 2001; Genereux *et al,* 2001; Muskens *et al,* 2006; Seward *et al,* 1974; Hasui *et al,* 2000; Matsuda *et al,* 2005). We can also remark that ion implantation and deformation of metallic nanoparticles allow obtaining a nanocomposite with a given organization of the nanoparticles therein and preventing their oxidation. On the other hand, chemical methods offer a wide variety of shapes and sizes of metallic nanoparticles, but organizing them into a matrix is not a trivial matter. A very ready-to-hand challenge is the combination of different methods to obtain an application tailored nanocomposite (Pérez-Juste *et al,* 2005).

4. Anisotropic nonlinear absorption with femtosecond pulses.

The nonlinear optical properties of the nano particle containing sample were probed using femtosecond pulses in the resonant regime. By using femtosecond pulses, we made sure that any thermal effects on the nonlinearity would be complete minimized (Rangel-Rojo *et al,* 2009). By going to the resonant regime, in this case with one of the surface plasmon resonances of the composite, we expect a large nonlinearity and anisotropy to be found. The z-scan technique was chosen since it is very sensitive to refractive index and absorption changes induced by light.

The light source employed in the experiments was an OPA (model Spectra Physics OPA-800) which was pumped by a regeneratively amplified Ti:Sapphire laser (Spectra Physics Spitfire) emitting 1 mJ pulses at a 800 nm wavelength with a repetition rate of 1 kHz, and a pulse width of 70 fs. The signal wavelength of the OPA was tuned to around 1290 nm, and the idler output oscillating at 2.1 μm was quadrupled in frequency by two consecutive second harmonic crystals to yield pulses at a 527 nm wavelength. The standard open-aperture z-scan set-up was used to study the nonlinear absorption of the sample, using a lens with focal length f=20 cm to focus the 233 fs pulses at 527 nm into the sample, resulting on a beam waist w=33 μm.

Fig. 7 shows experimental results obtained for *150 nJ* pulses, and two orthogonal polarizations. The data for $\theta = 0^o$ polarization shows the signature of saturable absorption, *i.e.* increased transmittance with higher irradiance, while the one at $\theta = 90^o$ shows no discernible change, indicating a much weaker nonlinearity, that is, a higher saturation irradiance I_s for this polarization. The nonlinear absorption of the sample is thus highly anisotropic, so we make a study of the nonlinearity for different input polarization angles and pulse energies to fully characterize the response. For all the input polarizations and pulse energies studied, the open aperture results showed a saturating nonlinearity, which can be modelled by an intensity dependent absorption coefficient $\alpha(I)$ given by:

$$\alpha(I) = \frac{\alpha_{lin}}{1 + I / I_s}, \tag{11}$$

where α_{lin} is the linear absorption coefficient, and I_s is the saturation irradiance. This model nonlinear absorption corresponds for example to a two-level system near resonance.

The z-scan results were analyzed using expression (11) to calculate transmission through the sample (by solving $dI/dz' = -\alpha(I)\, I$, integrated from $z'=0$ to $z'=L$, the sample thickness), and the input irradiance was considered as the gaussian at each sample position z, as described in references (Rangel-Rojo *et al*, 1995; Rangel-Rojo, *et al*, 1998) for example. Fig. 8 shows the experimental results obtained for a $\theta = 0^o$ input polarization, and a pulse energy of *30 nJ*. Also shown is the fit made using the procedure previously described. The same procedure was performed with all the z-scan traces obtained at the different polarization angles employed, and using the results at the lowest pulse energies where an effect was clearly seen. We used the results for the lowest energies possible, in order to make sure that the nonlinear absorption approximates to that in expression (11) as much as possible. For high irradiance values the nonlinear absorption can deviate considerable from that of a two-level saturable absorber, as it has been shown for other materials (Rangel-Rojo *et al*, 1998).

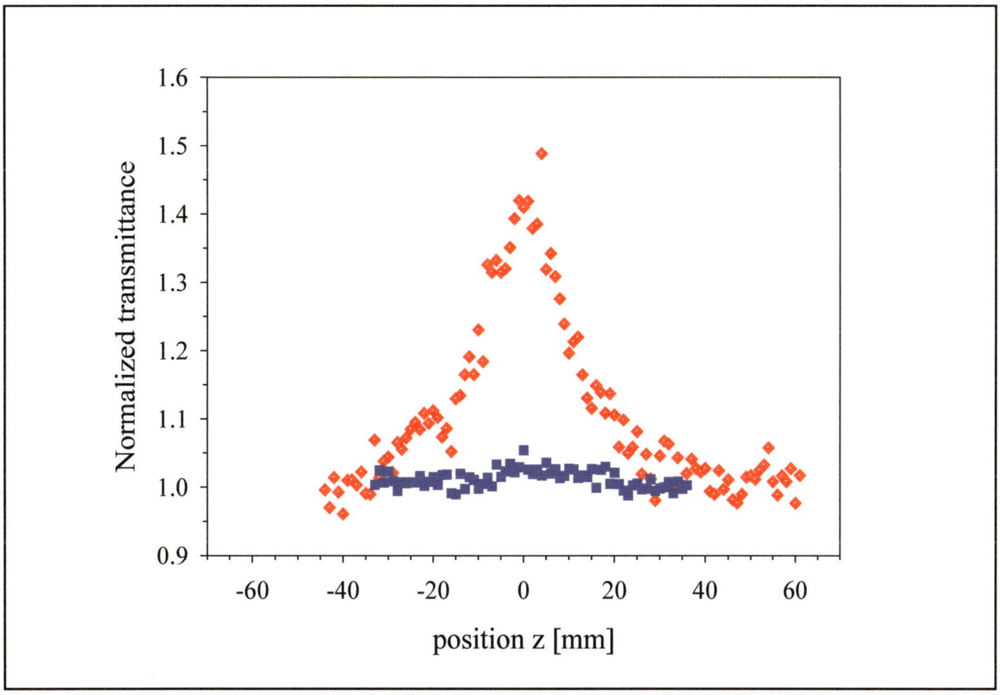

Fig. 7. Open aperture z-scan results for the prolate spheroidal nanoparticle sample with fs pulses at 527 nm. The results are shown for a pulse energy of 150 nJ, and for two different input polarizations, which are labelled $\theta=0^o$ (diamonds), and $\theta=90^o$ (squares).

For each input polarization angle θ, the fit to the open z-scan results yields a value for I_s. Fig. 9 shows the values of I_s extracted from the fits as a function of θ. The figure shows a rapidly growing I_s value for θ going to 90°. In fact, pulse energies as high as *300 nJ* were needed for θ =90° to actually obtain a measurable effect. In order to understand this dependence we notice that for a two-level saturable absorber, the saturation irradiance is given by:

$$I_s = \frac{\hbar\omega}{\sigma\tau} \tag{12}$$

where $\hbar\omega$ is the photon energy, σ is the absorption cross section of the transition, and τ is the lifetime of the excited state. Using the fact that the absorption coefficient can be written as $\alpha_{lin}= \sigma N$ with N the number density of the particles, together with expression (1) for $\alpha(\theta)$, we can write the polarization-angle dependent saturation irradiance $I_s(\theta)$ as:

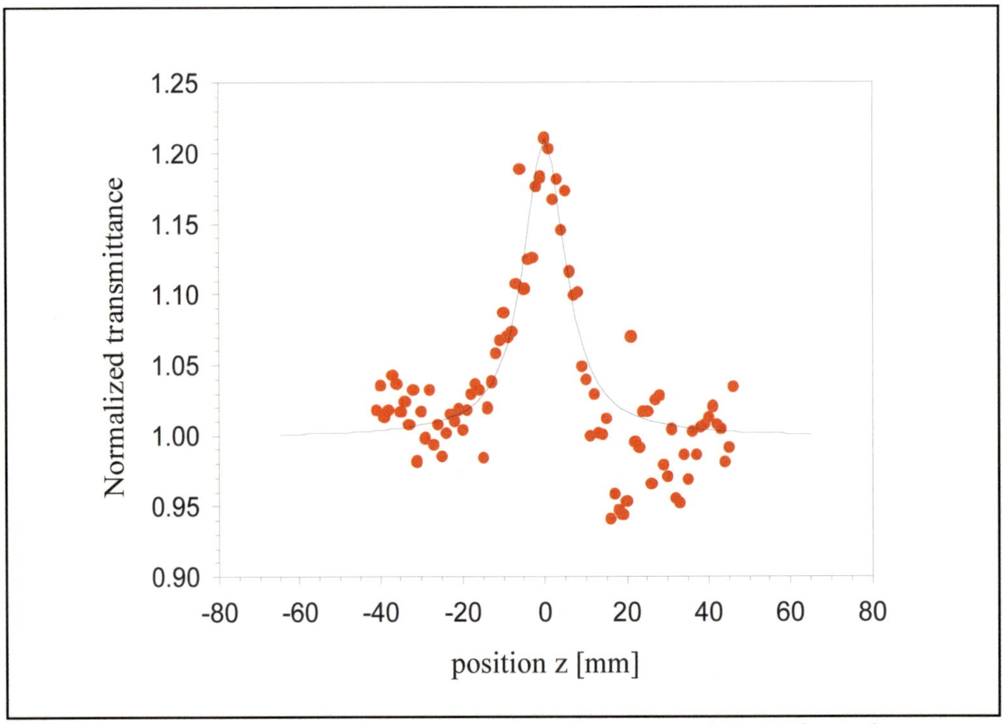

Fig. 8. Z-scan result for a $\theta=0^{o}$ polarization, and a pulse energy of *30 nJ*. The circles represent the experimental data, while the continuous line represents the fit made using the procedure described in the text.

$$I_s(\theta)=\frac{N\hbar\omega}{\tau[(\alpha_0-\alpha_{\pi/2})\cos^2\theta+\alpha_{\pi/2}]}. \tag{13}$$

Fig. 9 shows the fit obtained using this expression, together with the data extracted from the z-scan results. Given the fit for α (θ) shown in Fig. 4, the only fitting parameter is the constant $N\hbar\omega/\tau$. As it can be seen from the figure, the fit reproduces the observed angle dependence of the saturation irradiance reasonably well, for a $N\hbar\omega/\tau = 9.5 \times 10^5$ *GW/cm*^{3} value. To get a better fit, it is probably necessary to consider a more realistic nonlinear absorption model, such as the three-level model described in (Rangel-Rojo *et al*, 1998).

Although a saturating nonlinearity is not strictly a third order one, for small values of I/I_s expression (11) can be approximated as:

$$\alpha(I)=\frac{\alpha_{lin}}{1+I/I_s}\approx\alpha_{lin}\left[1-\frac{I}{I_s}\right]. \tag{14}$$

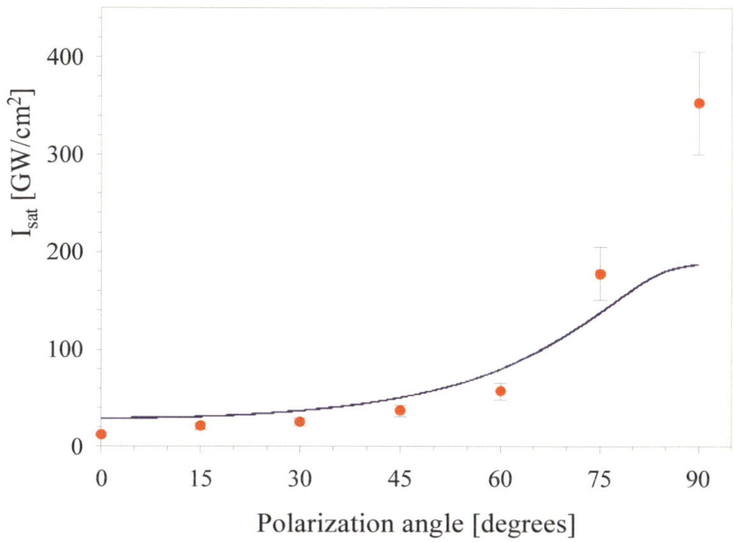

Fig. 9. Saturation irradiance extracted from the z-scan data as a function of polarization angle θ. The filled circles represent the I_s values extracted from the z-scan data, and the line represents a fit made using expression (4).

When this is compared with the usual expression for a third-order nonlinearity, $\alpha(I) = \alpha_{lin} + \beta I$, with β the two-photon absorption coefficient, we can make $\beta = -\alpha_{lin}/I_s$. Since β is related to $Im\ \chi^{(3)}$ through $Im\ \chi^{(3)} = \lambda \varepsilon_0 n_0^2 c\beta / 4\pi$ (in SI units), we can use the fitted $I_s(\theta)$ values to calculate angle dependent β and $Im\ \chi^{(3)}$ values, which are contained in table 2. From the table, it can be seen that $|Im\ \chi^{(3)}|$ takes values as large as $6.55 \times 10^{-18}\ m^2/V^2$ (4.7×10^{-10} *esu*).

θ [degrees]	α_{lin} [cm^{-1}]×10^4	I_s [GW/cm^2]	β [cm/GW]	Im $\chi^{(3)}$ [m^2/V^2]×10^{-18}
0	3.2	12.2	-2610	-6.55
15	3.16	21.0	-1510	-3.78
30	2.54	25.6	-994	-2.49
45	1.8	36.7	-489	-1.23
60	1.2	56.9	-211	-0.528
75	0.844	178	-47.5	-0.1119
90	0.648	353	-18.4	-0.046

Table 2. Fitted nonlinear optical coefficients.

We have shown results for the nonlinear absorption observed in a silica substrate containing a layer of aligned prolate spheroidal silver nanoparticles. The nonlinear absorption mechanism is saturable absorption from the broad surface plasmon resonance band, and we have found it to be highly anisotropic. We have also shown that the dependence of the linear absorption, and saturation irradiance with the polarization angle observed, can be reasonably well explained by a simple two-level saturation model.

5. Anisotropic nonlinear response to picosecond pulses. Self-diffraction studies.

In order to see how the nonlinear response would change for longer pulse durations, we also performed nonlinear optical measurements in the picosecond regime (Rodríguez-Iglesias et al, 2009; Reyes-Esqueda et al, 2009), using for these experiments a self-diffraction technique, together with 'P-scan' measurements of the transmitted intensities as a function of the incident one. The self-diffraction technique has proven to be very sensitive for measuring absorptive and refractive index changes induced by light (Torres-Torres et al, 2009). The advantages of this technique are related to its simplicity for obtaining absorptive and refractive coefficients, and for the possibility to identify the physical mechanisms responsible of the measured nonlinear absorption and refraction at the same time (Torres-Torres et al, 2008a). The main disadvantage is that sophisticated techniques of noise elimination could be required if the nonlinear response is small (Torres-Torres et al, 2008b). Therefore, in this section, we report the use of self-diffraction and P-scan (Banerjee et al, 1998) methods at 532 nm and 26 ps, to resolve the real and the imaginary parts of the linear independent components of $\chi^{(3)}$ for the anisotropic Ag nanocomposites. For nonlinear refraction, we obtained a pure electronic and self-focusing response, while for nonlinear absorption, as in the femtosecond regime, we observed saturable absorption.

5.1 Third order nonlinear polarization for anisotropic metallic nanocomposites

Since we are dealing with anisotropic systems, it is necessary to take into consideration the full tensorial character of the nonlinear response characterized by the third order nonlinear macroscopic susceptibility tensor (Reyes-Esqueda et al, 2009). In particular, we will write the proper contributions to the third order nonlinear polarization for metallic nanocomposites showing uniaxial symmetry, when measuring by means of a totally degenerate wave mixing setup. Such an expression will put this polarization in terms of the nonzero, independent components of the macroscopic susceptibility tensor and, most importantly, in terms of the angular position of the composite.

According to (Sutherland, 1996), the third order nonlinear polarization is written in general as

$$\mathbf{P}^{(3)} = \boldsymbol{\chi}^{(3)} \vdots \mathbf{EEE}\,, \tag{15}$$

which can be written for each Cartesian component as

$$P_i^{(3)}(\omega_4) = 6\sum_{jkl} \chi_{ijkl}^{(3)}(-\omega_4;\omega_1,\omega_2,\omega_3)\,E_j(\omega_1)E_k(\omega_2)E_l(\omega_3)\,, \tag{16}$$

where $\chi^{(3)}_{ijkl}(-\omega_4;\ \omega_1,\omega_2,\omega_3)$ is the macroscopic third order susceptibility of the material, with $\omega_4=\omega_1+\omega_2+\omega_3$, and ω_i, $i=1,2,3$, are the frequencies of the incident beams.

On the other hand, for an uniaxial system, aligned but not oriented (D_∞ symmetry) (Davis et al, 2008), the susceptibility tensor has only 11 nonzero elements, 10 of which are independent, for completely non-degenerate wave mixing. In the case of a single degeneracy, only 8 nonzero elements remain, 7 of which are independent, but in fully degenerate wave mixing, where $\omega_1=\omega_2=\omega$, and $\omega_3=-\omega$, only 3 nonzero independent components remain, which are $\chi^{(3)}_{1111}$, $\chi^{(3)}_{1133}$, and $\chi^{(3)}_{3333}$ (Davis, et al, 2008). In consequence, the nonlinear polarization of a general uniaxial system, for the fully degenerate case, may be written as

$$P_1(\omega)=\begin{bmatrix}\chi^{(3)}_{1111}\{3E_1(\omega)E_1(\omega)E_1^*(\omega)+2E_1(\omega)E_2(\omega)E_2^*(\omega)+E_2(\omega)E_2(\omega)E_1^*(\omega)\}+\\ \chi^{(3)}_{1133}\{6E_1(\omega)E_3^*(\omega)+3E_3(\omega)E_1^*(\omega)\}E_3(\omega)\end{bmatrix};$$

$$P_2(\omega)=\begin{bmatrix}\chi^{(3)}_{1111}\{3E_2(\omega)E_2(\omega)E_2^*(\omega)+2E_1(\omega)E_2(\omega)E_1^*(\omega)+E_1(\omega)E_1(\omega)E_2^*(\omega)\}+\\ \chi^{(3)}_{1133}\{6E_2(\omega)E_3^*(\omega)+3E_3(\omega)E_2^*(\omega)\}E_3(\omega)\end{bmatrix};$$

$$P_3(\omega)=\begin{bmatrix}\chi^{(3)}_{1133}\{3(E_1(\omega)E_1(\omega)+E_2(\omega)E_2(\omega))E_3^*(\omega)+6(E_1(\omega)E_1^*(\omega)+E_2(\omega)E_2^*(\omega))E_3(\omega)\}+\\ 3\chi^{(3)}_{3333}E_3(\omega)E_3(\omega)E_3^*(\omega)\end{bmatrix},$$

(17)

where $E_i^*(\omega)=E_i(-\omega)$, such that $E_i(\omega)E_i^*(\omega)=E_i(\omega)E_i(-\omega)=|E_i(\omega)|^2$, and $E_i(\omega)E_i(\omega)=E_i^2(\omega)$. Now, in order to determine the components of the susceptibility tensor for an uniaxial material by using the last expression and fully degenerate wave mixing, it is necessary to choose the laboratory coordinate system such that the *z*-axis is parallel to the optical axis of the nanocomposite. However, such a coincidence it is not obvious for the anisotropic metallic nanocomposites studied in this work, since the metallic nanoparticles are embedded into a SiO_2 matrix and then deformed in the direction of a Si ion beam, becoming an uniaxial system. In consequence, the electric field of the incident light beam will always make an angle θ with this optical axis. Therefore, the most convenient way of performing the calculations is giving preference to the deformed nanoparticle coordinate system, expressing the incident electric field in this frame and then coming back to the laboratory frame. To do this in the simplest manner, one can choose the laboratory and the particle systems such that their *x*-axes coincide, the *y*-axis of the laboratory frame is parallel to the wavevector of the incident light, the electric field is parallel to the *z*-axis of the laboratory, and that this last makes an angle θ to the nanoparticle *z*-axis. This choice is shown in detail in Fig. 10.

As it was established above, Eq. (17) gives the nonlinear polarization of the whole system in its main axes; consequently, for the anisotropic metallic nanocomposite, this polarization may be expressed in the nanoparticle frame by rewriting the electric field in that system, coming back later to the laboratory frame. Thus, Eq. (17) may be written in the *xyz*-frame of the nanoparticle, as

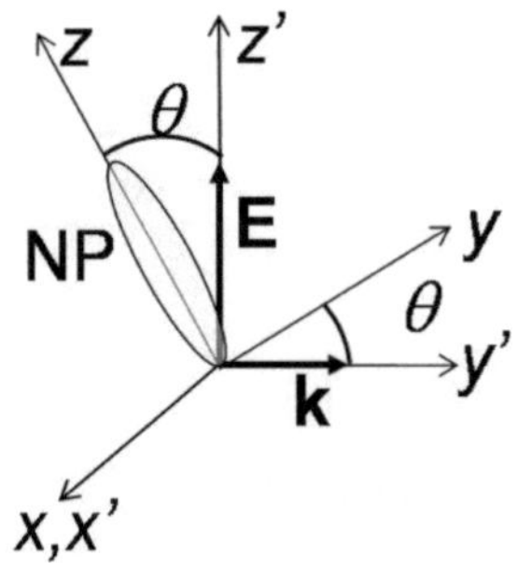

Fig. 10. Reference systems for the laboratory (primed) and the anisotropic metallic nanoparticle (unprimed).

$$\mathbf{P}_{\mathbf{NL}}^{(3)}(\omega)=\chi_{1111}^{(3)}\begin{bmatrix}\left\{3E_x(\omega)\left|E_x(\omega)\right|^2+2E_x(\omega)\left|E_y(\omega)\right|^2+E_y^2(\omega)E_x^*(\omega)\right\}\hat{\mathbf{i}}+\\ \left\{3E_y(\omega)\left|E_y(\omega)\right|^2+2E_y(\omega)\left|E_x(\omega)\right|^2+E_x^2(\omega)E_y^*(\omega)\right\}\hat{\mathbf{j}}\end{bmatrix}+$$
$$+\chi_{1133}^{(3)}\begin{bmatrix}\left\{6E_y(\omega)E_z^*(\omega)+3E_z(\omega)E_y^*(\omega)\right\}E_z(\omega)\hat{\mathbf{i}}+\\ \left\{6E_x(\omega)E_z^*(\omega)+3E_z(\omega)E_x^*(\omega)\right\}E_z(\omega)\hat{\mathbf{j}}+\\ \left\{3\left(E_x^2(\omega)+E_y^2(\omega)\right)E_z^*(\omega)+6\left(\left|E_x(\omega)\right|^2+\left|E_y(\omega)\right|^2\right)E_z(\omega)\right\}\hat{\mathbf{k}}\end{bmatrix}+, \quad (18)$$
$$+3\chi_{3333}^{(3)}E_z(\omega)\left|E_z(\omega)\right|^2\hat{\mathbf{k}}$$

where, by using Fig. 10, the components of the electric field can be written in such a frame as

$$E_x(\omega)=E_x^*(\omega)=0,\quad (19)$$
$$E_y(\omega)=E(\omega)\sin\theta\,,$$
$$E_z(\omega)=E(\omega)\cos\theta$$

This allows rewriting the third order nonlinear polarization for the anisotropic metallic nanocomposite in terms of the angle between the electric field and the axis of the nanoparticle, which defines the optical axis of the nanocomposite, as

$$\mathbf{P}_{\mathbf{NL}}^{(3)}(\theta;\omega)=3\left|E(\omega)\right|^2E(\omega)\left[\chi_{1111}^{(3)}\sin^3\theta\hat{\mathbf{j}}+\frac{3}{2}\chi_{1133}^{(3)}\sin2\theta\left\{\cos\theta\hat{\mathbf{j}}+\sin\theta\hat{\mathbf{k}}\right\}+\chi_{3333}^{(3)}\cos^3\theta\hat{\mathbf{k}}\right]. \quad (20)$$

When the incident electric field is parallel to the x-axis, the nonlinear polarization is trivially given by

$$\mathbf{P}_{\mathbf{NL}}^{(3)}(\omega) = \mathbf{P}_{\mathbf{NL,lab}}^{(3)}(\omega) = 3\left|E(\omega)\right|^2 E(\omega)\chi_{1111}^{(3)}\hat{\mathbf{i}}\,. \tag{21}$$

When projecting the polarization components given by Eq. (20) on the laboratory frame y',z', the nonlinear polarization is expressed as

$$\mathbf{P}_{\mathbf{NL,lab}}^{(3)}(\theta;\omega) = 3\left|E(\omega)\right|^2 E(\omega)\left[\chi_{1111}^{(3)}\sin^3\theta\left(\cos\theta\hat{\mathbf{j}}' - \sin\theta\hat{\mathbf{k}}'\right) + 3\chi_{1133}^{(3)}\sin\theta\cos\theta\hat{\mathbf{j}}' + \chi_{3333}^{(3)}\cos^3\theta\left(\sin\theta\hat{\mathbf{j}}' + \cos\theta\hat{\mathbf{k}}'\right)\right] \tag{22}$$

These last two equations, Eqs. (21) and (22), apply to an uniaxial, aligned but not ordered system, in the same way that the one typically found in literature (Boyd, 1992; Sutherland, 1996):

$$\mathbf{P}_{\mathbf{NL}}^{(3)}(\omega) = 6\mathbf{E}\cdot\mathbf{E}^*\mathbf{E}\chi_{1122}^{(3)} + 3\mathbf{E}\cdot\mathbf{E}\mathbf{E}^*\chi_{1221}^{(3)}, \tag{23}$$

applies to isotropic systems. Eq. (21) determines $\chi^{(3)}{}_{1111}$ directly when making measurements for the incident electric field perpendicular to the major axis of the nanoparticles, but all three components are present in Eq. (22). Therefore, in order to determine each of them separately, it becomes necessary to perform at least two other measurements as a function of the angle of incidence of the light with respect to the major axis of the particles, such that we obtain enough equations to determine the other two components of the nanocomposite tensor, $\chi^{(3)}{}_{1133}$ and $\chi^{(3)}{}_{3333}$.

Nevertheless, for the type of materials in consideration, it is necessary to take into account the refraction of light at the input face of the sample. Then, the simplest way is to perform one measurement of the wave mixing signal at normal incidence and other one such that the electric field is parallel to the major axis of the nanoparticle, which can be difficult due to refraction, such that the angle between them is the smallest possible. In conclusion, one has to perform at least three measurements in order to determine the three nonzero, independent components of the third order susceptibility tensor of an anisotropic metallic nanocomposite: one measurement should be performed at normal incidence with the electric field in the same plane of the major axis of the nanoparticle, making an angle between them given by the angle of deformation of the particles by the Si ion beam (Fig. 11(a)); a second one, at normal incidence too, but the electric field parallel to the minor axis of the nanoparticles, which is achieved by rotating the sample by 90° with respect to the previous case (Fig. 11(b)); finally, a last measurement such that the electric field and the major axis of the particle are parallel, or the angle between them is the minimum allowed by refraction (Fig. 11(c)). This last measurement is achieved by rotating the sample from case (a) with respect to an axis perpendicular to the plane of Fig. 11.

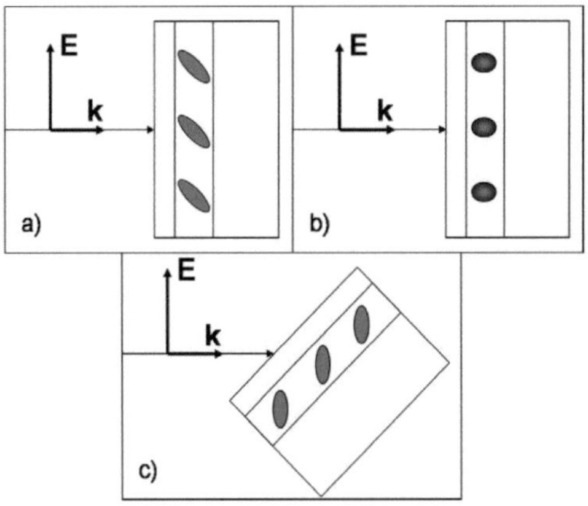

Fig. 11. Fully degenerate wave mixing measurements for anisotropic metallic nanocomposites. a) measurement at normal incidence generally implying all the three components of the tensor, b) determination of $\chi^{(3)}_{1111}$ at normal incidence, and c) determination, if possible due to the light refraction (not illustrated), of $\chi^{(3)}_{3333}$.

From what has been said above, one can rewrite Eq. (20) as

$$\mathbf{P}^{(3)}_{\mathbf{NL}}(\theta;\omega)=\left|E(\omega)\right|^2 E(\omega)\boldsymbol{\chi}^{(3)}_{\mathbf{eff}}, \tag{24}$$

where

$$\boldsymbol{\chi}^{(3)}_{\mathbf{eff}}=3\left[\left(\chi^{(3)}_{1111}\sin^3\theta+\frac{3}{2}\chi^{(3)}_{1133}\sin 2\theta\cos\theta\right)\hat{\mathbf{j}}+\left(\frac{3}{2}\chi^{(3)}_{1133}\sin 2\theta\sin\theta+\chi^{(3)}_{3333}\cos^3\theta\right)\hat{\mathbf{k}}\right], \tag{25}$$

is the effective third order susceptibility of the nanocomposite, when measuring at a given tilt of θ between the nanoparticle and the incident electric field, after considering light refraction. This last expression allows us finally to write

$$\left|\boldsymbol{\chi}^{(3)}_{\mathbf{eff}}\right|^2=9\left[\begin{array}{l}\left|\chi^{(3)}_{1111}\right|^2\sin^6\theta+\left|\chi^{(3)}_{3333}\right|^2\cos^6\theta \\ +\dfrac{3}{2}\sin^2 2\theta\left\{\begin{array}{l}\dfrac{3}{2}\left|\chi^{(3)}_{1133}\right|^2 \\ +\sin^2\theta\left(\operatorname{Re}\chi^{(3)}_{1111}\operatorname{Re}\chi^{(3)}_{1133}+\operatorname{Im}\chi^{(3)}_{1111}\operatorname{Im}\chi^{(3)}_{1133}\right) \\ +\cos^2\theta\left(\operatorname{Re}\chi^{(3)}_{1133}\operatorname{Re}\chi^{(3)}_{3333}+\operatorname{Im}\chi^{(3)}_{1133}\operatorname{Im}\chi^{(3)}_{3333}\right)\end{array}\right\}\end{array}\right]. \tag{26}$$

In consequence, when measuring according to what has been explained for Fig. 11, cases b) and c), we would obtain $|\chi^{(3)}_{eff}|^2=9|\chi^{(3)}_{1111}|^2$, and $|\chi^{(3)}_{eff}|^2=9|\chi^{(3)}_{3333}|^2$, respectively; while for case a), we would apply Eq. (26) fully.

5.2 Nonlinear optical measurements

The third order nonlinear optical response for a thin nonlinear optical media with strong absorptive response can be obtained by identifying the vectorial self-diffraction intensities generated by two incident waves (Torres-Torres et al, 2009). In this work, we first measured the nonlinear optical absorption using a P-scan technique (Banerjee et al, 1998), and later we used these results in order to fit the experimental data obtained by scalar self-diffraction experiments. The measurements were performed at IFUNAM's Nonlinear Optics laboratory using a Nd-YAG PL2143A EKSPLA laser system at λ=532 nm with a 26 ps pulse duration (FWHM) and linear polarization. The maximum pulse energy in the experiments was 0.1 mJ, while the intensity ratio between the two beams in the self-diffraction setup was 1:1. The radius of the beam waist at the focus in the sample was measured to be 0.1 mm. The results obtained are the average of single-pulse measurements widely spaced in time, well below the ablation threshold, in order to avoid thermal effects from accumulated pulses and assure reversible and reproducible nonlinear optical effects. Both setups are schematized in Fig. 12, where RPD represents a photodetector used for monitoring the laser stability; PD1 and PD2 are photodetectors for measuring the optical transmittance of the sample, while PD3 detects the self-diffraction signal. The mirrors were placed in order to obtain the same optical path for the two incident beams. We calibrated the self-diffraction measurements using a CS_2 sample, which is a well known nonlinear optical material with $|\chi^{(3)}{}_{eff}|=1.9\times10^{-12}$esu (Eisaman et al, 2005). For the single beam transmittance measurement in the P-scan experiments, we blocked one of the beams in the same experimental setup, as indicate in Fig. 12.

In order to perform both self-diffraction and P-scan measurements, as described at the end of Section 5.1, we started by positioning the sample as indicated in Fig. 12, which would correspond to Fig. 11(a). However, in the case of the Ag samples, due to the refraction of light, it was impossible to obtain $\chi^{(3)}{}_{3333}$ according to Fig. 11(c). Then, the measurements were performed for normal incidence and incidence at 45°. In the first case, it corresponds to Fig. 11(a) (normal incidence, vertical in Tables 3 and 4). In the second case (45°), the signal was measured for two positions, one such that the substrate containing the Ag nanocomposites was rotated 45° as indicated by the superior arrow in Fig. 12, but the incident optical polarization was kept parallel to the major axis of the nanoparticle (45° incidence, vertical in Tables 3 and 4, corresponding to Fig. 11(c)), addressing mainly $\chi^{(3)}{}_{3333}$. The other measurement was done in a similar way, but the incident optical polarization was kept parallel to the minor axis (45°, horizontal in Tables 3 and 4, substituting what it had been described for Fig. 11(b)), mainly related to $\chi^{(3)}{}_{1111}$. This has been achieved by rotating the sample 45° with respect to the axis normal to the setup, taking the first measurement and then, rotating the sample around its normal by 180°, measuring again.

5.3 Nonlinear optical results

According to what has been shown above and to our previous results (Oliver et al, 2006; Rodríguez-Iglesias, et al; Reyes-Esqueda et al, 2008), the nanocomposites thus fabricated show uniaxial symmetry. However, although the metallic anisotropic nanoparticles are all oriented in the same direction, as shown in Figs. 1 and 2, they do not exhibit polar order since they do not possess intrinsic dipolar moment. In consequence, as it has been shown in section 5.1, for nonlinear optical measurements using a fully degenerate setup, *i.e.* for $\chi^{(3)}(-\omega;\omega,\omega,-\omega)$, that the tensor has only three independent nonzero components: $\chi^{(3)}{}_{1111}$, $\chi^{(3)}{}_{1133}$ and $\chi^{(3)}{}_{3333}$. By following Fig. 11 and its discussion, we can perform the necessary measurements to obtain these three elements, show the anisotropy of the nonlinear optical properties of the nanocomposites, and therefore their usefulness for straightforward nonlinear applications, as we have already shown for linear optical properties (Oliver et al, 2006; Reyes-Esqueda et al, 2008). Then, we measured the P-scan transmitted and the scalar self-diffracted intensities while varying the angular position of the nanocomposites according to Fig. 11, for a fixed polarization of the incident beams. First, we measured the single beam optical transmittance in order to obtain the nonlinear optical absorption coefficient for each position of the sample. The fitting was made considering that absorption is described by the expression $\alpha=\alpha_0+\beta I$; where α_0 and β represent the linear, and nonlinear absorption coefficients, respectively; and I is the incident intensity. Then, the self-diffraction experiments were performed and the transmitted and self-diffracted intensities were measured for each case.

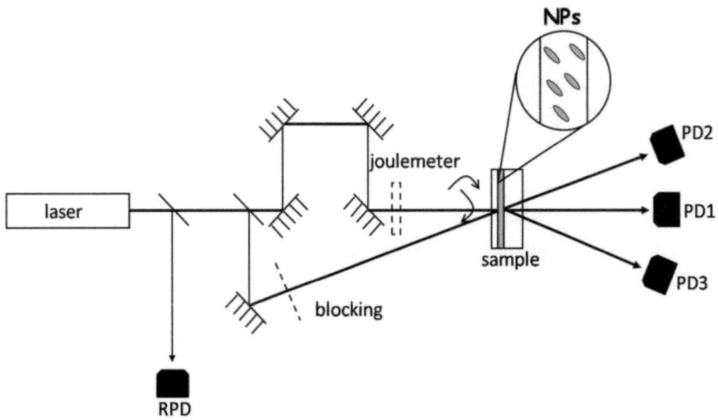

Fig. 12. Experimental setup used for self-diffraction and P-scan measurements (dashed components).

For the evaluation of the corresponding $|\chi^{(3)}_{eff}|$ value, we considered that nonlinear absorption takes place during the self-diffraction experiments. Thus, the numerical fitting was made by following the analysis of degenerate two-wave mixing with self-diffraction in the stationary regime for a thin nonlinear medium (Sánchez, 1992), but taking into account the results obtained for the nonlinear absorption for each case.

The $|\chi^{(3)}_{eff}|$, $\mathrm{Re}(\chi^{(3)}_{eff})$, and $\mathrm{Im}(\chi^{(3)}_{eff})$ values obtained for each sample, for each angular position, according to what has been described in Section 5.2, are shown in Tables 3 and 4. It is worth remarking that the self-diffraction and the P-scan signals were measured also for isotropic metallic nanocomposites, i.e. spherical-like Ag nanoparticles not deformed with Si, for each angular position as mentioned before, finding practically the same value for the nonlinear response.

Si fluence (×10^{15} ions/cm^2)	$\lvert\chi^{(3)}_{eff}\rvert$ normal incidence, vertical	$\lvert\chi^{(3)}_{eff}\rvert$ 45° incidence, vertical	$\lvert\chi^{(3)}_{eff}\rvert$ 45° incidence, horizontal
0.0		2.77	
0.5	2.73	3.42	1.46
0.8	2.29	3.09	1.80
1.0	2.28	3.26	1.33
2.0	1.99	2.12	1.44
5.0	3.36	2.79	2.05

Table 3. $|\chi^{(3)}_{eff}|$ *(×10^{-9} esu)* measurement for each angular position, Ag nanocomposites *(4.7×10^{16} ions/cm^2)*, $\Delta|\chi^{(3)}_{eff}| = \pm 10^{-11}$ *esu* .

From Table 3, the difference in the measured values for the anisotropic nanocomposites, for each position, is quite clear. Therefore, from the given association of them to the different components of the nanocomposite third-order susceptibility tensor $\chi^{(3)}$, *i.e.*, the vertical measurements are associated to the major axis component $\chi^{(3)}_{3333}$, while the horizontal ones to the minor axis component $\chi^{(3)}_{1111}$; for similar geometric parameters of the nanocomposites, we can establish the inequality $|\chi^{(3)}_{eff}|_{\text{minor axis}} < |\chi^{(3)}_{eff}|_{\text{isotropic}} < |\chi^{(3)}_{eff}|_{\text{major axis}}$. These results are valid considering that the magnitude of the nonlinear optical response seems to be strongly dependent on the size and shape of the particles (López-Suárez et al, 2009), and our deformation technique for obtaining nanoellipsoids is a direct way of modifying both. One obvious implication that we can sort out from this result, is that the smaller the nanoparticle that 'sees' the incident light, the smaller the third order nonlinear optical response. This could be related to the near-field enhancement associated with the shape of the nanoparticle, indicating that the largest enhancement is obtained for the major axis, as has been shown for Au nanorods (Chau *et al*, 2009).

From Table 4, the anisotropy of the nonlinear absorption and refraction is quite clear for these anisotropic nanocomposites. Furthermore, Re($\chi^{(3)}_{eff}$), is positive for Ag indicating a self-focusing response. On the contrary, the imaginary part is negative for both axes, although the anisotropy is quite clear when comparing the measurements related to the minor and the major axes (the last two columns of Table 4). It has been previously indicated that there are different contribution to the nonlinear response for different pulse durations, since hot electrons, or even thermal effects, can take place besides the electronic response of the material (Torres-Torres *et al*, 2008a; Torres-Torres *et al*, 2009; Torres-Torres *et al*, 2008c). In this case, if we compare the femtosecond and picosecond results, it is possible to observe that the magnitude of the absorptive nonlinearity in anisotropic Ag nanoparticles is approximately two orders of magnitude stronger for the femtosecond regime. We assume that the density of particles in the sample and the lifetime associated with the optical excitation in the absorption process cannot be fast enough to allow the linear optical interactions to take place. Apparently, optical saturation is more easily induced for femtosecond irradiation, even when similar mechanisms of nonlinear optical absorption are exhibited in both regimes. This behavior related with the temporal response and changes in the absorptive nonlinearity mechanisms have been previously studied in other nanostructured materials (López-Suárez *et al*, 2009). On the other hand, an important consequence that can be derived from the changes in the absorptive nonlinearities in the different temporal regimes is that the nonlinear refractive index is strongly stimulated with picoseconds pulses, while for the femtosecond experiments we do not obtain a significant refractive nonlinearity. We estimate that with the picosecond irradiation, strong intraband and interband transitions may take place according to the resonance of the major and the minor axis of the nanoparticle, respectively. As a result, changes in the magnitude of the real part of the third order nonlinearity are observed for picoseconds experiments as shown in Table 4.

Si fluence ($\times 10^{15}$ ions/cm^2)	normal incidence, vertical	45° incidence, vertical	45° incidence, horizontal
0.0		-3.5, 1.89	
0.5	-3.5, 1.84	-2.4, 2.34	-0.3, 1.00
0.8	-3.5, 1.56	-3.5, 2.12	-3.5, 1.22
1.0	-3.5, 1.56	-2.8, 2.23	-3.5, 0.89
2.0	-1.4, 1.34	-3.8, 1.45	-1.0, 1.00
5.0	-6.6, 2.28	-7.0, 1.89	-3.5, 1.39

Table 4. Im($\chi^{(3)}_{eff}$) (×10^{-12} *esu*), and Re($\chi^{(3)}_{eff}$) (×10^{-9} *esu*) measurement for each angular position, Ag nanocomposites (*4.7×10^16^ ions/cm^2^*).

6. Conclusions

We have shown a novel technique for producing composite materials containing deformed metallic nanoparticles that are placed at random but aligned in a given direction. This is achieved by a double ion-implantation technique; the first implantation produces the nanoparticles, while the second elongates them in a given direction. The resulting composite material has polarization dependent absorption spectra, due to a split of the surface plasmon resonance into two bands. Because of the alignment of the particles in a preferential direction, the material also presents strong birefringence, which has been characterized by a relatively simple ellipsometric technique.

We have also studied the nonlinear optical properties of these highly anisotropic nanocomposites in the femtosecond and picosecond regimes. In the femtosecond regime we find a mainly absorptive nonlinearity for wavelengths close to the surface plasmon resonance for polarization parallel to the long axis of the nanoparticles. The nonlinearity, studied using the z-scan technique, was found to consist of saturable absorption which was also polarization dependent, and this dependence was fitted reasonably well by a simple two-level model for anisotropic particles.

In the picosecond regime, the resonant nonlinear response was studied using the self-diffraction and 'P-scan' techniques. In this case the anisotropy of the nonlinear response was used to determine the different tensorial elements of $\chi^{(3)}_{ijkl}$, and to resolve the absorptive and refractive contributions to the nonlinearity. For these experiments, although significant nonlinear absorption was observed, the nonlinear response was shown to be dominated by nonlinear refraction, which was attributed to strong intra- and inter-band transitions induced by the resonant radiation.

The capability of tailoring their linear and nonlinear optical properties by the precise control of particle size, density, shape, and alignment, achievable with the implantation technique shown, makes these aligned nanostructured materials very interesting for applications in all-optical switching devices, which exploit their nonlinearities, and possibly for the emerging field of plasmonic circuits.

7. References

Banerjee, P.P.; Danileiko, A.Y.; Hudson, T. & McMillen, D. (1998). P-scan analysis of inhomogeneously induced optical nonlinearities. *J. Opt. Soc. Am. B* 15, 2446-2454.

Barnes, W.L.; Dereux, A.; and Ebbessen, T.W. (2003). Surface plasmon subwavelength optics, *Nature* 424, 824-830.

Boyd, R.W. (1992) *Nonlinear Optics*, Academic Press, San Diego, chapter 4.

Cheang-Wong, J.C.; Morales, U.; Oliver, A.; Rodríguez-Fernández, L. & Rickards, J. (2006). MeV ion beam deformation of colloidal silica particles. *Nuclear Instruments and Methods in Physical Research B* 242, 452-454.

Davis, R. P.; Moad, A.J.; Goeken, G.S.; Wampler, R.D. & Simpson, G.J. (2008). Selection rules and symmetry relations for four-wave mixing measurements of uniaxial assemblies. *J. Phys. Chem. B* 112, 5834-5848.

Eisaman, M. D.; André, A.; Massou, F.; Fleischhauer, M.; Zibrov, A. S. & Lukin, A.S. (2005). Electromagnetically induced transparency with tunable single-photon pulses. *Nature* 438, 837-841.

Elim, H.I.; Yang, J.; Lee, J.Y.; Mi, J. & Ji, W. (2006). Observation of saturable and reverse-saturable absorption at longitudinal surface plasmon resonance in gold nanorods, *Applied Physics Letters* 88, 083107.

Genereux, F.; Leonard, S.W.; van Driel, H. M.; Birner, A. & Gösele, U. (2001). Large birefringence in two dimensional silicon photonic crystals. *Phys. Rev. B* 63, 161101(R).

Gonzalez, A. L.; Reyes-Esqueda, J.A. & Noguez, C. (2008). Optical properties of elongated noble metal nanoparticles, *J. Phys. Chem C* 112, 7356-7362

Hasui, K.; Grossman, D.G.; Mann, L.G.; Takahashi, H. & Borrelli, N.F. (2000). A high performance dichroic glass polarizer with a thickness of 15-35 μm. *Jpn. J. Appl. Phys.* 39, 1494-1496.

Hoa, X. D.; Kirk, A.G. & Tabrizian, M. (2007). Towards integrated and sensitive surface plasmon resonance biosensors: a review of recent progress. *Biosens. Bioelectron.* 23, 151-160.

Inouye, H.; Tanaka, K.; Tanahashi, I.; Hattori, T.; Nakatsuka H., (2000). Ultrafast Optical Switching in a Silver Nanoparticle System, *Japanese Journal of Applied Physics* 39, 5132-5133 (2000).

Johnson, N.F. & Christy, R.W. (1972). Optical constants of the noble metals. *Phys. Rev. B* 6, 4370-4379.

Kasai, H.; Nalwa, H.S.; Oikawa, H.; Okada, S.; Minami, S.; Kakuta A.; Ono, K.; Mukoh A. & Nakanishi, H. (1992) . *Japanese Journal of Applied Physics* 31, L1132.

Künzner, N.; Kovalev, D.; Diener, J.; Gross, E.; Timoshenko, V. Yu.; Polisski, G.; Koch, F. & Fujii, F. (2001). Giant birefringence in anisotropically nanostructured silicon. *Opt. Lett.* 26, 1265-1267.

Kyoung, M. & Lee, M. (1999). Nonlinear absorption and refractive index measurements of silver nanorods by the Z-scan thecnique, *Optics Communications* 171, 145-148.

Lamarre, J.M.; Billard, F.; Kerboua, C.H.; Lequime, M.; Roorda, S. & Martinus, L. (2008). Anisotropic nonlinear absorption of gold nanorods in a silica matrix. *Opt. Commun.* 281, 331-340.

Lopez-Suarez, A.; Torres-Torres, C.; R. Rangel-Rojo, Reyes-Esqueda, J.A.; Santana, G.; Ortíz, A.; Alonso, J.C. & Oliver, A. (2009). Modification of the nonlinear optical absorption and optical Kerr response exhibited by nc-Si embedded in a silicon-nitride film. *Opt. Express* 17, 10056-10068

Maier, S.A.; Kik, P.G.; Atwater, H.A.; Meltzer, S.; Harel, E.; Koel, B.E. & Requicha, A.A.G. (2003). Local detection of electromagnetic energy transport below the diffraction limit in metal nanoparticle plasmon waveguides. *Nature Materials* 2, 229-232.

Matsuda, S.; Yasuda, Y. & Ando, S. (2005). Fabrication of polyimide-blend thin films containing uniformly oriented silver nanorods and their use as flexible, linear polarizers. *Adv. Mater.* 17, 2221-2224.

Muskens, O.L.; Borgström, M.T.; Bakkers, E.P.A.M. & Gómez-Rivas, J. (2006). Giant optical birefringence in ensembles of semiconductor nanowires. *Appl. Phys. Lett.* 89, 233117.

Noguez, C. (2007). Surface plasmons on metal nanoparticles: the influence of shape and physical environment. *J.Phys. Chem. C.* 111, 3806-3819.

Oliver, A.; Reyes-Esqueda, J.A.; Cheang-Wong, J.C.; Román-Velázquez, C.E.; Crespo-Sosa A.; Rodríguez-Fernández, L.; Seman, J.A.; & Noguez, C. (2006). Controlled anisotropic deformation of Ag nanoparticles by Si ion irradiation. *Phys. Rev. B* 74, 245425.

Pérez-Juste, J.; Pastoriza-Santos, I.; Liz-Marzán, L. M. & Mulvaney, P. (2005). Gold nanorods: synthesis, characterization and applications. *Coordin. Chem. Rev.* 249, 1870-1901

Rangel-Rojo, R.; Kar, A.K.; Wherrett, B.S.; Carroll, M.; Cross, G.H.; & Bloor, D. (1995). Third-order optical nonlinearities of a polymeric film doped with a novel zwitterion, DEMI-3CNQ, *Rev. Mex. Fis.* 41, 832--840

Rangel-Rojo, R.; Yamada, S.; Matsuda, H.; Kasai, H.; Nakanishi, H.; Kar, A.K. & Wherrett, B.S. (1998). Spectrally resolved third-order nonlinearities in polydiacetylene microcrystals: influence of particle size, *J. Opt. Soc. Am. B* 15, 2937-2945

Rangel-Rojo, R.; McCarthy, J.; Bookey, H.T.; Kar, A.K.; Rodriguez-Fernandez, L.; Cheang-Wong, J.C.; Crespo-Sosa, A.; Lopez-Suarez, A.; Oliver, A.; Rodriguez-Iglesias, V. ; Silva-Pereyra, H. (2009). Anisotropy in the nonlinear absorption of elongated silver nanoparticles in silica, probed by femtosecond pulses. *Optics Communications* 282, 1909-1912.

Reyes-Esqueda, J.A.; Torres-Torres, C.; Cheang-Wong, J.C.; Crespo-Sosa, A.; Rodríguez-Fernández, L.; Noguez, C. & Oliver, A. (2008). Large optical birefringence by anisotropic silver nanocomposites, *Opt. Express* 16, 710-717.

Reyes-Esqueda, J.A.; Rodríguez-Iglesias, V.; Silva-Pereyra, H.G.; Torres-Torres, C.; Santiago-Ramírez, A.L.; Cheang-Wong, J.C.; Crespo-Sosa, A.; Rodriguez-Fernandez, L.; Lopez-Suarez, A. & Oliver, A. (2009). Anisotropic linear and nonlinear optical properties from anisotropy-controlled metallic nanocomposites. *Opt. Express* 17, 12849-12868.

Rodríguez-Iglesias, V.; Silva-Pereyra, H.G.; Cheang-Wong, J.C.; Reyes-Esqueda, J.A.; Rodríguez-Fernández, L.; Crespo-Sosa, A.; Kellerman, G. & Oliver, A. (2008). MeV Si ion irradiation effects on the optical absorption properties of metallic nanoparticles embedded in silica. *Nucl. Instrum. Methods B* 266, 3138-3142

Rodríguez-Iglesias, V.; Silva-Pereyra, H.G.; Torres-Torres, C.; Reyes-Esqueda, J.A.; Cheang-Wong, J.C.; Crespo-Sosa, A.; Rodriguez-Fernandez, L.; Lopez-Suarez, A. & Oliver, A. (2009). Large and anisotropic third-order nonlinear optical response from anisotropy-controlled metallic nanocomposites. *Opt. Commun.* 282, 4157-4161

Ruda, H.E. & Shik A. (2007). Nonlinear optical phenomena in nanowires. *Journal of Applied Physics* 101, 034312.

Saleh, B.E.A. & Teich, M.C. (1991). *Fundamental of Photonics*. Wiley-Interscience, John Wiley & sons, Inc. This equation differs from to the usual one in the reference because our analysis considers the angle between the wavevector and the normal to the particle, and not with its optical axis as it is usual. Anyway, both angles are related since their sum gives $\pi/2$, which explains the difference in this equation.

Sanchez, F. (1992). Two-wave mixing in thin nonlinear local-response media: a simple theoretical model. *J. Opt. Soc. Am. B* 9, 2196-2205

Seward III, T. P. (1974). Elongation and spheroidization of phase-separated particles in glass. *J. Non-Crystalline Solids* 15, 487-504.

Sutherland, R.L. (1996). Handbook of Nonlinear Optics. Marcel Dekker Inc, New York,.

Torres-Torres, C.; Reyes-Esqueda, J.A.; Cheang-Wong, J.C.; Crespo-Sosa, A.; Rodríguez-Fernández, L. & Oliver, A. (2008a). Optical third order nonlinearity by nanosecond and picosecond pulses in Cu nanoparticles in ion-implanted silica. Journal of Applied Physics, 104, 014306 1-0143065.

Torres-Torres, C.; Khomenko, A.V.; Tamayo-Rivera, L.; Rangel-Rojo, R.; Mao, Y. & Watson, W. H. (2008b). Measurements of nonlinear optical refraction and absorption in an amino-triazole push-pull derivative by a vectorial self-diffraction method. *Opt. Commun.* 281 3369–3374.

Torres-Torres, C.; Lopez-Suarez, A.; Tamayo-Rivera, L.; Rangel-Rojo, R.; Crespo-Sosa, A.; Alonso, J.C. & Oliver, A. (2008c). Thermo-optic effect ant third order nonlinearity in nc-Si embedded in a silicon nitride film. *Opt. Express* 16, 18390-18396.

Torres-Torres, C.; Trejo-Valdez, M.; Sobral, H.; Santiago-Jacinto, P. & Reyes-Esqueda, J.A. (2009). Stimulated emission and optical third order nonlinearity in Li-doped nanorods. *J. Phys. Chem C* 113, 13515-13521

Yildirim, H. & Bulutay, C. (2008). Enhancement of optical switching parameter and third-order optical nonlinearities in embedded Si nanocrystals: a theoretical assessment. *Optics Communications* 281, 4118-4120.

3

The applicability of global and surface sensitive techniques to characterization of silver nanoparticles for Ink-Jet printing technology

M. Puchalski[1], P.J. Kowalczyk[1,2], Z. Klusek[1] and W. Olejniczak[1]
[1]*Division of Physics and Technology of Nanometer Structures, Solid State Physics Department, University of Lodz, Pomorska 149/153, 90-236 Lodz, Poland*
[2]*MacDiarmid Institute for Advanced Materials and Nanotechnology, Department of Physics and Astronomy, University of Canterbury Private Bag 4800, Christchurch 8140, New Zealand*

1. Introduction

In recent years, silver nanoparticles (AgNPs) have been investigated by many research groups all over the world (Klabunde, 2001). It is mainly due to potential applications of this new nanomaterial in modern electronic devices (Xue et. al., 2006), medicine (Shahverdi et. al. 2007), optics (Mc Farlandand et. al., 2003) and biology (Huangand et. al., 2004). Among others it is promising to use AgNPs in electronic industry to create printed circuits due to good electric and thermal conductivity of silver. This is commonly performed by the use of so called Ink-Jet printing technology.

Ink-Jet printing is based on throwing onto substrate picolitters portions of ink containing nanoparticles of metal (especially silver) which could be formed in micrometer dots and tracks (Fuller et. al., 2002). In consequence it is possible to print any matrices consisting micrometer elements which after further treatment can be used as a electronic tracks and devices. This technology requires well defined and stable metal nanoparticles which remain on surface after evaporation of solvent which forms stable colloid (ink). Nanoparticles usually contain a metal (nanocrystalline) core and an outer shell - surfactants or ligands, which typically are alkyl (Jiang et. al., 2002) or polymer chains (Tsuji et. al., 2008). The presence of organic molecules stabilizes nanoparticles, prevents their aggregation and facilitates creation of colloid. Unfortunately, it is also responsible for insulating character of printed structures. To make them conducting thermal sintering process is required. It results in ligands decomposition and in the crystallization of silver cores (Smith et. al., 2006).

We present our results related to physical and chemical characterization of ligand-stabilized silver nanoparticles (Amepox MC) using different experimental methods. Particularly, we focus on problems associated with estimation of size distribution of silver nanoparticles using: high resolution transmission electron microscopy (HRTEM), atomic force microscopy (AFM), dynamic light scattering (DLS) and ultraviolet and visible spectroscopy (UV-VIS) - the similarity and the differences between these techniques will be briefly outlined. The global chemical analysis by using energodispersive X-ray spectroscopy (EDX), and the surface chemical analysis by using X-ray photoelectron spectroscopy (XPS) of AgNPs deposited on different substrates will be presented and the applicability of the used techniques described. Finally, the global ultraviolet photoelectron spectroscopy (UPS) and local scanning tunnelling spectroscopy STS techniques, will be presented as a valuable techniques in electronic structure investigations of AgNPs. This is especially important in the case of studies of quantum size effects on nanoparticles.

1.1 AgNPs for Ink-Jet printing

The most popular nanomaterials for Ink-Jet application are silver nanoparticles synthesized by thermal decomposition method (Khanna et. al., 2008). This method is based on decomposition of silver salt of fatty acid at high temperature in inert atmosphere (Fig. 1).
Silver salt (silver alkanoate) is obtained as the result of the reaction of silver nitrate and fatty acid dissolved in sodium hydroxide water solution. The proportions of compounds are very important. For example using of: 52g of $AgNO_3$, 75g of myristic acid, 12g of NaOH and 1L of high-purity water (Nagasawa et al., 2002) seams to form good proportions to prepare initial solution. The next step is a decomposition that received silver alkanoate in inert atmosphere. Such process could be realised in two ways. According to Nagasawa et. al. 2002, stable silver nanopartiles with narrow size distribution could be obtained as a result of thermal decomposition of silver alkanoate in nitrogen atmosphere at 250°C. Recently Shim et. al. 2008, suggest reaction at 210°C in liquid solution - fatty acid dissolve in non-polar and slowly evaporating solvent (for example 1-octadecene). The both reaction give silver powder composed of strongly aggregated ligand-stabilized AgNPs with narrow size distribution. However, the efficiency of the reaction, the size of the final product and stabilization of nanoparticles in solvent strongly depend of annealing temperature.

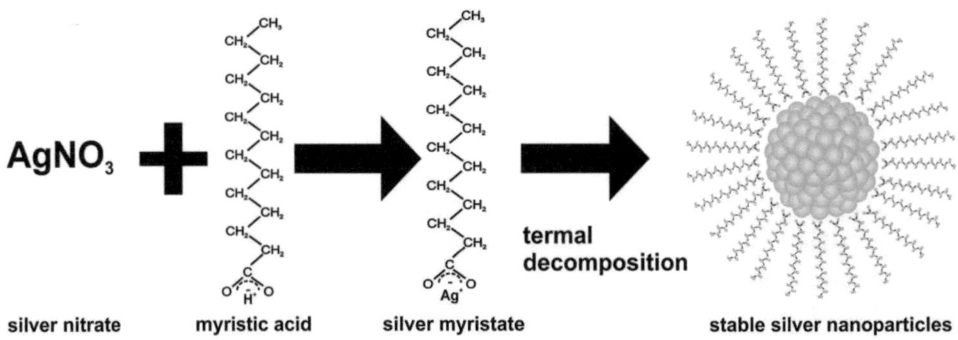

Fig. 1. Schematic of the thermal decomposition synthesis of the stable silver nanoparticles with fatty acid coated.

Preparation silver ink requires dissolving of the synthesised silver powder in non-polar organic solvent (usually *n*-tetradecane). Tracks printed using such ink become conducive after sintering at 240 - 250°C for 30 min.
In our experiments we investigated commercially available silver powder synthesized using thermal decomposition method in Amepox Microelectronics laboratory (Moscicki et. al.,

2007). For characterisation we re-dissolve that material in cyclohexane (Aldrich) what give good quality colloid with low concentration of AgNPs.

2. AgNPs characterization

The quality of the ink and its properties depend mainly on size distribution and chemical composition of nanoparticles. There are many techniques which could be used to perform such measurement however among others a few are most popular and widely used:

- high resolution transmission electron microscopy (HRTEM);
- atomic force microscopy (AFM);
- dynamic light scattering (DLS);
- ultraviolet and visible spectroscopy (UV-VIS).

Each of these techniques are based on different physical property and because of that they return slightly different distributions. Two of them are global techniques which give information about average distribution in whole sample (DLS, UV-VIS). Other two are local which allow to measure shape and diameter of individual nanoparticles (HRTEM, AFM). Beside that some of them measurement is performed in solvent (DLS, UV-VIS) while other require earlier deposition of nanoparticles on the substrate (HRTEM, AFM) which could result in distortion of their diameter. Because of these properties more than one technique of diameter estimation should be used.

Beside of size distribution measurements chemical analysis of nanoparticles is often performed using mainly three techniques:

- X-ray photoelectron spectroscopy (XPS);
- energodispersive X-ray spectroscopy (EDX);

Due to different probing depth XPS could be qualified as a surface technique while EDX is rather bulk like. XPS and EDX could give information about chemical composition and XPS could be used to estimate type of bonds formed between surfactant and metallic core as well as in surfactant itself.

2.1 Size distribution measurement

The first technique which we used, was high resolution transmission electron microscopy (JEOL JEM2000EX HRTEM). Fig. 2(a) shows a typical electron micrograph recorded on silver nanoparticles. It is clearly seen that the their spatial distribution is random which is a typical result in case of drop cast method. More controlled sample preparation and better spatial distribution of nanoparticles could be achieved using Langmuir-Blodgett (LB) which results in high quality monolayers of nanoparticles (Lu et. al., 2005). Despite the preparation method the estimation of AgNPs size distribution are similarly. By measuring of about 250 individual particles we were able to created a histogram which is shown in Fig. 2(b). It is clearly seen, that the obtained histogram is not really symmetrical due to a low amount of small AgNPs and almost constant number of counts in the range of 5 - 6 nm. Nevertheless, we decided to fit our experimental data with normal distribution. Fitted distribution (shown in Fig 2 (b)) is characterised by the mean value and standard deviation equal 5.5 nm and 1.5 nm respectively. This result is consistent with reports of other research groups (Abe et. al. 1998). It is worth noting again that HRTEM is only sensitive to silver cores (crystalline) size. The organic shell is not observed on electron micrographs, however distances between cores could be used to estimate ligands lengths.

Another microscopic technique which was used in our experiments was atomic force microscopy (home build (Klusek et. al., 1998)). Due to is local character and huge magnifications this technique found its application in nanoparticles investigations world wide. Especially AFM is used to estimate their diameter. Commonly tapping mode is used which is believed to be less interacting with NPs than contact mode. In our experiments we prepared samples in similar way like in case of HRTEM studies. Silver nanoparticles were deposited on atomically flat Muscivite (001) surfaces by drop cast method.

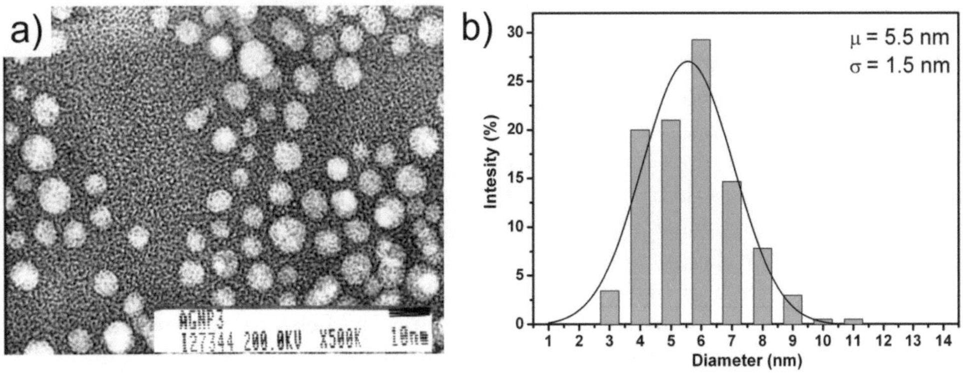

Fig. 2. The measurement of nanoparticles diameter by means of HRTEM: (a) electron micrograph, (b) estimated histogram.

In Fig. 3(a) we present the 1400×1400nm^2 AFM topography of AgNPs decorated the Muscovite (001) substrate. The image evidently shows the nanoparticles of different size dispersed randomly on the whole surface. Although, it is easy to observe the single isolated nanoparticles one can also notice that nanoparticles tend to form bigger structures, conglomerates as a result of coagulation process. The nanoparticles were not chemically bonded to the substrate, however, we found out that the sample was stable during investigation in the air which allowed us to measure size distribution of nanoparticles. In order to minimize the effect of AFM tip radius in the measurements, a height of the particles has been measured (Ebenstein et. al., 2002). In Fig. 3(b) we present a histogram of the particle height distribution. The mean value and the standard deviation are 4.3 nm and 1.2 nm, respectively.

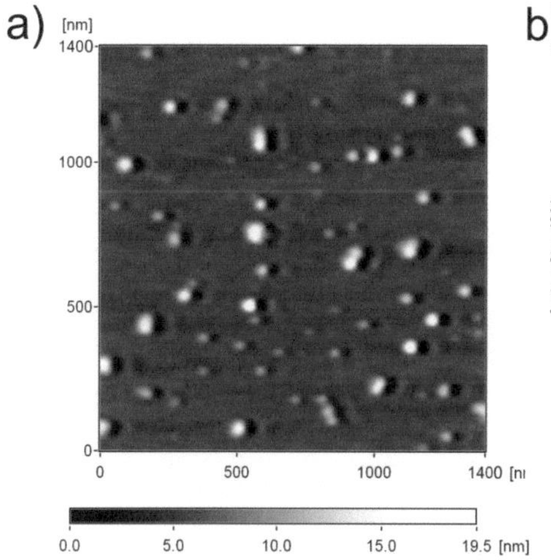

Fig. 3. AFM measurement of nanoparticles diameter: (a) topography image, (b) estimated histogram of high distribution.

This results which we recorded are different than these obtained using high resolution transmission electron microscopy. Because we expect that AFM should be sensitive not only to metallic cores like HRTEM is but also to ligands. Thus, we expected to obtain mean value at the level of 8 nm (5 nm silver core and 2×1.4 nm long alkyl chain). Much lower value could be related with specific properties of investigated system. In case of homogeneous samples, height measurement is quite precisions. However, nanoparticles are composed of 5 nm silver cores capped by 1.4 nm long alkyl chain which form highly not homogenous system. Such system could behave in various ways after depositing on substrate. At least three different behaviours are expected: (i) ligands forms spherical shell around core and AFM does not interacts strongly with the NPs, (ii) alkyl chains are distorted or even removed at the substrate core interface but interaction with the cantilever is weak and (iii) there is strong interaction wit the substrate and also alkyl chains bend as a result of presence of the cantilever. Each of these situations would result in different apparent height starting from 8 nm and finishing on 5 nm. We think that in our case we dealt with situation (iii) which could explain height distribution which would reflect real core diameters. However situation is much more complicated because different values of height could be obtained after deposition of nanoparticles on different substrates. For that reason, AFM seams to be not a good technique to precisely estimate diameter of ligand-stabilized small nanoparticles or clusters (<2 nm) (Hornyak et. al. 1998). However, this effect can be ignored when diameter of nanoparticles are 20 nm or bigger.
In the next step, we used techniques based on optical properties of nanoparticles: UV-Vis (Hitachi U-1900) and DLS (Zetasizer 1000HS by Malvern Instruments). Ultraviolet and visible light spectrometer is an instrument mainly used to characterise the optical properties of materials, especially to estimate the surface plasmon resonance peak. It could be used to measure the mean value of nanoparticles core radius. It is possible using Gustav Mie theory of light scattering by spherical particle (Mishchenko et. al., 2002). This theory is described in terms of two parameters:

- normalized refractive index n_r:

$$n_r = \frac{n_p}{n_m}, \tag{1}$$

where n_p is refractive index of particles and n_m of surrounding medium

- size parameter x:

$$x = \frac{2\pi r n_m}{\lambda}, \tag{2}$$

where r is particles radius and the light wavelength λ.

According to the Mie theory, efficiency of scattering is expressed by:

$$\sigma_s = C_s A, \tag{3}$$

where A is a geometrical area of particles and C_s - scattering coefficient.

The scattering (C_s) and extinction (C_e)coefficients are expressed by:

$$C_s = \frac{2}{x^2}\sum_{n=1}^{\infty}(2n+1)\left(|a_n|^2 + |b_n|^2\right), \tag{4}$$

$$C_e = \frac{2}{x^2}\sum_{n=1}^{\infty}(2n+1)\mathbb{R}(a_n + b_n), \tag{5}$$

where a_n and b_n are Mie coefficients. The expressions of these coefficients are discussed further by Mishchenko et. al., 2002.

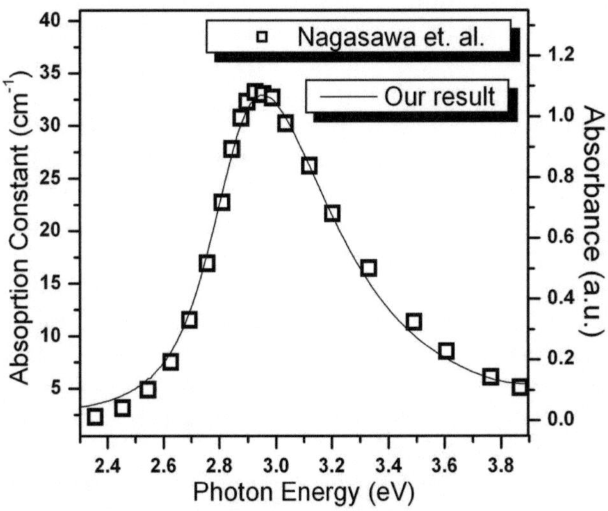

Fig. 4. Absorption spectra of 5 nm ligand-stabilized silver nanoparticles.

The mean value of nanoparticle by analyse of light absorption could calculate by the using presented Mie equations for example using implementation in MiePlot software by Philip Laven (www.philiplaven.com/mieplot). This software estimate the absorption, scattering and extinction spectra for silver, gold and copper nanoparticles.

In the our experiments, we measured absorption for ligand-stabilized silver nanoparticles in cyclohexane. In Fig 4 we show our spectrum and this measured by Nagasawa et. al. 2002.

Perfect agreement between both measurements is seen. The both spectra have maximum about 2.9 eV and similar shape which additionally confirms that we dealt with nanoparticles with mean diameter of 5 nm.

In the recent years, dynamic light scattering was widely used to measure nanoparticles sizes distribution. The DLS is capable to make measurements of particles as small as 10 nm. The advantage of this technique is fast measurement of diameter distribution. In Fig. 5 result of our investigation using DLS is presented. It is characterized by the mean value and standard deviation equal 5.1 nm and 1.1 nm respectively. Results of our measurements confirm that the thermal decomposition method yields narrow distribution of nanoparticles with an average diameter equal to 5 - 6 nm, which is consistent with HRTEM, UV-Vis investigation.

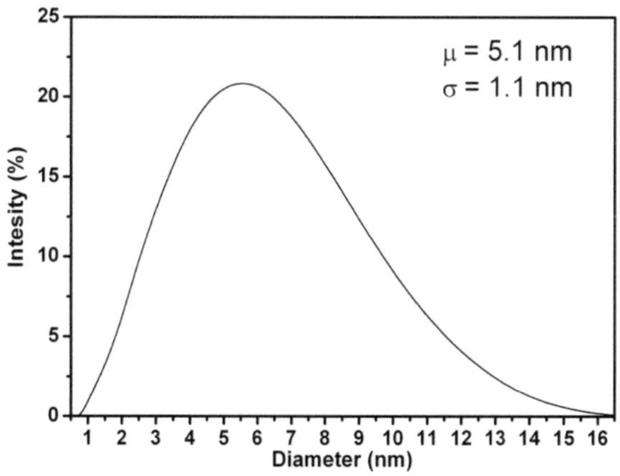

Fig. 5. Ligand-stabilized AgNPs diameter distribution by DLS.

Measurement of size distribution seams to be very important problem. All described techniques returned similar results which is surprising because AFM should measure cores and ligands. From our observations it is clear that it is also not sensitive to the presence of surfactants. We suspect that it is the result of distortion of ligands especially at the core substrate interface. However main advantage of that technique is its locality. Similar advantage could be found in case of HRTEM but here the main problem is related with suspending nanoparticles on special substrates. It in fact allows to measure diameter of the core but this diameter could be completely different after deposition on substrate typical in case of ink jet printing. Here one finds advantages of AFM which can be particularly performed on any substrate.

The two other techniques allows for measurements in solution. Unfortunately interpretation of UV-VIS spectra is not straightforward which makes this measuring technique rather difficult. In contrary the results generated by DLS are in interpretation and the time required for measurement is short which makes this technique the most popular. However it have to be remembered that in case of such measurement results are usually overestimated due to coalescence of the nanoparticles in colloid and also due to influence of not bonded ligands which could increase apparent diameter of the nanoparticles. Therefore, none of the techniques is best nor worst. Each of them could be used to estimate size distribution however experimentalist should always keep in mind limitations related with each of them.

2.2 Analysis of chemical composition

The analysis of chemical composition of nanoparticles seams to be very important task mainly due to two reasons. First of all structure of the surfactants (ligands) could be studied as well as processes related with its desorption (Puchalski et. al. 2009). The other reason is more straightforward - using chemical composition information about quality of the NPs could be deduced and their stability could be predicted. Finally, information about chemical composition is used as a feedback for chemist making synthesis o nanoparticles. In our experiments we have used two most popular and in some sense complementary techniques: energodispersive X-ray spectroscopy (EDX or EDS) and X-ray photoelectron spectroscopy (XPS).

EDX could be characterized as a global technique which returns qualitative and quantitative chemical information about investigated sample. Due to relatively large mean free path of X-rays in solid it belongs rather to techniques which could be used to characterize bulk than a surface. However it could be also useful in investigation of nanomaterials especially when they could be prepared in form of a powder. It is the case of AgNPs which could be prepared as a powder. In Fig 6a a EDX spectrum recorded on silver powder is presented. In the middle part of the spectrum five peaks located between 2 keV and 4 keV could be seen. These maxima are related with silver K and L core levels. The maximum which is located at 0.2 keV is related with carbon. The small maximum located at 0.5 keV is related to oxygen 1s characteristic line. The carbon and oxygen signals recorded in our sample indicate the presence of alkyl chains. Beside identification of the elements present in the sample by the use of EDX it is also possible to estimate their concentration. Such analysis is shown in Fig 6b. To obtain quantitative information we used SEMQuant software with implemented ZAF procedure. Analysis of recorded spectra (Fig. 6a) proved high silver contents in silver powder. That contents was evaluated for 74%. Beside that we also estimated amount of carbon and oxygen which contents was estimated to 22% and 4% respectively. This results seams to describe the chemical composition of ligand-stabilised silver nanoparticles with high precision. It is because the 5 nm nanoparticles consist of about 4200 silver atoms (Medasani et. al., 2007) and single ligand (derivative of myristic acid) consist 2 oxygen and 14 carbons atom.

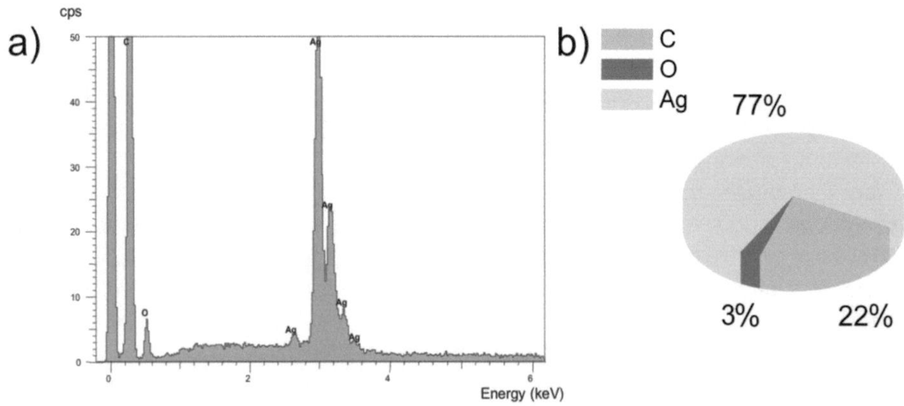

Fig. 6. The results EDX analysis of silver nanoparticles: (a) survey spectrum, (b) quantitative analysis (Puchalski et. al. 2007).

Unfortunately the main problem related with using EDX is related with it global character. To investigate samples with higher precision and in some sense more locally XPS have to be used. This technique is classified as a surface sensitive because a typical analysis depth is limited to the thickness of ~7 nm. Because of such attenuation depth it seems to be perfect for performing investigations on clusters and nanoparticles. In Fig. 7a survey XPS spectrum of AgNPs powder is shown. One can see that the most intense lines correspond to the core levels of carbon, silver and oxygen similarly like in case of EDX. This technique could also be used to obtain quantitative information about sample composition (Fig 7b).

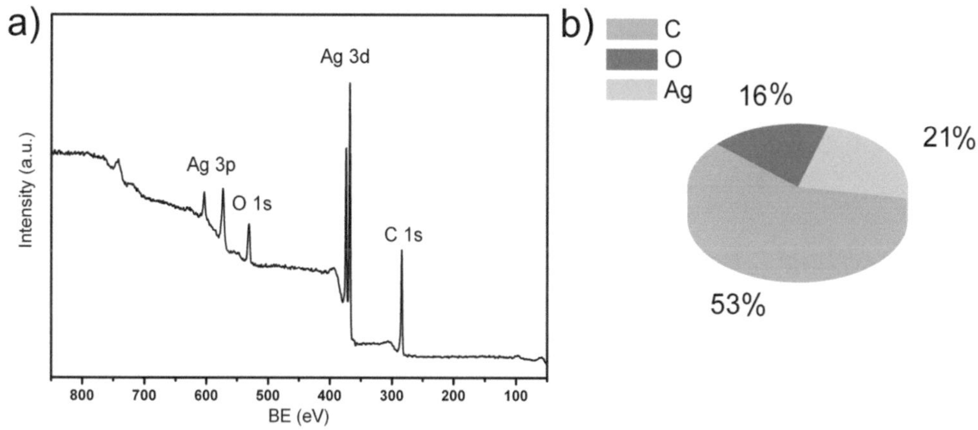

Fig. 7. The results XPS analysis of silver nanoparticles: (a) survey spectrum, (b) quantitative analysis

Or results suggest that sample is composed of 21% of sliver, 53% of carbon and 16% of oxygen. These numbers do not agree with similar estimation made using EDX. The main reason for this discrepancy is related with surface character of XPS. It have to be remembered that samples prior loading to UHV were kept in air which resulted in formation of water layer on top of our sample. Such layer is responsible for increase of amount of oxygen in our results. Relatively low amount of silver recorded during our experiment is misleading. Silver is buried under the layer of surfactants which results in screening of the photoelectrons emitted from cores by ligands. As a consequence we are dealing with decreasing of the Ag signal.

The biggest advantage of the XPS is related with its possibility of detecting of chemical bonds which are in the sample. In Fig 8 a C 1s and O 1s core level lines are shown. Vertical lines were added as a guide for an eye and denote location of main components. Carbon C 1s line is dominated by the C-C and C-H signal located at 284.9 eV. The presence of this type of bonds is related to alkyl chains used in our case as ligands. Additionally, carbon-oxygen bonds are detectable: C-O (285.9 eV), C=O (287.3 eV), O-C=O (289.1 eV), which are typical for fatty acids. The oxygen O 1s peak consists of C-O or H_2O (533.2 eV) and C=O (532.2 eV) components. Furthermore, an additional component located at 531.3 eV is seen and in our opinion this could be related to Ag-O-C (531.3 eV) (Gerenser L.J. 1988).

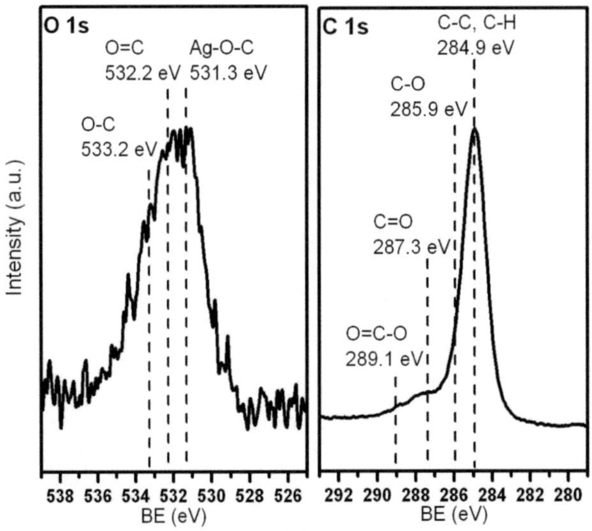

Fig. 8. XPS spectra of O 1s and C 1s core level lines.

2.3 UPS/STS investigations of AgNPs electronic structure

Another techniques which could be used to investigate silver nanoparticles are UPS and STS. Both these techniques allows to investigate electronic structure of materials in vicinity to the Fermi level. Knowledge of electronic structure is very important especially in case of application of materials in electronics. In our investigations we were mainly interested in using of the nanoparticles as a building blocks for conductive tracks formation. Therefore it is reasonable to use these techniques especially to study sintering process which should transfer initially insulating material into conductor. It have to be also pointed out here that both techniques were successively used in studies of quantum size effects which took place in small nanoparticles.

As it was mentioned above we are mainly concentrated on application of silver nanoparticles in Ink-Jet printing in nanoscale electronic. We have decided to investigate the sintering process using STM/STS and UPS. Thus we deposited AgNPs on HOPG (highly oriented pyrolytic graphite) substrate and sintered them in order to create nanometer conductive track. We have decided to use HOPG in these measurements due to two reasons. First of all it is easy to prepare, atomically flat and inert surface. The other reason was related with its semimetallic character. We have expected to observe transition in case of our nanoparticles from insulating state into conducting one. To observe such transition insulating substrate would be the best, unfortunately such substrates are not suitable for UPS and STM measurements thus we decided to select the one with low amount of electronic states located at the Fermi level. The best one in our opinion was HOPG. In Fig. 8 results of our UPS measurements are shown.

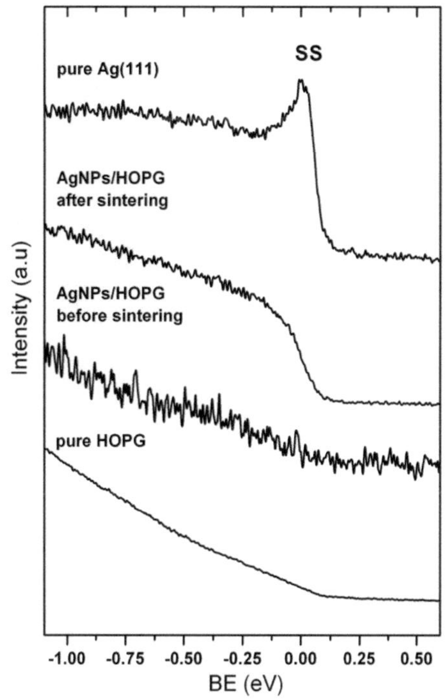

Fig. 8. The comparison of UPS spectra: pure HOPG, AgNPs/HOPG before and after sintering and pure Ag(111).

All spectra were recorded in vicinity to the Fermi level. The pure graphite surface is characterized by soft Fermi edge which resembles its semimetallic character. General shape of the spectrum recorded after deposition of AgNPs is similar to the one recorded on HOPG. The min difference is related with large drop of the intensity recorded on the AgNPs due to insulating character of deposited nanoparticles.

After sintering of NPs at 250°C (after ligand decomposition) the shape of the recorded spectra changed considerably. The distinct Fermi edge could be found which means that surface is covered with metal. It is also clear that obtained spectrum is featureless (see for example the Ag(111) spectrum which has maximum at the Fermi level related with the surface state) which is a clear indication that obtained silver film have polycrystalline character. In order of confirmation of our UPS results we carried out STM measurements on AgNPs deposited on HOPG. Results of these investigations are shown in Fig. 9. We have found out that imaging of NPs deposited on HOPG is extremely hard. In fact during our studies we have been able to record any image which would confirm that we are dealing with nanoparticles on the surface. It is because initially nanoparticles are insulating and STM was designed to investigate conducting or semiconducting materials. However, sintering of the AgNPs changed situation drastically. We have been able to observe large fractal structures composed form small clusters (possibly cores). What is more we measured diameter of individual clusters and it is equal ~5.5 nm which would correspond to presented average diameter of silver core.

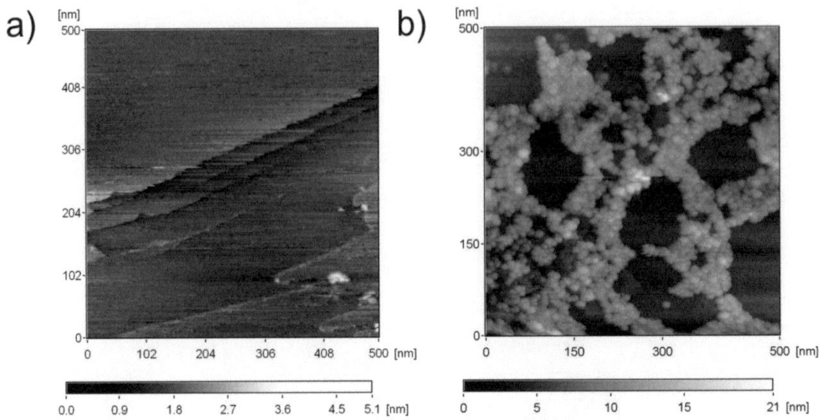

Fig. 9. The STM topography of AgNPs deposited on HOPG before (a) and after (b) sintering at 250°C.

Finally, we decided to use complementary to UPS technique to check electronic structure of obtained Ag film. We performed measurements using scanning tunnelling spectroscopy (STS). Results of these investigations are shown in Fig. 10. The red curve corresponds to a typical I(V) spectrum recorded on HOPG while the blue one on the silver film. It is clear that the curve recorded on Ag film is much steep close to the Fermi level (0 V–bias voltage) which is a measure of number of states at the Fermi level. Thus, we conclude that film which we obtained after sintering of AgNPs is more conductive than HOPG which confirms its metallic character. Furthermore in the Fig. 10b we showed tunnelling conductance maps in which lighter colour corresponds to the region with more electronic states. In some sense, these maps could be interpreted in straightforward way that the lighter areas correspond to more conductive regions of the sample. It is clearly seen that in general Ag is more conductive than surrounding graphite. However, there are still regions with considerably lower number of states than in surrounding regions. It could be related with some amount of ligands which have not been removed completely from the sample. It has to be pointed out here that STS in contrary to UPS allows to investigate sample extremely locally. Because of that these two techniques should be used together to correctly interpret data.

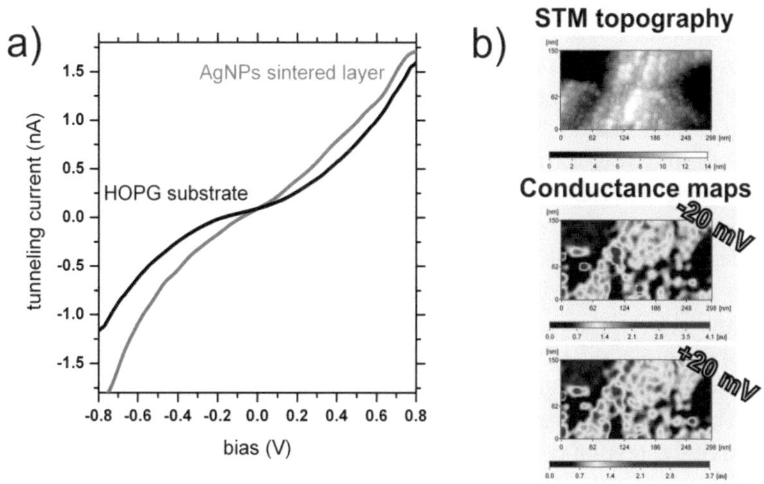

Fig. 10. STS results of AgNPs sintered on HOPG electronic properties investigation. I(V) curve (a) and conductance maps recorded on AgNPs sintered layer on HOPG (b) testify of well conductivity of porous silver fractal aggregates.

3. Summary

The detailed characterisation of nanoparticles is a crucial point in their further application. Without knowledge about their sizes, composition and lengths of ligands nanoparticles could not be used effectively. For example without information about diameters engineers are not able to design jet which with high efficiency would spread nanoparticles on substrate during printing. Without knowledge about composition and length of the ligands it would be difficult to predict what temperature to use during sintering. To learn more about nanoparticles which we are using (commercially available ligand stabilised silver nanoparticles) we have decided to perform series of experiments in order to better understand their properties. We have concentrated here on analysis of AgNPs morphology using HRTEM, AFM, DLS and UV-Vis. Chemical composition was checked using XPS and EDX, and finally electronic structure was investigated using UPS and STS.

The size distribution measured using all four techniques resulted in very similar distribution with average diameter equal 5 nm. This result is especially surprising in case of AFM which should be sensitive also to ligands, therefore expected diameter was equal ~ 8 nm. In our opinion, this discrepancy is the result of distortion of ligands especially at the core substrate interface. Such distortion of ligands could change from one to another substrate which could result in completely different size distributions. From the other point of view the main advantage of that technique is possibility of imaging of individual nanparticles deposited on different substrates. What is more, the imaging process does not necessary have to be carried out in laboratory but by the use of new portable AFMs could be done in any place particularly in the factory immediately after printing. In contrary HRTEM could be used only in laboratory and investigated material have to be deposited on special substrates. However, no distortion of nanoparticles is expected in case of this technique. Thus, it seams to be most precise one which could be used in diameter estimation. What is more, extreme magnifications allow to get information about crystallographic structure of prepared nanoparticles. The two other techniques (DLS and UV-Vis) allow for measurements in solution which is their biggest advantage. Unfortunately, interpretation of UV-VIS spectra is not straightforward which makes this measuring technique rather difficult. In contrary the results generated by DLS are simple in interpretation and the time required for measurement is short which makes this technique the most popular. However, it have to be remembered that in case of such measurement results are usually overestimated due to coalescence of the nanoparticles in the colloid. For all these reasons, none of the presented techniques is best nor worst. Each of them could be used to estimate size distribution however experimentalist should always keep in mind limitations related with each of them.

For chemical characterisation of ligand-stabilized silver nanoparticles we have used EDX and XPS. The both techniques give the same results of qualitative analysis of the studied silver powder. The maxima of carbon, oxygen and silver are clearly observed on recorded spectra. Unfortunately, results of quantitative analysis are completely different. Using EDX we recorded relative concentration of main components at the level of 22%, 3% and 75% respectively while in case of XPS 16%, 22% and 53% respectively. The difference originates from large attenuation depth in case of EDX and small one in case of XPS. Thus, EDX could be called bulk sensitive while XPS is surface sensitive. In case of AgNPs powder EDX is a technique which returns expected values of elements concentration. But it would completely fail in case of thin layers of NPs. In contrary XPS does not return correct values but could be successively used for small amounts of NPs. The wrong concentrations of elements returned by XPS are related with shielding of the cores by the ligands as well as by presence of water layer which additionally screens lower deposited layers. Thus, from this point of view XPS is not a good technique for such investigations. But if one knows mean free path of photoelectrons in ligands and cores than is able to calculate length of ligands. This could be the biggest advantage of that technique. Finally, XPS is sensitive to chemical bonds which is another advantage. For example our results confirmed ligand stabilisations of silver core by detection of Ag-O-C bonds (531.3 eV)

Finally we have presented result of our investigations by the use of UPS and STS. Both techniques are used to investigate density of electron states (DOS) in valence band and conduction band (STS only). Interpretation of UPS results is straightforward - recorded signal intensity is measure of DOS while in case of STS it is more difficult due to convolution of DOS with tunneling transmission coefficient. It have to be stressed here that STS is local and UPS global which sometimes disallow for straightforward comparisons of obtained results. In our experiments we have concentrated on investigation of sintering process. Both techniques successively showed that after sintering nanoparticles become conductive. Because local and global character of these two techniques should to be used together to interpret correctly data.

4. Acknowledgments

The part of works was supported by the Ministry of Science and Higher Education, Poland, grant No R02 021 03.

5. References

Abe, K., Hanada, T., Yoshida, Y., Tanigaki, N., Takiguchi, H., Nagasawa, H., Nakamoto, M., Yamaguchi, T. & Yase K. (1998). Two-dimensional array of silver nanoparticles. *Thin Solid Films*, Vol. 327–329, August 1998) 524–527, ISSN 0040-6090

Ebenstein, Y., Nahum, E. & Banin U. (2002). Tapping Mode Atomic Force Microscopy for Nanoparticle Sizing: Tip–Sample Interaction Effects. *NanoLetters*, Vol. 2, No. 9, (July 2002) 945-950, ISSN 1530-6984

Fuller, S.B., Wilhelm, E.J. & Jacobson J.M. (2002). Ink-Jet Printed Nanoparticle Microelectromechanical Systems. *Journal of Microelectromechanical Systems*, Vol. 11, No. 1, (February 2002) 54-60, ISSN 1057-7157

Gerenser L.J., (1988) An x-ray photoemission spectroscopy study of chemical interactions at silver/plasma modified polyethylene interfaces: Correlations with adhesion. *Journal of Vacuum Science & Technology A*, Vol. 6, (September 1988) 2897-2903, ISSN 0734-2101

Hornyak, G.L., Peschel, S., Sawitowski T. & Schmid G. (1998). TEM, STM and AFM as Tools to Study Clusters and Colloids. *Micron* Vol. 29, No. 2/3, (January 1998) 183-190, ISSN 0968-4328

Huangand, H. & Yang, X. (2004). Synthesis of polysaccharide-stabilized gold and silver nanoparticles: a green method. *Carbohydrate Research*, Vol. 339, (September 2004) 2627–2631. ISSN 0008-6215

Jiang, P., Xie, S.-S., Peng, S.-J. & Gao, H.-J. (2002). The combining analysis of height and phase images in tapping-mode atomic force microscopy: a new route for the characterization of tiol-coated gold nanoparticles on solid substrate. *Applied Surface Science*, Vol. 191, (March 2002) 240-246, ISSN 0169-4332

Khanna, P.K., Kulkarni, D. & Beri R.K. (2008). Synthesis and characterization of myristic acid capped silver nanoparticles. *Journal of Nanoparticles Research*, Vol. 10, (March 2008) 1059–1062, ISSN 1530-6984

Klabunde, K.J. (2001). *Nanoscale Materials in Chemistry*, John Wiley and Sons, Inc., ISBN 0471-38395-3, New York

Klusek, Z., Olejniczak, W., Pawlowski, S. & Kobierski, P. (1998) Modular Scanning System. *Elektron Technology*, Vol. 31, No. 3/4, (March 1998) 508-511, ISSN 0070-9816

Mc Farlandand, A.D. & Van Duyne, R.P. (2003). Single Silver Nanoparticles as Real-Time Optical Sensors with Zeptomole Sensitivity. *NanoLetters*, Vol. 3, No. 8, (November 2003) 1057-1062, ISSN 1530-6984

Medasani, B., Park, Y.H. & Vasiliev I. (2007). Theoretical study of the surface energy, stress, and lattice contraction of silver nanoparticles. *Physical Review B*, Vol. 75, (June 2007) 235436-1-235436-6, ISSN 1098-0121

Mishchenko, M.I., Travis, L.D. & Lacis, A.A. (2002). *Scattering, Absorption, and Emission of Light by Small Particles*, Cambridge University Press, ISBN 0521-78252, Cambridge

Moscicki, A., Felba, J., Gwiazdzinski, P. & Puchalski, M. (2007). Conductivity improvement of microstructures made by nano-size-silver filled formulations, *Proceedings of Polytronic 2007*, pp. 14-17, ISBN, Japan, January 2007, IEEE, Tokyo

Nagasawa, H., Maruyama, M., Komatsu, T., Isoda, S. & Kobayashi T. (2002). Physical Characteristics of Stabilized Silver Nanoparticles Formed Using a New Thermal-Decomposition Method. *physica status solidi (a)*, Vol. 191, No. 1, (October 2002) 67–76, ISSN 0031-8965

Lu, Y., Liu, G.L. & Lee, L.P. (2005). High-Density Silver Nanoparticle Film with Temperature-Controllable Interparticle Spacing for a Tunable Surface Enhanced Raman Scattering Substrate. *NanoLetters*, Vol. 5, No. 1, (November 2005) 5-9, ISSN 1530-6984

Puchalski, M., Dabrowski, P., Olejniczak, W., Krukowski, P., Kowalczyk, P.J. & Polanski, K. (2007). The study of silver nanoparticles by scanning electron microscopy, energy dispersive X-ray analysis and scanning tunnelling microscopy. *Materials Science-Poland*, Vol. 25, No. 2, (March 2007) 473-478, ISSN 0137-1339

Puchalski, M., Kowalczyk, P.J., Zasada, I., Krukowski, P. & Olejniczak, W. (2009). Alloying process at the interface of silver nanoparticles deposited on Au(1 1 1) substrate due to the high-temperature treatments. *Journal of Alloys and Compounds*, Vol. 481, (March 2009) 486–491, ISSN 0925-8388

Shahverdi, A.R., Fakhimi, A., Shahverdi, H.R. & Minaian S. (2007). Synthesis and effect of silver nanoparticles on the antibacterial activity of different antibiotics against Staphylococcus aureus and Escherichia coli. *Nanomedicine*, Vol. 3, (February 2007) 168–171, ISSN 1549-9634

Shim I.-K., Lee, Y., Lee, K.J. & Joung J. (2008). An organometallic route to highly monodispersed silver nanoparticles and their application to ink-jet printing. *Materials Chemistry and Physics*, Vol. 110, (August 2008) 316–321, ISSN 0254-0584

Smith, P.J., Shin, D.-Y., Stringer, J.E., Derby, B. & Reis, N. (2006). Direct ink-jet printing and low temperature conversion of conductive silver patterns. *Journal of Materials Science*, Vol. 41, (May 2006) 4153–4158, ISSN 0022-2461

Tsuji, T., Thang, D.-H., Okazaki, Y., Nakanishi, M., Tsuboi, Y. & Tsuji, M. (2008). Preparation of silver nanoparticles by laser ablation in polyvinylpyrrolidone solutions. *Applied Surface Science*, Vol. 254, (February 2008) 5224–5230, ISSN 0169-4332

Xue, F., Liu, Z., Su, Y. & Varahramyan, K. (2006). Inkjet printed silver source/drain electrodes for low-cost polymer thin film transistors. *Microelectronic Engineering*, Vol. 83, (September 2006) 298–302, ISSN 0167-9317

4

In situ photochemically assisted synthesis of silver nanoparticles in polymer matrixes

Lavinia Balan, Jean-Pierre Malval and Daniel-Joseph Lougnot
Institut de Science des Matériaux de Mulhouse LRC 7228,
15 rue Jean Starcky, 68093 Mulhouse
France

1. Introduction

The synthesis of dispersed nanoparticles is essential for many advanced applications because of their novel properties that are greatly different from those of corresponding bulk substances. During the past years, a whole bunch of synthetic methods of metal nanoparticles (NPs) have been developed: chemical, photochemical and thermal. Amongst them, the photochemical method has attracted much attention due to it being a versatile and convenient process with distinguishing advantages such as space-selective fabrication (Sakamoto et al., 2009). This method was the key of the development of silver photography but recent advances in the chemistry of metal NPs and nanomaterials gave it a new lease of life. One of the main interests of metal NPs stems from their unique physical properties, which can be addressed by the chemical control of their shape and size (Burda et al., 2005).
For instance, silver nanoparticles with spherical shape and nanometer size exhibit a very intense absorption band in the visible region due to the surface plasmon resonance. The absorption coefficient can be orders of magnitude larger than strongly absorbing organic chromophores. Besides, the enhanced electromagnetic fields generated in the close-proximity of the metal surface have a strong influence on the local environment which is illustrated by surface-enhanced Raman scattering (SERS) or by metal-enhanced fluorescence process. Nanocomposite materials combine the different properties of the components. However, in a bulk medium, the benefit of these unique performances mainly relies on the homogeneous dispersion of uniformly shaped and sized particles into the matrix. Generally, metal-polymer nanocomposites are obtained via multi-step methods. Thus, silver nanoparticles can be produced beforehand, and then dispersed into a polymerizable formulation to obtain self-assembly functionalized structures. However, besides the specific hazards related to handling dry nanoparticules, this *"ex-situ"* method is limited by the difficulty to control their monodispersity over a large scale (Balan et al, 2006, 2008).
In the *"in-situ"* approach, the metal nanoparticles are generated in a polymerizable medium from cationic precursors that exhibit better dispersion ability and that undergo facile chemical or photochemical reduction.
This study focuses on the *in-situ* synthesis of silver nanoparticles in polymer matrix through photo-assisted processes. *In-situ* photochemical fabrication is one of the most powerful

approach to synthesize metal NPs/polymer nanocomposites. Furthermore, embedding nano-sized metal NPs into polymer matrix has attracted much interest because these materials open new perspectives; they combine properties from both inorganic and organic systems. Thus, metal NPs homogeneously dispersed in polymer matrixes are widely investigated. They are already used as sensors (Freeman et al., 1995; McConnell et al., 2000; Duan et al., 2001), as materials with solvent switchable electronic properties (Holmes et al. 2000), as optical limiters or filters (Jin et al., 2001; Biswas et al., 2004), as optical data storage (Ouyang et al., 2004, 2005), surface plasmon enhanced random lasing media (Hao et al., 2007), catalytic applications (Boyd et al., 2006) , or for antimicrobial coatings (Sambhy et al., 2006; Anyaogu et al., 2008) Up to now, the fabrication of metal/polymer nanocomposites in poly(methyl methacrylate) (PMMA), poly(vinyl alcohol) (PVA), poly(vinyl acetate) (PVAc), poly(vinyl carbazole) (PVK), polyimide films and N-isopropylacrylamide (NIPAM), diallyldimethylammonium chloride (DADMAC) was reported (Weaver et al.1996; Malone et al., 2002; Hirose et al., 2004; Sakamoto et al., 2006, 2007).
Similarly to what is widely-known in homogeneous solution, direct photoreduction and photosensitization are powerful approaches for the *in-situ* synthesis in polymer matrixes (Balan et al., 2008, 2009; Yagci et al., 2008). The heart of the photochemical approach is the generation of M^0 in such conditions that their precipitation is thwarted. M^0 can be formed through direct photoreduction of a silver source, silver salt or complex, or reduction of silver ions using photochemically generated intermediates, such as radicals.
The photoreduction is often promoted by dyes dispersed or dissolved in the polymer or present in the chemical structure of the matrix. In this one-step approach, we report a strategy involving the photoinduced formation of homogeneous silver nanoparticles in an acrylate polymer stemming from a crosslinking photopolymerization of an acrylate monomer (Scheme 1).

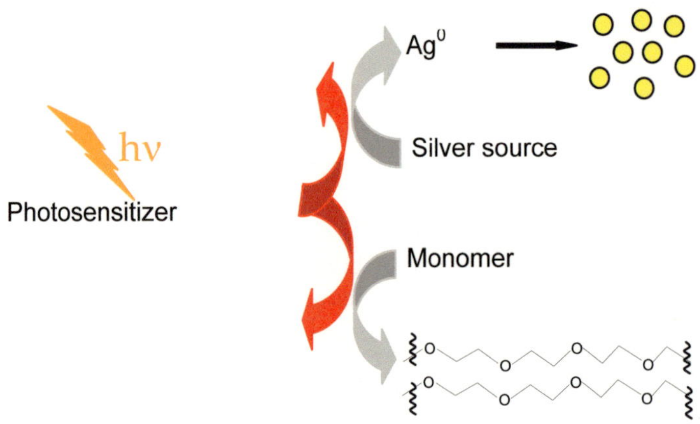

Scheme 1. Scheme of photochemical synthesis

Finally, no mean advantage of this *in situ* synthetic route to NPs and polymer/metal nanocomposite is its high flexibility in terms of actinic wavelengths used to trigger the photochemically assisted reduction of metal precursors (Scheme 2). Several formulations using laser sources with emissions ranging from the near-UV to the near-IR will exemplify this interesting feature.

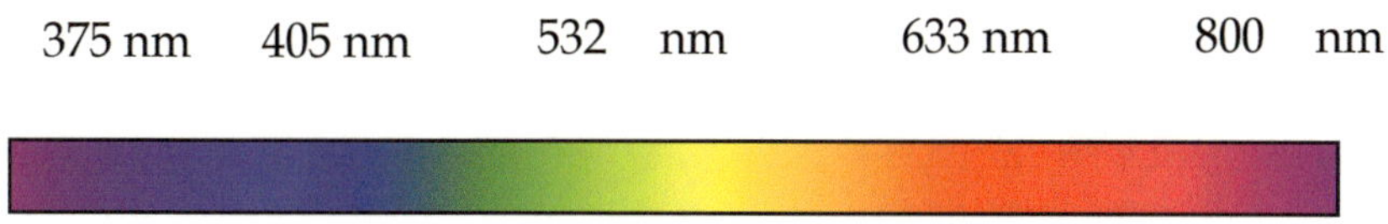

Scheme 2. The laser lines used to generate silver NPs *in situ* in polymerizable formulations

2. General Techniques

The absorption spectra and kinetic follow-ups were recorded with a Perkin Elmer Lambda 2 spectrometer. A FluoroMax 4 Luminescence Spectrometer was used for the fluorescence and time-gated phosphorescence measurements. Low temperature experiments were carried out in a glassy matrix of isopropanol using a 5-mm diameter quartz tube inside a Dewar filled with liquid nitrogen. The emission spectra were spectrally corrected in all cases.
The reference photopolymerizable formulation contained Eosin Y (0.1 wt %), Methyldiethanolamine - MDEA (3 % wt) and $AgNO_3$ (1 wt %) in acrylate monomer. It was sandwiched between two glass plates with a calibrated thickness wedge setting the optical path length to ca 30 µm. Photochemical reactions were carried out at 532 nm with a cw Verdi laser from Coherent. The progress of the reaction was monitored via UV-Vis absorption spectroscopy. The other formulations with various spectral sensitivity windows were formulated according to the same principle.
The photopolymerization was followed up in situ by real-time Fourier transformed infrared spectroscopy with an AVATAR 360 FTIR spectrometer from Nicolet. The laminated formulation (typ. 25 µm thick), deposited on a BaF_2 pellet, was irradiated at 532 nm with a green laser diode module from Crystalaser. The conversion rates were deduced from the disappearance of the vinyl C=C stretching vibration band at 1630 cm^{-1}.

Transmission electron microscopy (TEM) was used to characterize the size and shape of Ag nanoparticles. The nanocomposites were cut by means of a microtome (LKB model 8800) and placed onto the observation grid to get their TEM images. Transmission electron microscopy measurements were carried out at 200 kV using a Philips CM20 instrument with Lab6 cathode.

3. Results and discussion

The key step of the process is the reaction of silver cations with photogenerated transient species that are able to both reduce them to silver metal atoms and initiate the polymerization of the host medium. Two classes of photoinduced reactions were used to produce these primary radicals. The first one is based on the reaction of an electron rich molecule (amine, thiol, ether…) with the highly oxidant triplet state of a sensitizer excited upon absorption of the actinic photons. The second one involves the direct homolytic photocleavage of a sigma bond (mainly C-C bonds adjacent to a carbonyl).

3.1 Photo-oxidation – Activation at 532 nm

A dual system based on Eosin Y (E) and N-methyl diethanolamine (MDEA) was used to simultaneously photogenerate silver nanoparticles and photoinitiate the free radical polymerization. After a mechanistic analysis of the photoprocess by steady state spectroscopy, the elementary steps leading to the metal nanocomposite are correlated to structural properties of the reactants.

3.1.1 Mechanistic aspect of Ag nanoparticles photogeneration in solution

Figure 1 shows the lowest energy absorption band of EO^{2-} localized in the 450-575 nm region with a maximum at 530 nm in acetonitrile. The fluorescence spectrum of EO^{2-} is mirror image of its absorption band with a slight Stockes shift (535 cm^{-1}) suggesting a weak geometrical relaxation in the singlet state.

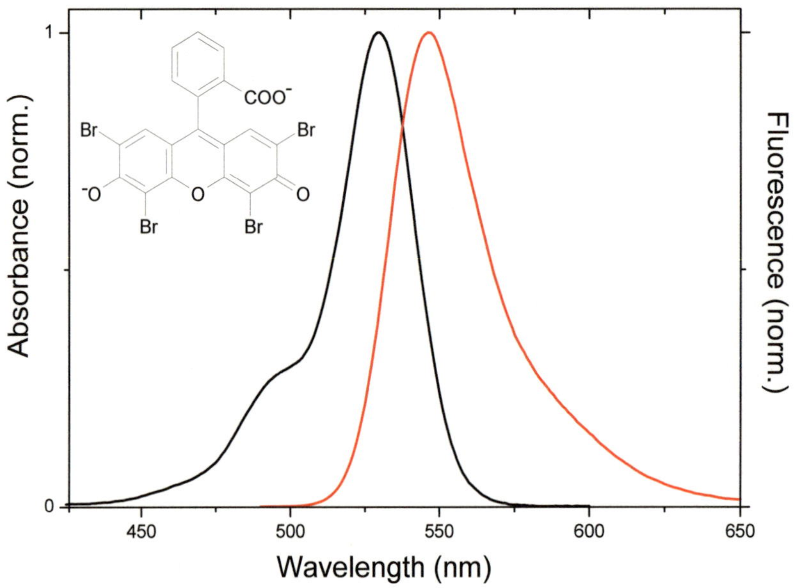

Fig. 1. Normalized absorption and fluorescence spectra of Eosin Y (solvent : acetonitrile)

Figure 2 shows the evolution of the absorption spectrum of EO^{2-} upon addition of increasing amounts of $AgNO_3$ in an alkaline solution of acetonitrile. The last absorption band is progressively shifted to the red ($\Delta\lambda_{max}$ = + 4 nm) and the presence of two isosbestic points located at 498 nm and 530 nm indicates the presence of an equilibrium in the ground state. The presence of two isosbestic points located at 498 nm and 530 nm clearly indicates the occurrence of an equilibrium in the ground state. Moreover, the addition of tetra-N-butylammonium hexafluoroborate (3 x 10^{-2} M) excludes any variation of the ionic strength during the reaction. Therefore, these observations can be ascribed to a complexation reaction of Ag^{+} by EO^{2-} which leads to the formation of an ion-pair complex [EO-Ag]. Under the same conditions, fluorescence emission shifts to the red ($\Delta\lambda_{max}$ = + 5 nm) with a slight decrease in intensity while the fluorescence lifetime decreases from 4.05 ns to 3.80 ns.

Because the acidic character of the hydroxylic group is stronger than that of the carboxylic group (Levillain & Fompeydie, 1985; Moser & Grätzel, 1984), such a cation exchange reaction should mainly involve the chelation of silver cation by the 'phenolate' oxygen. This is also in line with the Pearson's hard-soft acid-base (HSAB) principle (Pearson, 1963, 1968) which presumes a better stabilizing interaction of Ag^+ (Soft Lewis acid) with the phenolate group than with the carboxylate function. However, the participation of the carboxylate function in the coordination reaction cannot be excluded. The inset of Figure 3 shows the best-fitting of the experimental data using the method of least squares; log K1:1 exhibits a high value of ca. 4.7 ± 0.3 whereas log K1:2 has a value of ca. 5.0 ± 0.4. Such a slight difference suggests that the formation of the [1:2] complex can be reasonably neglected under the experimental conditions of this work (i.e. $[Ag^+]/[EO^{2-}] < 50$).

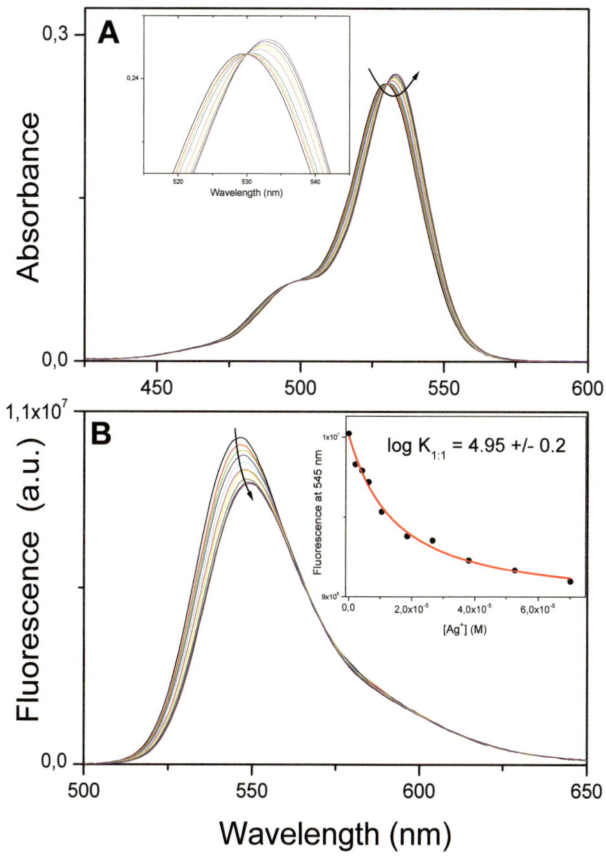

Fig. 2. (A) Evolution of the absorption spectrum of Eosin upon addition of $AgNO_3$ (solvent: acetonitrile), Inset: Isosbestic point at 530 nm. (B) Evolution of the fluorescence spectrum of Eosin upon addition of $AgNO_3$ (solvent acetonitrile), Inset: Dependence of fluorescence intensity at 545 nm vs. concentration of silver cations (circles). Best fitting curve (red line)

Results of the steady state photolysis experiments are shown on figure 3. The changes in the spectral features observed upon laser excitation of Eosin Y and $AgNO_3$ (50 eq.) solution in acetonitrile are reported on Figure 3. Upon increasing the exposure, the absorption band of EO^{2-} was progressively fading away while no trace of the formation of silver nanoparticles could be detected. The same photolysis experiment was then performed in an acetonitrile solution containing Eosin Y, $AgNO_3$ and MDEA. In this case, a new band centred around 435 nm and characteristic of the surface plasmon (SP) of silver nanoparticles was clearly developing (Figure 3). The only species capable of reducing Ag^+ in the presence of MDEA and under visible irradiation is a transient photoproduct deriving from Eosin. This suggests that the formation of silver nanoparticles was promoted by the excited state of EO^{2-} and confirms the mediating role of MDEA. Thus, the sequence of reaction would involve first, an electron transfer from the amine to $^3EO^{2-}$, and then a proton transfer within the ion pair formed between amine radical cation and Eosin radical anion (Jones & Chatterjee, 1988; Kepka & Grossweiner, 1971; Rele et al., 2004; Janata et al., 1994; Burget et al., 1999). Hence, the reaction should initially produce an α-aminoalkyl radical and the conjugated acid of semi-reduced Eosin ($EOH^{\bullet 2-}$). The photogenerated α-aminoalkyl that is known as a strongly reductive species can convert efficiently Ag^+ to Ag^0 as follows:

$$^3[EO\text{-}Ag]^- + MDEA \rightarrow EO^{\bullet 3-} + Ag^+ + MDEA^{\bullet +}$$
$$EO^{\bullet 3-} + Ag^+ + MDEA^{\bullet +} \rightarrow EOH^{\bullet 2-} + Ag^+ + MDEA^{\bullet}$$
$$Ag^+ + MDEA^{\bullet} \rightarrow Ag^0 + MDEA^+$$

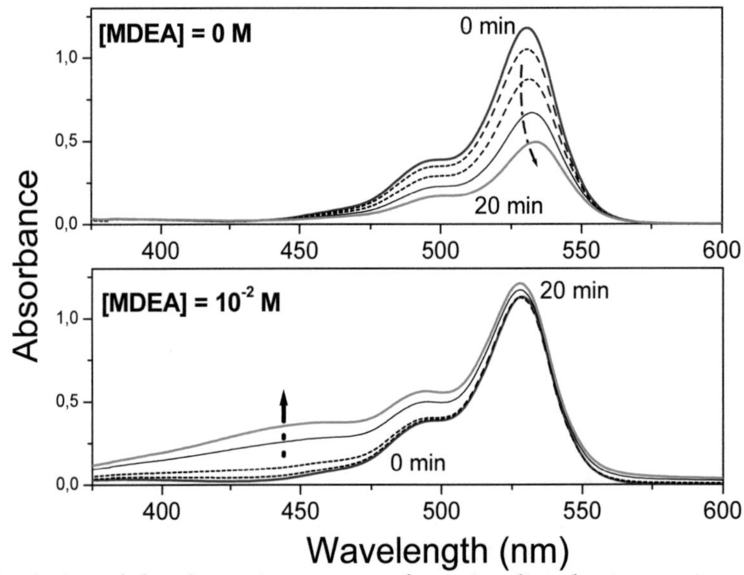

Fig. 3. Evolution of the absorption spectra of two irradiated mixtures in acetonitrile (λ_{irr} = 532 nm) (a) without MDEA and (b) with MDEA

3.1.2 In situ generation of Ag^0 embedded in a crosslinked polymer.

An acrylate formulation which contained EO^{2-} (0.1 wt %) and MDEA (3 % wt) (Espanet et al., 1999; Rathore et al., 2005).was mixed with $AgNO_3$ (1 wt %) and then photopolymerized at 532 nm (2.5 $mW.cm^{-2}$). The conversion of the acrylate double bonds was followed up by real-time FTIR at 1630 cm^{-1} and was compared with the conversion rate of a reference formulation without $AgNO_3$ (Figure 4).

Basically, addition of Ag^+ does not perturb the polymerization kinetics. After a 10-min exposure, the reference sample turned from pink to colourless whereas the sample with Ag^+ turned from pink to brown-yellowish. The visible absorption band of EO^{2-} decreased progressively whereas the plasmon band developed in the 350-500 nm region with a maximum at 437 nm and a FWHM of 115 nm. MDEA acts both as an electron donor in the photoinitiation process (Fouassier & Chesneau, 1991) and as a basic agent that quantitatively converts Eosin into its dianionic form. Figure 5 shows the photobleaching of Eosin Y which goes concomitantly with the growth of the surface plasmon band. The existence of an isosbestic point at 480 nm strongly suggests a simple reaction between EO* and Ag^+ leading to reduced EO^{2-} and Ag^0. This assumption was corroborated by the linear correlation obtained when plotting the absorbance at 532 nm *vs.* absorbance at 432 nm (inset Figure 5). This latter species is then involved into the photo-initiating process in the presence of MDEA.

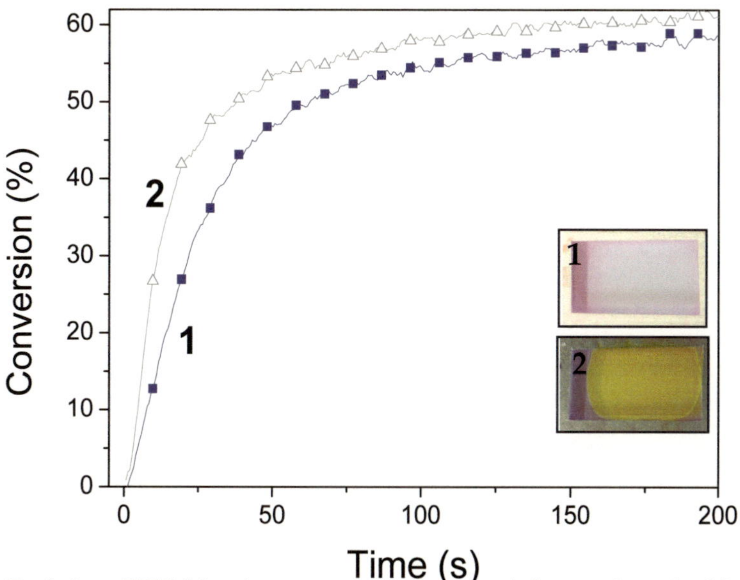

Fig. 4. Real-time FTIR kinetics conversion curves of the acrylate double bonds for visible curing at 532 nm: (1) EO^{2-}/ MDEA, 0.1 wt % / 3 wt % and (2) EO^{2-}/ MDEA / $AgNO_3$, 0.1 wt % / 3 wt % / 1 wt %. Inset: View of 30 μm thick samples (1) and (2) after curing.

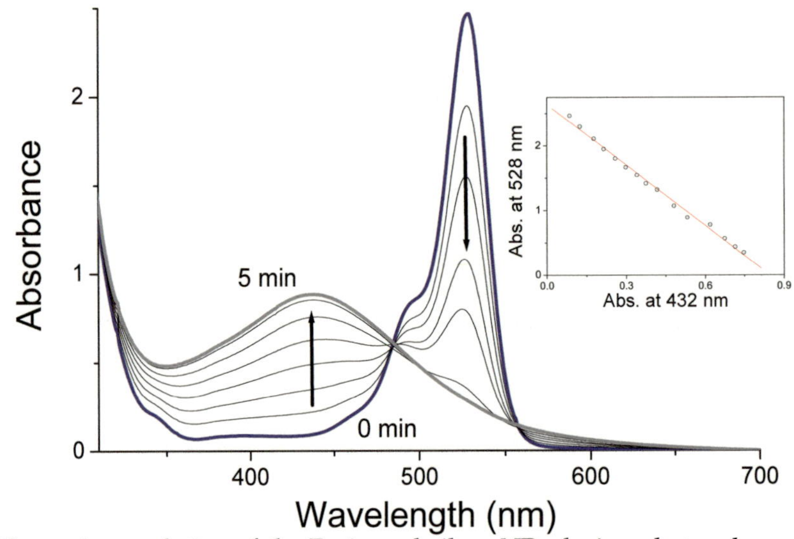

Fig. 5. Absorption evolution of the Eosin and silver NPs during photopolymerization, EO (0.1 wt %) / MDEA (3 wt %) / $AgNO_3$ (1 wt %). Inset: Linear correlation between absorptions at 532 nm and 432 nm.

Transmission electron microscopy analysis of the sample indicated the formation of monodisperse spherical particles whose diameters are in the 3 to 7 nanometer range. Analysis of a population of ca. hundred silver nanoparticles from a portion of the grid indicated that their average diameter was 5.0 ± 0.7 nm (Figure 6). The particles were homogeneous in size and no agglomeration was observed.

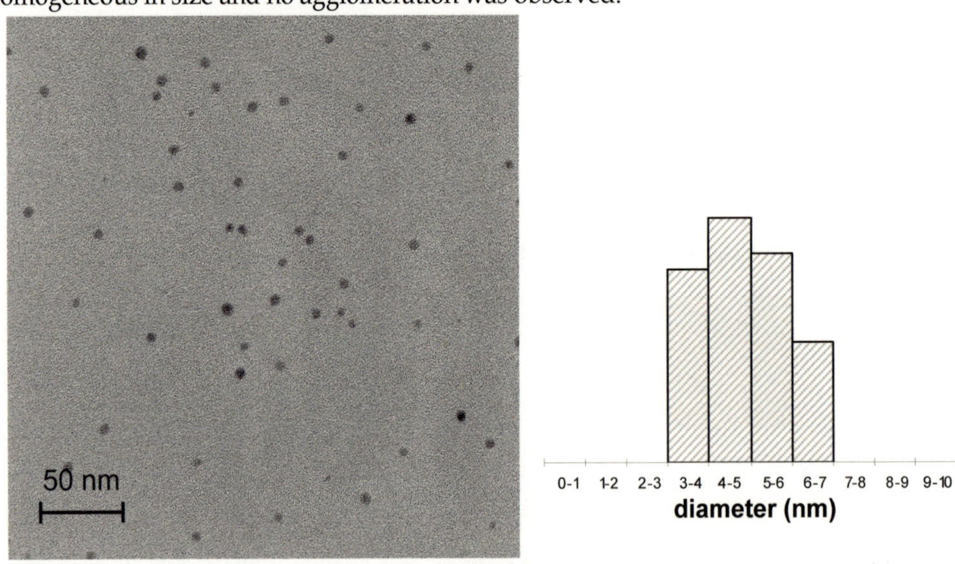

Fig. 6. TEM image of silver nanoparticles embedded in a polyacrylate matrix and histogram of the diameter distribution obtained from this TEM image

3.2 Photocleavage – Activation at 405 nm

An acylphosphine oxide was used as free radical source upon irradiation with near-UV light. Usually, this photoinitiator serves to initiate the photopolymerization of top-coatings formulations by UV A and deep blue sources. In its triplet state, bis-(2,4,6-trimethylbenzoyl)-phenylphosphine oxide (Irgacure 819) is known to undergo α-cleavage and generate a pair of radicals as shown on scheme 3. Indeed, the homolytic photocleavage of the C–P bond generates 2,4,6-trimethylbenzoyl and phenylphosphonyl radicals. In the present application, this photoinitiator simultaneously induces the formation of silver nanoparticles through reduction of $AgNO_3$ and initiates the radical polymerization of the acrylic resin. Thus, irradiation of Irgacure 819 in polyethylene glycol diacrylate monomer (SR 344, Mw = 508 g/mol) in the presence of $AgNO_3$ led to its reduction with rapid generation of both metallic silver and radical initiation of the crosslinking polymerization without any spurious side reactions. The formation of silver particles during the polymerization was confirmed by UV-Vis absorption (the color of the sample turned to brown-yellow).

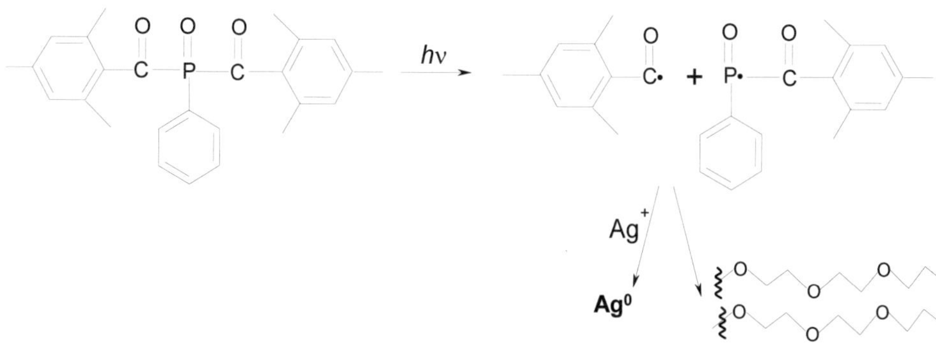

Scheme 3. Photocleavage mechanism of Irgacure 819.

Figure 7 shows the typical evolution of the UV-Visible spectrum of the Irgacure 819/Ag+ solution upon increasing the incident dose of actinic light. Before irradiation, Irgacure 819 exhibited a peak at about 375 nm and a strong absorption at 405 nm. As the photolysis proceeded, the absorption of Irgacure 819 decreased; the photo-generated radicals reduced Ag^+ to Ag^0 while a new band grew up at around 410 nm that corresponds to the surface plasmon resonance of silver metal particles. It must be emphasized that no metal NPs are generated when $AgNO_3$ was irradiated under the same photonic conditions in isopropanol in the absence of Irgacure 819. This observation thus, excluded the intervention of any photo-thermal effect in the process generating NPs.

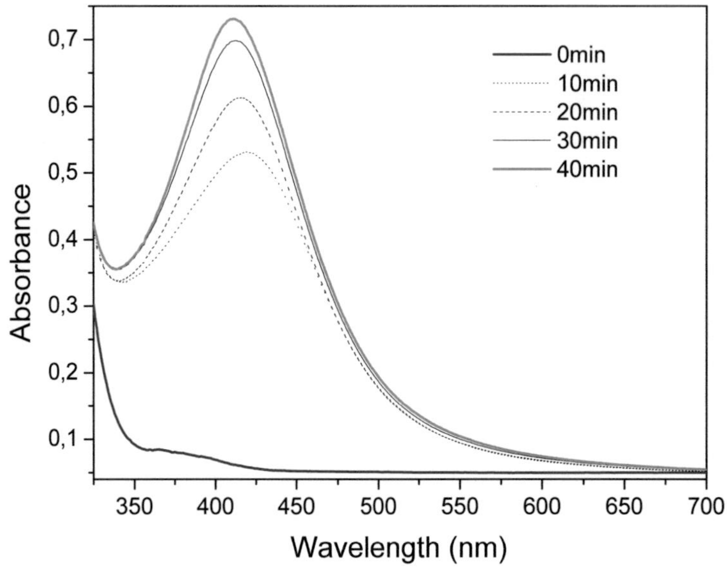

Fig. 7. Evolution of Irgacure 819 and silver absorptions during exposure of a photosensitive silver salt

The bright-field TEM micrograph (Figure 8) of the UV-cured films confirmed the synthesis of spherical particles and showed the well dispersed silver nanoparticles without macroscopic aggregation. The average diameter of the silver nanoparticules was 4.3 ± 0.4 nm.

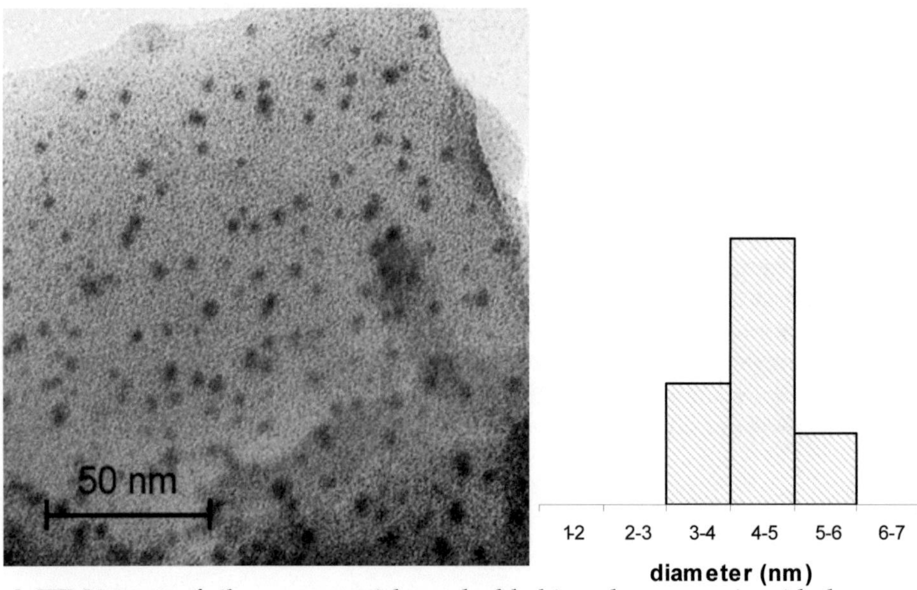

Fig. 8. TEM images of silver nanoparticles embedded in polymer matrix with the respective size distribution.

4. Conclusion

The photo-induced synthesis of silver nanoparticles was carried out in a wide range of experimental conditions: variety of actinic wavelengths (table 1), variety of photoreduction media (solution and acrylic monomer).

Molecule name	Structure	Wavelength (nm)	Co-initiator
2,7-diaminofluorene derivative	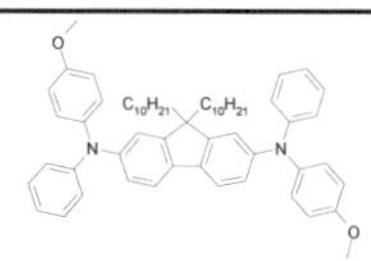	375 nm	*
Bis (2,4,6-trimethylbenzoy) phenylphosphine oxide	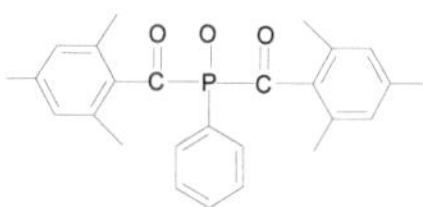	405 nm	-
Eosin Y	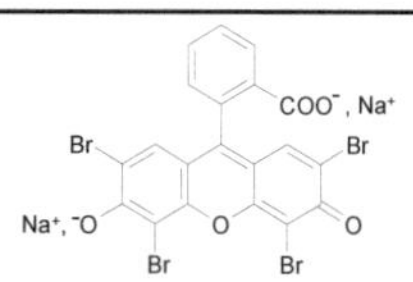	532 nm	*
Methylene blue	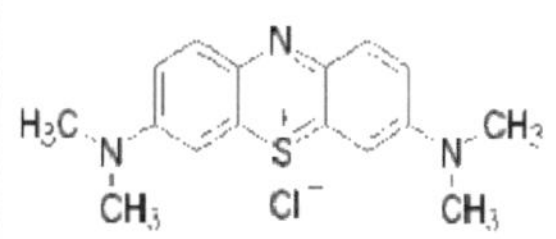	633 nm	*
Cyanine dye	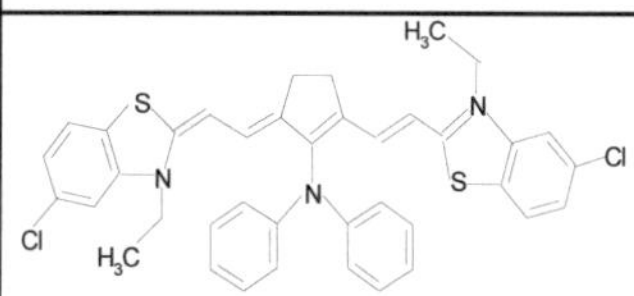	800 nm	*

Table 1. Names and chemical structures of the photoinitiators used

When the dye and silver cations are mixed together in the monomer resin, both photoreduction and photoinitiation occur in parallel without interaction. Silver nanoparticles are homogenously distributed within the polymer network without macroscopic agglomeration and they do not affect the photopolymerization process. Polymer/metal nanocomposite materials are thus, produced within seconds from a liquid formulation.

The one-pot and one-step photochemical process turns out to be a verily innovative route to synthesize silver nanoparticles especially in polymer matrixes that, to top it all, affords high spatial resolution and temporal controllability.

A few reports mentioning the photoinduced synthesis of other geometries than spheres (tetraedrons, sticks...) appeared recently in the literature. However, obtaining particles with well-defined geometries and narrow dispersity remains a challenging issue.

5. References

Anyaogu, K. C.; Fedorov, A. V. & Neckers, D. C. (2008). Synthesis, Characterization, and Antifouling Potential of Functionalized Copper Nanoparticles. Langmuir, 24, 4340-4346

Balan, L. & Burget. D. (2006). Synthesis of metal /polymer nanocomposite by UV-radiation curing. European Polymer Journal, 42, 3180-3189

Balan, L.; Schneider, R. & Lougnot, D. J. (2008). A new and convenient route to polyacrylate/silver nanocomposites by light-induced cross-linking polymerization. Progress in Organic Coatings, 62, 351-357

Balan, L.; Jin, M.; Malval, J-P.; Chaumeil, H.; Defoin, A. & Vidal, L. (2008). Fabrication of silver nanoparticle-embedded polymer promoted by combined photochemical properties of 2,7-diaminofluorene derivative dye. Macromolecules, 41, 9359-9365

Balan, L.; Malval, J-P.; Schneider, R.; Le Nouen, D. & Lougnot, D. J. (2009). In situ fabrication of polyacrylate-silver nanocomposite through photoinduced tandem reactions involving eosin dye. Polymer, DOI: 10.1016/j.polymer.2009.05.003.

Boyd, D. A.; Greengard, L.; Brongersma, M.; El-Naggar, M. Y. & Goodwin, D. G. (2006). Plasmon-assisted chemical vapor deposition. Nanolett., 6, 2592-2597

Burda, C.; Chen, X. B.; Narayanan, R. & El-Sayed, M. A. (2005). Chemistry and properties of nanocrystals of different shapes. Chem. Rev., 105(4), 1025-1102

Burget, D.; Fouassier, J. P.; Amat-Guerri, F.; Mallavia, R. & Sastre, R. (1999). Enhanced activity as polymerization photoinitiators of Rose Bengal and Eosin esters with an O-benzoyl-α-oxooxime group : The role of the excited state reactivity. Acta Polym., 50, 337-346

Duan, X.; Huang, Y.; Cui, Y.; Wang, J. & Lieer, C. M. (2001). In situ fabrication of silver nanoparticles in a photopolymer matrix. Nature, 409, 66-69

Espanet, A.; Ecoffet, C. & Lougnot, D. J. (1999). PEW: Photopolymerization by evanescent waves . II - Revealing dramatic inhibiting effects of oxygen at submicrometer scale. J Polym Sci A: Polym Chem., 37, 2075-2085.

Fouassier, J-P. & Chesneau, E. (1991). Polymérisation induite sous irradiation laser visible, Le système éosine/photoamorceur ultra violet/amine. Makromol. Chem., 192, 245-260

Freeman, R. G.; Grabar, K. C.; Allison, K. J.; Bright, R. M.; Davis, J. A.; Guthrie, A. P.; Hommer, M. B.; Jackson, M. A.; Smith, P. C.; Walter, D. G. & Natan, M. J. (1995) Self-assembled metal colloid monolayers: An approach to SERS substrates. Science, 267, 1629

Hao, F.; Nehl, C. L.; Hafner, J. H. & Nordlander, P. (2007). Plasmon resonances of a gold nanostar. NanoLett., 7(3), 729-732

Hirose, T.; Omatsu, T.; Sugiyama, M.; Inasawa, S. & Koda, S. (2004). Au-nano-particles production by pico-second ultra-violet laser deposition in Au-ion doped PMMA film. Chem. Phys. Lett., 390, 166-169

Holmes, J. D.; Johnston, K. P.; Doty, R. C. & Korgel, B. A. (2000). Control of thickness and orientation of solution-grown silicon nanowires. Science, 287, 1471-1473

Janata, E.; Henglein, A. & Ershovt, B. G. (1994). First clusters of Ag+ ion reduction in aqueous solution. J. Phys. Chem., 98, 10888-10890

Jones, G. & Chatterjee, S. (1988). Steric control of distrace parameters and the yield of charge- carriers in photochemical electron-transfer the quenching of eosin-Y by hindered phenols. J. Phys. Chem., 92, 6862-6864

Kepka, A. G. & Grossweiner, L. I. (1971). Photodynamic oxidation of iodide ion and aromatic amino acids by eosin. Photochem. Photobiol., 14, 621-639

Levillain, P. & Fompeydie, D. (1985). Determination of equilibrium constants by derivative spectrophotometry. Application to the pKas of Eosin..Anal. Chem., 57, 2561-2563

Malone, K, Weaver, S.; Taylor, D.; Cheng, H.; Sarathy, K. P. & Mills, G. (2002). Formation Kinetics of Small Gold Crystallites in Photoresponsive Polymer Gels. J. Phys.Chem. B, 106, 7422-7431

McConnell, W. P.; Novak, J. P.; Brousseau, L. C.; Fuierer, R. R.; Tenent, R.C. & Feldheim, D. L. (2000). Electronic and Optical Properties of Chemically Modified Metal Nanoparticles and Molecularly Bridged Nanoparticle Arrays. J. Phys. Chem. B, 104, 8925-8930

Moser, J.; & Grätzel, M. (1984). Photosensitized electron injection in colloidal semiconductors. J. Am. Chem. Soc., 106, 6557-6564

Ouyang, J.; Chu, C.-W.; Szmanda, C. R.; Ma, L. & Yang, Y. (2004). Programmable polymer thin film and non-volatile memory device. Nat. Mat., 3, 918-922

Ouyang, J.; Chu, C.-W.; Sieves, D.; Yang, Y. (2005). Electric-field-induced charge transfer between gold nanoparticle and capping 2-naphthalenethiol and organic memory cells. Appl. Phys. Lett., 86, 123507-123513

Pearson, R. G. (1963). Hard and Soft Acids and Bases. J. Am. Chem. Soc., 85, 3533-3539

Pearson, R. G. (1968). Hard and soft acids and bases. HSAB, Part I, Fundamental principles. J. Chem. Educ., 45, 581-587

Rathore, R.; Chebny, V. J. & Abdelwahed, S. H. (2005). A versatile and conformationally adaptable fluorene-based receptor for efficient binding of silver cation. J. Am. Chem. Soc., 127, 8012-8013

Rele, M.; Kapoor, S.; Sharma, G. & Mukherjee, T. (2004). Reduction and aggregation of silver and thallium ions in viscous media. Phys. Chem. Chem. Phys., 6, 590-595

Sakamoto, M.; Tachikawa, T.; Fujitsuka, M. & Majima, T. (2000). Acceleration of Laser-Induced Formation of Gold Nanoparticles in a Poly(vinyl alcohol) Film. Langmuir, 22, 6361-6366

Sakamoto, M.; Tachikawa, T.; Fujitsuka, M. & Majima, T. (2007). Photochemical Formation of Au/Cu Bimetallic Nanoparticles with Different Shapes and Sizes in a PVA Film. Adv. Funct. Mater., 17, 857-862

Sakamoto, M.; Fujistuka, M. & Majima, T. (2009). Light as a construction tool of metal nanoparticles: Synthesis and mechanism. J. Photochem and Photobiol C: Photochem Reviews 10(1) 33-56

Sambhy, V.; MacBride, M. M.; Peterson, B. R. & Sen, A. Silver Bromide NanoparticlePolymer Composites: Dual Action Tunable Antimicrobial Materials. J. Am. Chem. Soc., (2006). 128, 9798-9808

Shipway, A N.; Katz, E.; Willner, I.; (2000). Nanoparticle Arrays on Surfaces for Electronic, Optical, and Sensor Applications. Chem Phys Chem, 1, 18-52

Weaver, S.; Taylor, D.; Gale, W. & Mills, G. (1996). Photoinitiated Reversible Formation of Small Gold Crystallites in Polymer Gels. Langmuir, 12, 4618-4620

Linear and nonlinear optical properties of silver nanoparticles synthesized in dielectrics by ion implantation and laser annealing

Andrey L. Stepanov
Kazan Physical-Technical Institute, Russian Academy of Sciences
Russian Federation
Laser Zentrum Hannover
Germany

1. Introduction

Composite metamaterials containing metallic nanoparticles (MNPs) are now as considered a basis for designing new photonic media for optoelectronics and nonlinear optics (Sarychev & Shalaev, 2007). Simultaneously with the search for and development of modern technologies intended for nanoparticle synthesis, substantial practical attention has been devoted to designing techniques for controlling the MNP size. This is caused by the fact that the properties of MNPs, such as the quantum size effect, single-electron conduction, etc., which are required for various applications, take place up to a certain MNP size. An example of their application in optoelectronics is a prototype of integrated electronic circuit - chip that combines metallic wires as conductors of electric signals with fibers as guides of optical signals. In practice, light guides are frequently made of synthetic sapphire or silicon oxide, which are deposited on or buried in semiconductor substrates. In this case, electrooptic emitters and that accomplish electric-to-optic signal conversion are fabricated inside the dielectric layer. This light signal from a microlaser is focused in a light guide and then transmitted through the optoelectronic chip to a high-speed photodetector, which converts the photon flux to the flux of electrons. It is expected that light guides used instead of metallic conductors will improve the data rate by at least two orders of magnitude. Moreover, there is good reason to believe that optical guide elements will reduce the energy consumption and heat dissipation, since metallic or semiconductor components of the circuits may be replaced by dielectric ones in this case. Prototype optoelectronic chips currently available are capable of handling data streams with a rate of 1 Gbit/s, with improvement until 10 Gbit/s in future.

Key elements of dielectric waveguides used for light propagation are nonlinear optical switches, which must provide conversion of laser signal for pulse duration as short as pico- or femtoseconds. The nonlinear optical properties of MNP-containing dielectrics stem from the dependence of their refractive index and nonlinear absorption on incident light intensity

(Flytzanis et al., 1991). This effect is associated with MNPs, which exhibit an enhancement of local electromagnetic field in a composite and, as consequence, a high value of the third order nonlinear susceptibility when exposed to ultrashort laser pulses. Therefore, such MNP-containing dielectric materials may be used to advantage in integrated optoelectronic devices, for example, as shown in Fig. 1.

It is well known (Kreibig & Vollmer, 1995) that a local field enhancement in MNPs stimulates a strong linear optical absorption called as surface plasmon resonance (SPR). The electron transitions responsible for plasmon absorption in MNPs cause also a generation of an optical nonlinearity of a composite in the same spectral range. As a result, the manifestation of nonlinear optical properties is most efficient for wavelengths near the position of a SPR maximum. In practice, to reach the strong linear absorption of a composite in the SPR spectral region, attempts are made to increase the concentration (filling factor) of MNPs. Systems with a higher filling factor offer a higher nonlinear susceptibility, when all other parameters of composites being the same. Usually noble metals and copper are used to fabricate nonlinear optical materials with high values of third order susceptibility. There are variety ways to synthesis MNPs in dielectrics, such as magnetron sputtering, the convective method, ion exchange, sol–gel deposition, etc. One of the most promising enhanced fabrication methods is ion implantation (Townsend et al., 1994, Stepanov, 2005a) because it allows reaching a high metal filling factor in an irradiated matrix beyond the equilibrium limit of metal solubility and provides controllable synthesis of MNPs at various

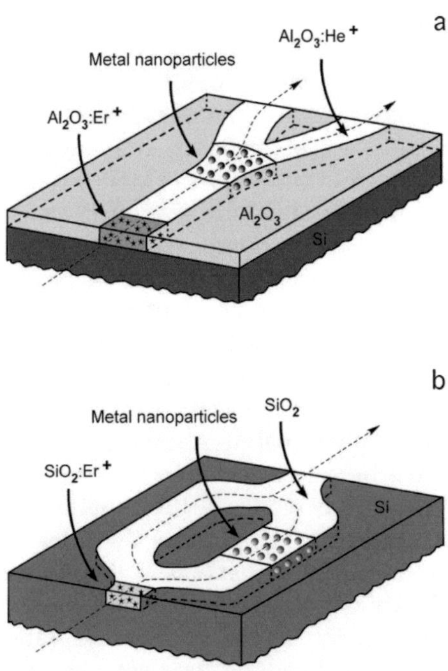

Fig. 1. Prototypes of optoelectronic chip with a dielectric waveguide combined with silicone substrate. Ion implantation can be applied to fabricate selective area doped by rear metal ions (marked by stars) to work as microlaser and to illuminate in waveguide, created by rear-gas ion radiation with MNPs to form an optical switcher.

depths under the substrate surface. Nearly any metal-dielectric composition may be produced using ion implantation. This method allows for strict control of the doping ion beam position on the sample surface with implant dose as, for example, in the case of electron- and ion-beam lithography. Ion implantation is widely used in industrial semiconductor chip fabrication. Therefore, the combination of MNP-containing dielectrics with semiconductor substrates by same technological approach as ion implantation could be reached quite effective. Moreover, ion implantation can be applied for different steps in optoelectronic material fabrication such as creation of optical waveguides by implantation with rear gas ions (H^+, He^+ etc.), a designing of electric-to-optic signal convectors and microlaser by irradiation of dialectics waveguides with rear metal ions (Er^+, Eu^+ etc.) and a synthesis of MNPs (Fig. 1).

The history of MNP synthesis in dielectrics by ion implantation dates back to 1973, when a team of researchers at the Lyons University in France (Davenas et al., 1973, Treilleux et al., 1978) pioneered this method to create particles of various metals (sodium, calcium, etc.) in LiF and MgO ionic crystals. First work on ion-synthesis of noble nanoparticles was done in study of Au- and Ag-irradiated lithia-alumina-silica glasses (Arnold et al., 1975, Arnold & Borders, 1977). Later developments have expanded from the metal implants to the use of many ions and the active formation of compounds, including metal alloys and totally different composition precipitate inclusions. In ion implantation practice MNPs were fabricated in various materials, such as polymers, glass, artificial crystals, and minerals. By implantation, one can produce almost any metal–dielectric composite metamaterials, as follows from Table 1, which gives a comprehensive list of references of various dielectrics with implanted silver nanoparticles with conditions for their fabrications. This chapter focuses on recent advantages in fabrication of silver nanoparticles by low-energy in implantation in various inorganic matrixes. Also some examples of nonlinear optical repose in such composites are presented and discussed.

2. Ion synthesis of metal nanoparticles

Ion implantation is an effective technological tool for introducing single impurities into the surface layer of the substrate to a depth of several micrometers. The degree of surface modification of the materials depends on their individual chemical and structural properties, as well as on variations of implantation parameters, such as the type and energy of an implant, current density in ion beam, substrate temperature, etc. A most critical parameter is ion dose F_0, which determines the implant amount. Depending on the modification of dielectrics by irradiation, ion implantation can be conventionally divided into low-dose and high-dose processes.

In the case of low-dose irradiation ($\sim F_0 \leq 5.0 \cdot 10^{14}$ ion/cm^2), the Ag ions implanted, after stopping and thermalization, are dispersed throughout the volume of the dielectrics and are well separated from each other. The energy of the implant is transferred to the matrix via electron shell excitation (ionization) and nuclear collisions. This causes radiation-induced defects, which, in turn, may reversibly or irreversibly modify the material structure (Townsend et al., 1994). Various types of crystal structure damage have been observed in practice : extended and point defects, amorphization and local crystallization, precipitation of a new phase made up of host atoms or implanted ions, etc.

Matrix type	Ion energy, keV	Ion dose, ion/cm^2	Current density, μA/см^2	Post-implantation heat treatment	Authors
Al_2O_3 crystal <1010>	50 360	$4.0\cdot10^{16}$ $5.0\cdot10^{16}$ $8.0\cdot10^{16}$	1-5	Annealing in air at 650°C, 30 min	Rahmani et al., 1988, Rahmani & Townsend 1989
Al_2O_3 crystal <0001>	$1.5\cdot10^3$ $1.8\cdot10^3$	$(0.2\text{-}2.0)\cdot10^{17}$		Annealing in air at 1100°C, 2 h or in Ar+4%H_2 at 500°C, 2 h	White et al., 1993
Al_2O_3 crystal	25 30	$(0.2\text{-}2.0)\cdot10^{17}$			Steiner et al., 1998
Al_2O_3 crystal	30	$3.8\cdot10^{17}$	3, 6, 10		Ganeev et al., 2005, 2006
Al_2O_3 crystal	160	$(0.1\text{-}1.0)\cdot10^{17}$			Marques et al., 2006
ABSG, BPYR glasses	270	$1.5\cdot10^{16}$			Mazzoldi et al., 1993
MgO crystal (100)	$1.5\cdot10^3$	$1.2\cdot10^{17}$	2-3	Annealing in air at 550 and 1100°C	Qian et al., 1997, Zimmerman et al., 1997
MgO crystal (100)	600	$1.0\cdot10^{16}$		Annealing in air at 1200°C, 22 h	van Huis et al., 2002
MgO crystal (100)	200	$2.0\cdot10^{17}$	2	Some samples annealed in air, Ar, O_2 or 70%N_2+30%H_2 at 300-900°C, 1h	Xiao et al., 2008
$MgOP_2O_5$ glass	150	$(0.1\text{-}1.0)\cdot10^{17}$	0.5-3		Matsunami & Hosono, 1993a
Lithia-alumina-silica glass	275-285	$1.0\cdot10^{16}$	1-2		Arnold & Borders, 1977
$LiNbO_3$ crystal.	50 360	$(4.0\text{-}0.8)\cdot10^{16}$	1-4	Annealing in at 650°C,30min	Rahmani et al., 1988, 1989
$LiNbO_3$ crystal.	20, 25 $3\cdot10^3$ $4.2\cdot10^3$	$(0.5\text{-}8.0)\cdot10^{16}$		Some samples annealed in air at 200 -600°C, 1-3 h	Deying et al., 1994 Shang et al., 1996 Saito & Kitahara, 2000

$LiNbO_3$ crystal.	160 $1.5 \cdot 10^3$	$2.0 \cdot 10^{16}$ $4.0 \cdot 10^{16}$ $1.7 \cdot 10^{17}$		As implanted and annealing in air at 500 - 800°C, 1 h	Sarkisov et al. 1998a, 1998b, 1998c, 1998d, 1999, 2000, Williams et al., 1998a, 1998b, 1999
$LiNbO_3$ crystal.	$1.5 \cdot 10^3$	$2.0 \cdot 10^{16}$		Annealing in Ar at 100 - 1100°C, 30 min	Amolo et al.; 2006
SiO_2 crystal.	200	$(2.3\text{-}9.0) \cdot 10^{16}$	1-5	Annealing in air at 300-500°C, 30 min	Rahmani & Townsend, 1989
SiO_2	65 130 270	$(1.5\text{-}5.0) \cdot 10^{16}$	1 1.5	Some samples annealed in air or 4%H_2	Mazzoldi et al., 1993, Mazzoldi & Mattei, 2007, Antonello et al., 1998, Battaglin et al., 1998, 2001, Bertoncello et al., 1998, Caccavale, 1998, Cattaruzza et al., 1999, Gonella et al., 1999, Osborne, 1998
SiO_2	150	$(0.1\text{-}6.0) \cdot 10^{17}$	1.5-14		Matsunami & Hosono, 1993b
SiO_2	305	$(3.0\text{-}9.0) \cdot \cdot 10^{16}$	2		Magruder III et al., 1994, 1995, 1996, 2009, Anderson et al., 1996, 1997, 1998, 2000, Zuhr et al., 1998
SiO_2	20-58, 130	$(0.4\text{-}2.0) \cdot 10^{17}$	0.6		Pham et al., 1997
SiO_2 crystal.	200	$(2.3\text{-}9.0) \cdot 10^{16}$			Liu et al., 1998a, 1998b, 1998c, 2000
SiO_2	$1.5 \cdot 10^3$	$2.0 \cdot 10^{16}$ $4.0 \cdot 10^{16}$ $1.4 \cdot 10^{17}$	2	Annealing in Ar gas at 500 - 1000°C, 1 h	Ila et al., 1998
SiO_2	65	$5.0 \cdot 10^{16}$			D'Acapito & Zontone, 1999
SiO_2	60	$4.0 \cdot 10^{16}$	10		Stepanov et al. 2000b, 2003b
SiO_2	43	(0.06-	1-2.5	Some samples	Jiang et al., 2000,

	90 150 200 300	2.0)·10^{17}		annealed in air, Ar, O_2 or 70%N_2+30%H_2 at 300-800°C, 1 h	Ren et al., 2004a, 2004b, 2005a, 2005b, 2005c, 2006, 2007, 2008, 2009, Liu et al., 2005, Xiao et al., 2006, 2007a, 2007b, Wang et al. 2007, 2008, Zhang et al., 2004, Cai et al., 2008
SiO_2 sol-gel films	5-100	(5.0-6.0)·10^{16}	1.5		Armelao et al., 2002
SiO_2 on Si	10 30 40	(1.0-5.0)·10^{15} (1.0-5.0)·10^{16} 1.0·10^{17}	2	Annealing in Ar at 500-900°C, 1 h.	Ishikawa et al., 2002, Tsuji et al. 2002a, 2002b, 2003a, 2004, 2005a, 2005b, Arai et al., 2003, 2005, 2006, 2007a, 2007b
SiO_2	2.0·10^3	(0.4-1.0)·10^{17}	2	Some samples annealed in 50%N_2+ 50%H_2 or in air at 230-800°C, 1 h	Roiz et al., 2004, Peña et al., 2007, Cheang-Wong et al., 2007, Rodrigues-Iglesias et al., 2008
SiO_2 on Si	40	0.3·10^{15}		Annealing in vacuum at 550°C, 20 min	Romanyuk et al., 2006
SiO_2	60	(0.3-1.0)·10^{17}	3		Takeda et al., 2006
SiO_2	32-40 1.7·10^3 2.4·10^3	(0.1-1.0)·10^{17}	3-5	Some samples annealed in air at 500°C, 1 h	Joseph et al., 2007a, 2007b, Sahu et al., 2009
SiO_2	0.7, 1.5, 3	(1.2-4.7)·10^{15}	3-5		Carles et al., 2009
SiO_2	200	(0.1-2.0)·10^{17}	< 2.5		Wang et al., 2009a, 2009b
SiO_2+ TiO	305	6.0·10^{16}	7		Magruder III et al., 2007
Si_3N_4	20 130	4.0·10^{16}	0.6		Pham et al., 1997
BPYR glass	270	1.5·10^{16}			Mazzoldi et al., 1993
SLSG	60	4.0·10^{16} 2.0·10^{16}	-	-	Nistor et al., 1993 Wood et al., 1993

SLSG	200	$(0.5\text{-}4.0)\cdot 10^{16}$	0.5-2	-	Dubiel et al., 1997, 2000, 2003, 2008
SLSG	60	$(2.0\text{ -}7.0)\cdot 10^{16}$	10	Some samples annealed in at 350°C, 3 h	Stepanov et al., 1998, 1999a, 1999b, 1999c, 2000a, 2000b, 2000c, 2001a, 2001b, 2001c, 2002a, 2002b, 2002c, 2002b, 2003a, 2003b, 2003c, 2004a, 2004b,2005b, 2008a, 2008b, 2009a, 2009 Hole et al. 1999 Ganeev at al. 2003a, 2003b,2004a, 2004b, 2004c
Ta_2O_5	80-130	$6.0\cdot 10^{16}$	0.6 – 6.4		Pham et al., 1997
TiO_2 crystal	50 65	$(0.3\text{-}1.0)\cdot 10^{17}$	2	Annealing in Ar at 400°C, 1h	Tsuji et al., 2002c, 2003b
TiO_2 sol-gel films	30 65	$(0.1\text{-}0.5)\cdot 10^{17}$	2	Annealing in Ar at 300-600°C, 1 h	Tsuji et al., 2005c, 2006

Table 1. Types of dielectric inorganic matrix with silver nanoparticles fabricated by ion implantation combined with post-implantation heat treatment. (Abbreviations - ANZT glass, $2Ag_2O{\cdot}3Na_2O{\cdot}25ZnO{\cdot}70TeO_2$; ABSG, alkali-borosilicate glass, $0.2K2O{\cdot}0.8B_2O_3{\cdot}3SiO_2$; BPYR – borosilicate Pyrex glass; SLSG, soda-lime silicate glass).

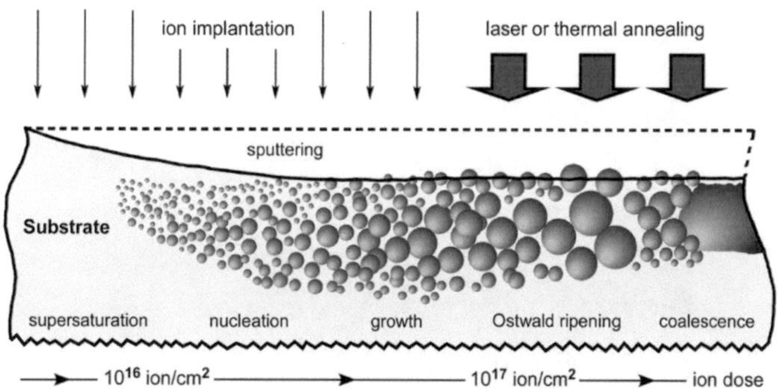

Fig. 2. Basic physical processes (from left to right) involved in the formation of nanoparticle from an implant vs. the ion dose with regard to surface sputtering under irradiation.

The range of high-dose implantation may be divided into two characteristic dose sub-ranges. In the range $10^{15} \le \sim F_0 \le 10^{17}$ ion/cm^2, the concentration of Ag ions exceeds the solubility limit of metal atoms in matrices and the system relaxes by nucleation and growth of MNPs (Fig. 2), as illustrated in plane (Carles, et al. 2009) and cross-section (Ren, et al., 2007) transmission electron microscopy (TEM) views of SiO2 glass with ion-synthesized Ag particles (Fig. 3 and 4). The threshold dose value (at which MNPs nucleate) depends on the type of the dielectric matrix and implant. For example, for 25-keV Ag+-ion implantation into $LiNbO_3$, the threshold dose was found to be $F_0 \sim 5.0 \cdot 10^{15}$ ion/cm^2 (Deying, et al. 1994), for 30-keV silver ions embedded in epoxy glass, F0 ~ 1016 ion/cm2 (Stepanov et al., 1995). The next subrange of high-dose implantation,

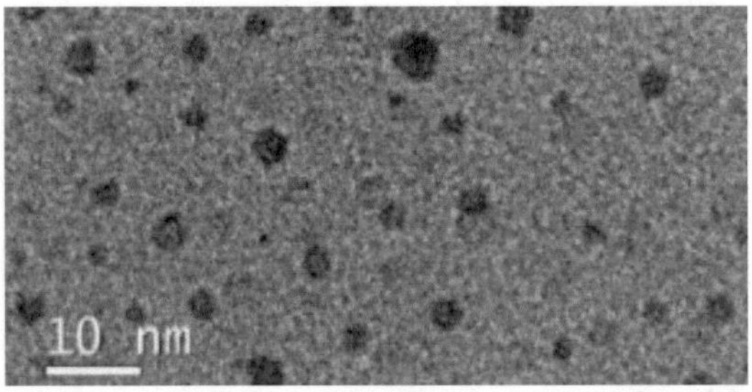

Fig. 3. Plan-view TEM image of SiO_2 with Ag nanoparticles fabricated at a dose of $6.0 \cdot 10^{16}$ ion/cm^2 and an energy of 3 keV. Fragment from (Carles, et al. 2009).

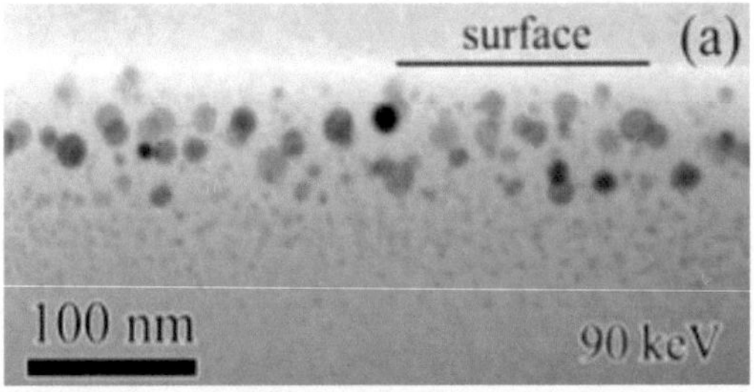

Fig. 4. Cross-section TEM image of SiO_2 with Ag nanoparticles fabricated at a dose of $5.0 \cdot 10^{16}$ ion/cm^2 and an energy of 90 keV. Fragment from (Ren, et al., 2007).

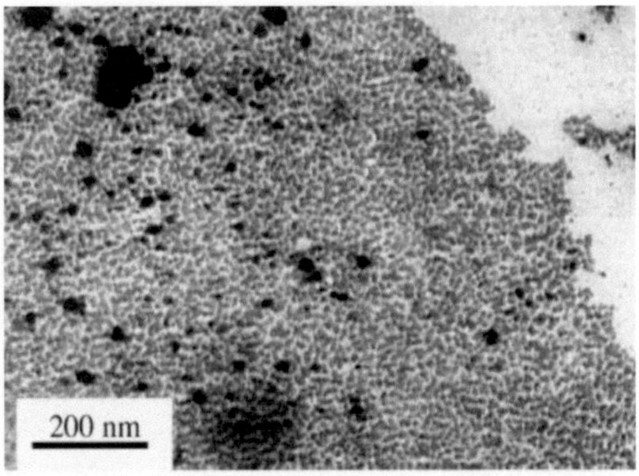

Fig. 5. Plan-view TEM image of silicone polymer-glass with Ag nanoparticles fabricated at a dose of $3.0 \cdot 10^{16}$ ion/cm^2 and an energy of 30 keV. Fragment from (Khaibullin, et al. 1999).

$\sim F_0 \geq 10^{17}$ ion/cm^2, leads to the coalescence of already existing MNPs with the formation of either MNP aggregates or thin quasi-continuous films at the dielectric surface. For instance, the irradiation of silicone polymer-glass by 30-keV Ag ions at higher-than threshold-nucleation doses favors the formation of aggregate structures (Fig. 5) (Khaibullin, et al.

1999). The MNP distribution established in the dielectrics after coalescence or Ostwald ripening may be dramatically disturbed by postimplantation thermal or laser annealing.

3. Laser annealing of glasses with silver nanoparticles

Despite advantages the use of ion implantation for nanoparticle synthesis there have not yet emerged clear mechanisms which allow precisely controlled particles sizes and depth distributions. Latter has a certain drawback, which is the statistically nonuniform depth of penetration of implanted ions into a material (Stepanov et al., 2000d). This leads to a wide size distribution of synthesized nanoparticles not only in the plane parallel to the irradiated surface but to a great extent also over the depth of the sample (Nistor et al., 1993). Dispersion of nanoparticles with respect to sizes leads to a broadening of the SPR optical absorption band accompanied by a decrease it in the intensity (Kreibig & Vollmer, 1995). This is also attributable to the dependence of the SPR spectral position on the particle size, i.e. the absorption spectrum in real sample is a superposition of several overlapping less intense bands that corresponding to particles of various sizes. The concern of the modern task is to increase the uniformity of the size distribution of MNPs synthesized by ion implantation using an approach of high-power pulse laser annealing with sequential furnace one. Experience gained from using the laser annealing techniques for various purposes allowed MNPs to be modified in various dielectrics. The main feature most of all the experiments with laser annealing of composites with MNPs is that the laser light was applied directly into the spectral region of the transparency of the dielectric matrix, and consequently, the intense laser pulses were primarily absorbed by the metal particles. Contrary to that, a new approach for annealing was demonstrated, when SLSG with Ag particles was irradiated by a laser light at wavelengths of glass absorption in the ultraviolet region (Wood et al., 1993). When applying high-power excimer ArF (193 nm) laser pulses, a decrease of the reflectance intensity of composite samples was observed. It was suggested that the implanted silver particles in glass can be dissolved and the glass matrix can be modified to be a silver rich metastable new glass phase. If this is correct then the new phase will be the potential to be destabilized to precipitate out the new silver particles in a controlled fashion by furnace.

As seen in Fig. 6 at low energy implantation the silver depth concentration in the SLSG derived from experimental Rutherford backscattering spectrometry (RBS) spattering shows a maximum near the implanted surface of the sample with some penetration to about 60-65 nm (Stepanov et al., 1999a).

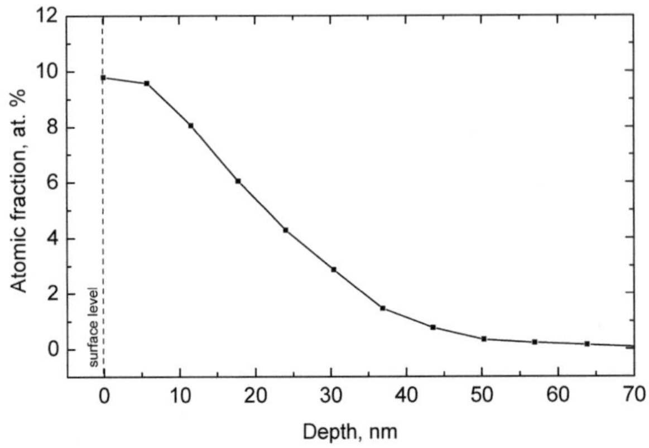

Fig. 6. The depth distribution of silver derived from the RBS spectrum for ion implantation with a dose of $7 \cdot 10^{16}$ ion/cm^2 and an energy of 60 keV into the SLSG.

An excess of metal atoms in the glass, above the solubility limit, causes nucleation of the nanoparticles. As usually the size distribution of MNPs appears. Additional to this distribution in the implanted sample there is distribution particles in the depth. Since the increase of metal concentration in the depth profile and the sputtering yield depend on implantation time, then the metal particle nucleation and growth will also vary with time and depth. It is obvious that during the implantation the size of the growthing particles with depth is "proportional" to the metal filling factor, because they are both determined by the ion concentration profile. Consequently, according to the Fig. 6 the large silver nanoparticles (or/and the higher filling factor) in the glass are close to the implanted surface with small ones in the interior of the implant zone. These features can be recognized in optical spectra of dielectrics with implanted nanoparticles. As example, transmittance and reflectance data of the non-implanted and implanted glass are presented in Fig. 7 (Stepanov et al., 1999a).

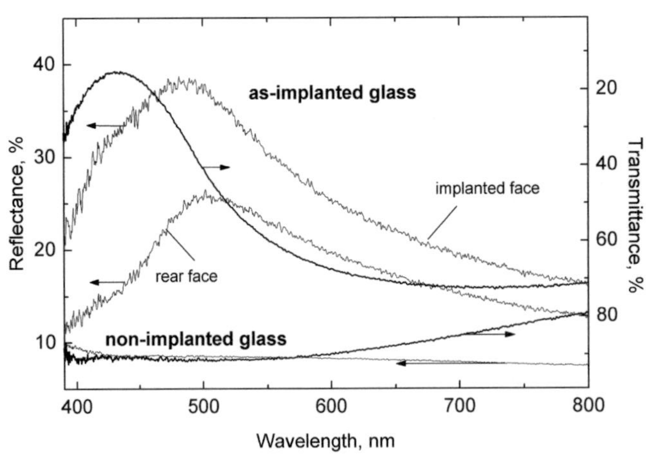

Fig. 7. Optical transmittance and reflectance of the silver implanted and original SLSG with a dose of $7 \cdot 10^{16}$ ion/cm^2 and an energy of 60 keV. Reflectance was measured from both the implanted and rear faces of the sample.

The transmission spectrum of implanted glass is maximized near 430 nm and the shape of the spectral curve is almost symmetrical. The reflectance spectra of same sample are more complex and, although the transmission must be the same whether the glass is viewed from the implanted or the reverse face, the shapes of the reflectivity curves differ. Overlapping peaks of reflectance spectra measured from the implant face of the samples exhibit a shoulder at about 430 nm, on the side of a clearly determined maximum at 490 nm, whereas reflectivity from the rear face appears to have a simpler peak at longer wavelengths near 500 nm. The contrast between the spectrum available from transmittance and reflectance of the implanted glass is recognized as coming from the non-uniform silver profile and size distribution of the nanoparticles in the implanted glass (Fig. 6). Since the distribution pattern is not symmetric, the reflectivity determined from the ion implanted and rear faces of the sample differ (Fig. 7), and the reflected intensity from a particular layer depends on the their local absorption in the depth. The atomic force microscope (AFM) images implanted surfaces of the same samples shown in Fig. 8 (Stepanov et al., 1998). As seen from the figure the implanted surface is smoother (roughness) and there are many hemispherical hills on this surface with an average diameter of approximately 100-150 nm. There are no such protrusions on the unimplanted sample. The reason for the existence of

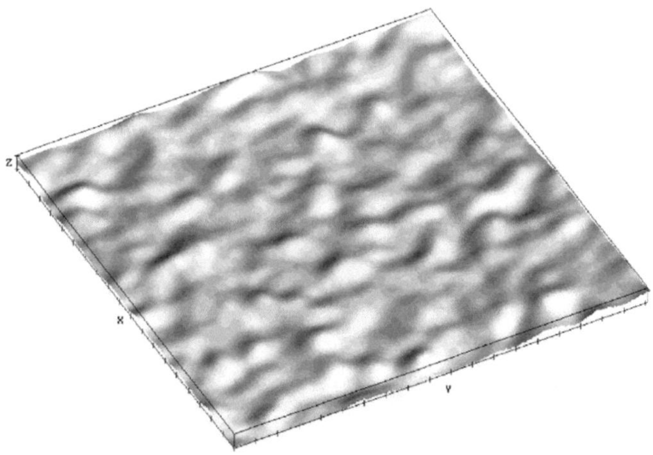

Fig. 8. An AFM image as a top view under lateral illumination of the surface of SLSG Ag-implanted with a dose of $7 \cdot 10^{16}$ ion/cm^2 and an energy of 60 keV. The step along the X and Y axes id 100 nm, and the step along the Z axis is 3 nm.

surface hills is assumed to be from a sputtering of irradiated glass during implantation, which leads to unequal ejection of ions of different elements from the surface, exposing the synthesized nanoparticles in the sub-surface glass. The sputtered glass thickness is typically of the order of tens of nanometers for the present ion dose (Hole et al., 1998), and hence, the synthesized buried MNPs appear near to the glass surface in the implanted sample.
Consequences of the excimer laser pulse with nanosecond pulse width and high beam intensity are heating, melting and/or vaporisation (ablation) of material on a time scale of nanoseconds to microseconds. The excimer laser treatment has been applied to many glasses, but there is less information on high-power pulse laser interaction with dielectrics containing metal nanoparticles. In the present case the energy density is lower than the value of the ablation threshold for the SLSG, which was determined to be about 5 J/cm^2

(Buerhop et al., 1990). Also, the excimer laser is characterised by a UV-wavelength, which is much longer than the typical sizes of the nanoparticles formed by ion implantation. Hence, present metal/glass composites may be considered similar to be a homogeneous material for light propagation (Kreibig & Vollmer, 1995). This is a simplification, which is true generally for low intensity light, but it gives an estimate of the optical penetration depth as (α^{-1}) of the laser pulses into the composite material, where α is the linear absorption coefficient. An intense laser pulse is absorbed and relaxed into heat into the surface SLSG layer of a thickness of α^{-1}, which is several microns (Townsend & Olivares, 1997), i.e., deeper than the thickness of the implanted layer (Fig. 6).
The optical spectral result of pulse laser treatment by 5 pulses of a KrF excimer laser with pulse length of 25 ns full-width at half-maximum at a wavelength of 248 nm with the total released energy of 0.2 J/cm^2 on the optical spectra of the Ag-implanted glass is presented in Fig. 9 (Stepanov et al., 2001). Applied laser pulses did not change the reflectance and transmittance spectra of the non-implanted SLSG, but for implanted sample the location of the transmittance minimum shifts slightly towards shorter wavelengths, and the transmittance in peak position increases from 16 to 23% (Fig. 9a). Remarkable change was found in the reflectance spectra, where in the case of the implanted surface, the peak of the overlapping bands shifts continuously from 490 to 450 nm with modifications in the

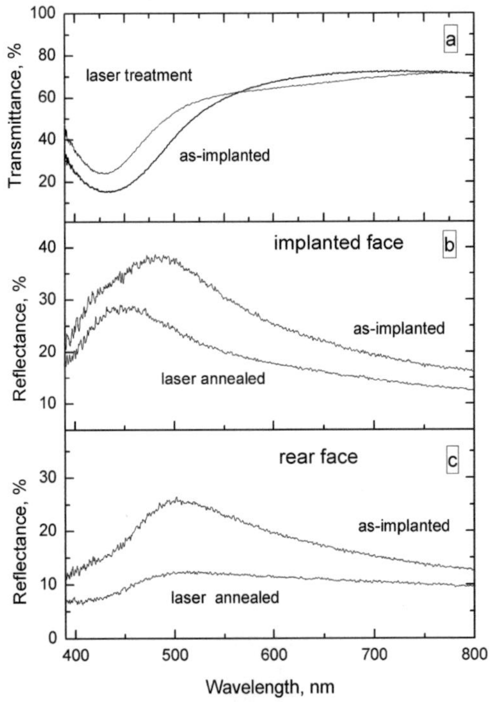

Fig. 9. Optical spectra of the SLSG after silver implantation as in Fig. 7 and the implantation followed with laser treatment (0.2 J/cm^2): (a) transmittance; (b) reflectance measured from the implanted face; (c) reflectance measured from rear faces of the sample.

shape of the envelope of the bands, which become narrower (Fig. 9b). The reflectance intensity falls from 38 to 27 %. However, for reflectance from the rear face of the same sample (Fig. 9c) there is only a decrease of the intensity to a 13 % maximum, which is at the same initial wavelength as in the implanted sample.

For composite materials with MNPs effective-medium theories (Maxwell-Garnett, Bruggeman, etc) can be considered (Kreibig & Vollmer, 1995). According to this approach, the measured optical intensities corresponding to the implanted glass (Figs. 7 and 9) are determined mainly by the near-surface layer, where there is the largest metal filling factor, and also the largest nanoparticles. The smaller filling factors (and smaller nanoparticles) that exist at the other depths effectively only influence a smearing of the optical spectra. Moreover, consistent with effective-medium considerations, a decrease of the filling factor is accompanied by movement of the optical peak position to shorter wavelengths, as is observed in Fig. 9. Thus, a decrease of the filling factor in the composite resulting from laser treatment suggests there is a size reduction of all the silver nanoparticles, and again the biggest nanoparticles are localized near the SLSG surface. The difference between reflectance spectra from implanted and rear faces of laser treated sample (Fig. 9 b, c) emphasizes the existence of a non-symmetrical depth distribution of the Ag nanoparticles as being similar to the distribution in the implanted sample.

As presented in Fig. 10, (note an increased magnification of ten times in the direction perpendicular to plane of the figure), it is seen that there are a lot of clearly defined hills

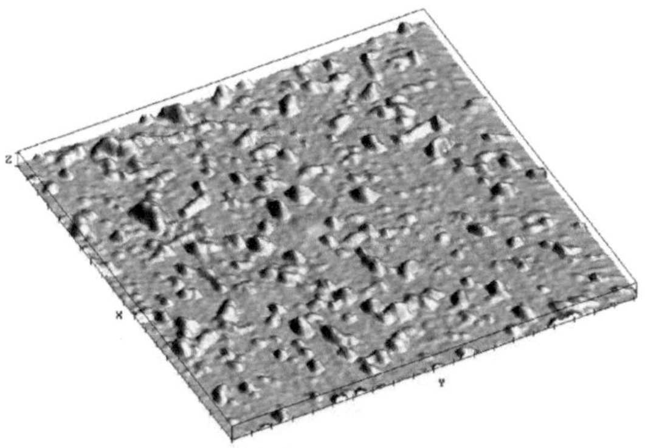

Fig. 10. An AFM image as a top view under lateral illumination of the surface of SLSG Ag-implanted with a dose of $7 \cdot 10^{16}$ ion/cm^2 and an energy of 60 keV followed with irradiation with an excimer laser (0.2 J/cm^2). The step along the X and Y axes id 100 nm, and the step along the Z axis is 40 nm.

(believed to be silver particles) on the glass surface whose size is one order smaller in contrast to large hills on the implanted glass surface in Fig. 8 (Stepanov et al., 1998). The Ag MNPs accumulate effect on the laser irradiated surface of glass is result of melting implanted composite layer and some desorption of glass material under the laser pulses, which exposes the melted metal particles after their solidification. Hence, the first of the conclusions from the present data on excimer laser treatment of implanted glass is the reduction of the size of the silver nanoparticles, and second is the existence of some asymmetry in their depth distribution (Fig 9b, c). To recognize the mechanisms by which the changes occur for strongly absorbed excimer laser pulses the thermal propagation after the laser-irradiation must be considered. The laser heating is traditionally characterized by the heat diffusion length, $l(t) = (D \cdot t)^{1/2}$, where D is the heat diffusivity, and t is the laser pulse duration. In the present experiment with laser pulses of 25 ns, the heat propagation is approximately $l(t)$ = 115 nm, that is shorter than the α^{-1}, i.e., the temperature rise is no longer controlled by the diffusion of the heat. However, the $l(t)$ surpasses the depth of the implanted Ag nanoparticles. As was estimated earlier (Townsend & Olivares, 1997), the temperature at the surface of laser treated SLSG reaches values exceeding ~700°C, which is equivalent to the SLSG melting temperature. Under these temperature conditions there is also a possibility for melting small silver particles, because their melting temperature is drastically decreased, for example, to ~400°C for sizes < 30 nm, compared with the bulk metal melting temperature of 960°C (Castro et al., 1990). The time scale of electronic relaxation and energy transfer to the lattice vibrations in the metal particles is several orders faster than in the surrounding glass medium. Therefore, during the interaction of the excimer laser pulse with the metal/glass composite, the Ag nanoparticles are heated and melted more quickly than solidification of the melted glass can occur. Atomically dispersed Ag released from nanoparticles enters into the glass melt, and immediately diffuse throughout all the heated thickness of the laser treated substrate. In principle in time this could lead to a uniform metal distribution, where the silver atom concentration exceeds the solubility value in the solid glass. However, following glass solidification spreading from the depth to the surface, as heat from the laser pulse penetrates into the depth of the sample, the cooling part of the annealing cycle will stimulate new nucleation and regrowth of metal particles. In this case, the possibility of regrowth of metal particles will depend on competition between regrowth and the cooling speed of the moving solidification front, resulting in a new non-uniform size distribution of new MNPs over a depth scale similar to that after ion implantation. Obviously, under some conditions the metal particles may be dispersed into separate metal ions and/or into such small units that they cannot display nanoparticle type optical properties. As shown here, subsequent high power excimer laser pulse treatments of the ion implanted layer may be used to melt, and/or regrow, the MNPs within the insulator medium. Overall this results in a tighter distribution of small particles. The laser treatments have slightly reduced, but not completely removed evidence for a non-symmetric depth distribution of the silver particles. The Ag-insulator composite material is complex, and so a much wider range of laser pulse conditions, and more data on the cooling rates are required to fully model the changes in the size distributions, which can occur.

4. Nonlinear optical properties of ion-synthesized silver nanoparticles

The Ag nanoparticles doped in different dielectrics demonstrate variable nonlinear optical properties (Palpant, 2006). The interest on such structures is based on the prospects of the elaboration of optical switchers with ultrafast response, optical limiters, and intracavity elements for mode locking. Ag nanoparticles have an advantage over another metal nanoparticles (i.e., gold and copper) from the point of view that the surface plasmon resonance energy of Ag is far from the interband transition energy. So, in the silver nanoparticle system it is possible to investigate the nonlinear optical processes caused solely by SPP contribution. It should be noted that previous studies of nonlinear optical parameters of silver nanoparticles-doped glasses were mostly focused on determination of third-order nonlinear susceptibility $\chi^{(3)}$. The saturated absorption in silicate glasses doped with Ag nanoparticles at wavelength of 532 nm and their dependence on laser radiation intensity are considered at present review.

The Ag nanoparticles ion-synthesized in SLSG (Ag:SLSG) and (Ag:SiO_2) demonstrate the SPP band with minimum transmission in the range of 410–440 nm (Fig. 11).

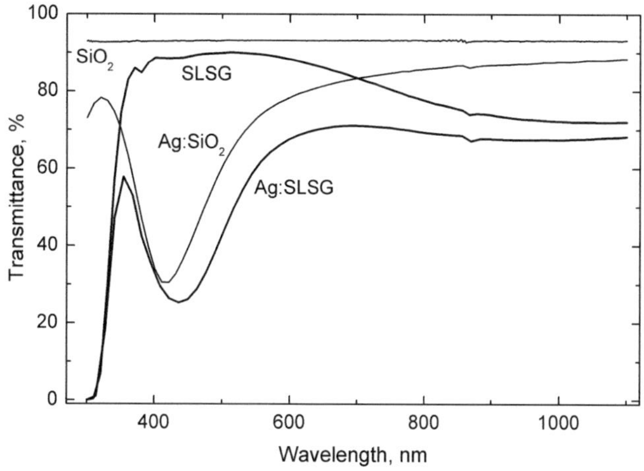

Fig. 11. Optical transmittance of the Ag nanoparticles formed in SLSG and SiO_2 by implantation with a dose of $4 \cdot 10^{16}$ ion/cm^2 and an energy of 60 keV.

It was previously predicted that silver-doped glasses possess by saturated absorption. The spectral dispersion of the imaginary part of third-order susceptibility Im$\chi^{(3)}$ of silver-doped glass matrices was analyzed and it was shown that Im$\chi^{(3)}$ was negative in the spectral range of 385 - 436 nm (Hamanaka et al., 2000). The nonlinear absorption coefficient β is also negative in the case of saturated absorption.

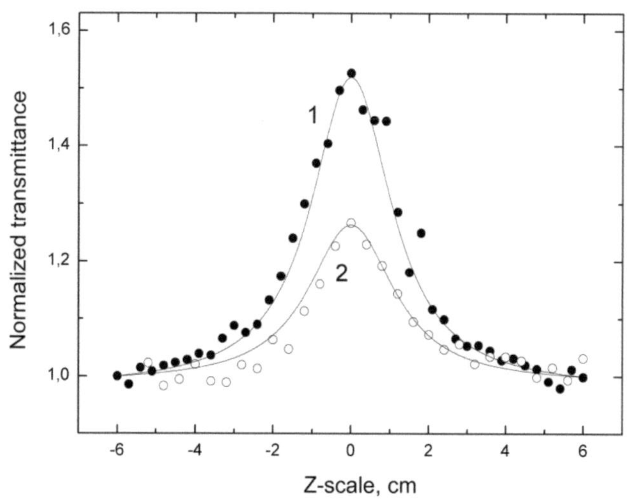

Fig. 12. Normalised transmittance Ag:SLSG (1) and Ag:SiO_2 (2) samples at laser radiation intensity of $I_0 = 2.5 \cdot 10^9$ W/cm^2. Solis lines show theoretical fittings.

The normalized transmittance dependences of Ag:SLSG and Ag:SiO_2 samples measured using open aperture Z-scan scheme (Sheik-Bahae et al., 1990) at laser radiation intensity of

$I_0 = 2.5 \cdot 10^9$ W/cm^2 and pulse duration of 55 ps is presented in Fig. 12. The transmission of samples was increased due to saturated absorption as they approached close to the focal plane. The theoretical fitting was done by the method described early (Ganeev et al., 2004b). After fitting of experimental data the β are -6.7 $\cdot 10^{-5}$ cm/W in Ag:SLSG and -3.6 $\cdot 10^{-5}$ cm/W in Ag:SiO_2. Coefficient β can be presented as $\beta = \alpha / I_s$ where is I_s saturated intensity. The values of I_s are 1.1 $\cdot 10^9$ and 1.4 $\cdot 10^9$ W/cm^{-2}, also the Im$\chi^{(3)}$ are -2.4 $\cdot 10^{-8}$ and -1.3 $\cdot 10^{-8}$ esu in Ag:SLSG and Ag:SiO_2, respectively.

In Figs. 13 and 14 values of β in dependence of laser intensity varied from 10^9 to $2 \cdot 10^{10}$ W/cm^2 are presented. As seen from the figures there are a decrease β of for higher intensities. In particularly, a 21- and 12-fold decrease of β was measured at $I_0 = 1.15 \cdot 10^{10}$ W/cm^2 for Ag:SLSG and Ag:SG, respectively, compared to β detected at $I_0 = 1 \cdot 10^9$ W/cm^2.

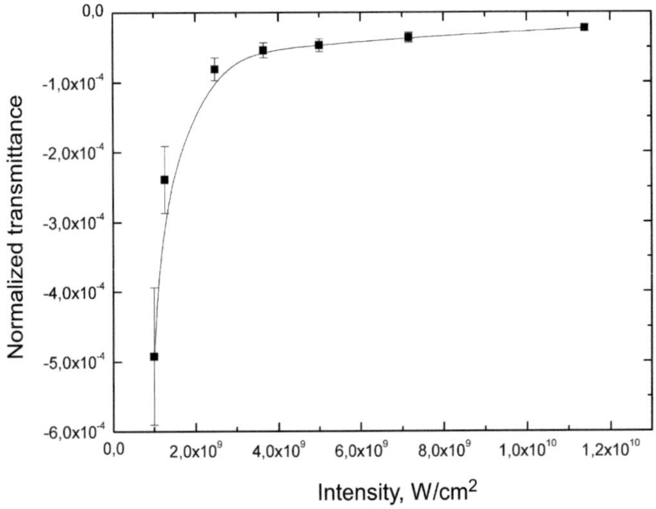

Fig. 13. Coefficient β of Ag:SLSG in dependence of laser intensity.

The variations of transmission in similar MNP structures were attributed in some cases to the fragmentation (Kurata et al., 1998; Link et al. 1999; Mafune et al., 2002), or fusion (Chandrasekharan et al., 2000) of nanoparticles following the photothermal melting. It was reported about the alteration of the sign of nonlinear refractive index of small Ag clusters embedded in SLSG (Osborne et al., 1998). They noted that thermal effects could change the properties of nanoclusters. The transparency in these samples was associated with oxidation of Ag nanoparticles. However, no irreversible changes of transmittance were observed in present experiments.

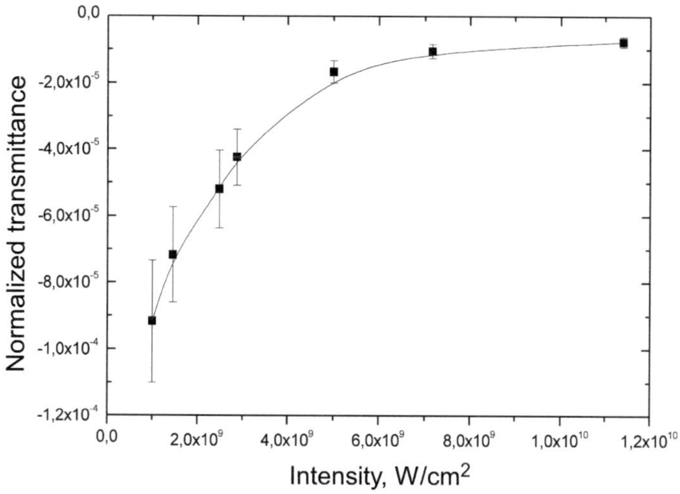

Fig. 14. Coefficient β of Ag:SiO_2 in dependence of laser intensity.

The reverse saturated absorption can be responsible for the decrease of negative nonlinear absorption of Ag nanoparticles and it could be assume that in the case of picosecond pulses the reverse saturated absorption starting to play an important role in the overall dynamics of nonlinear optical transmittance of metal nanoparticles contained compounds, taking into account the saturation of intermediate transitions responsible for saturated absorption. Thus, saturated absorption in Ag:SLSG and Ag:SiO_2 was dominated at small intensities and decreased with the growth of intensity due to influence of competing effects, whereas the selfdefocusing at low intensities was changed to self-focusing at high intensities. The possible mechanism of the decrease of Im$\chi^{(3)}$ is the influence of nonlinear optical processes with opposite dependences on laser intensity, also such as two-photon absorption (Sheik-Bahae et al., 1990). The wavelength range corresponded to the interband transitions in Ag is located below 320 nm, so the two-photon absorption connected with interband transitions can be involved in the case of 532 nm radiation. The possibility of two-photon absorption due to interband transition of photoexcited electrons was previously demonstrated for Ag partilces (Magruder et al., 1994). The three-photon absorption connected with interband transition for Ag nanoparticles was analysed in (Kyoung et al., 1999). Thus, saturated absorption in Ag:SLSG and Ag:SiO_2 was dominated at small intensities and decreased with the growth of intensity due to influence of competing effects, whereas the selfdefocusing at low intensities was changed to self-focusing at high intensities.

5. Acknowledgements

Author is grateful to the Alexander Humboldt Foundation for the financial support of this work in Germany and Austrian Scientific Foundation in the frame of Lisa Meitner Fellowship. I wish to thank my partners and co-authors D. Hole, P.D. Townsend, I.B. Khaibullin, V.I. Nugdin, V.F. Valeev, R.I. Khaibullin, V.N. Bazarov, Yu.N. Osin, S.N. Abdullin, V.A. Zikharev, I.A. Faizrakhmanov, A.A. Bukharaev, V.N. Popok, U. Kreibig, A.I. Ryasnyansky, R.A. Ganeev, E. Alves.

6. References

Amolo, G.O.; Comins, J.D.; Naidoo, S.R.; Connell, S.H.; Witcomb, M.J. & Derry, T.E. (1999). Effects of Ag^+ and Au^{3+} ion implantation of lithium niobate, *Nucl. Inst. Meth. Phys. Res. B* 250, 233-237

Anderson, T.S.; Magruder III, R.H.; Zuhr, R.A. & Wittig, J.E. (1996). Optical properties of multicomponent nanometer dimension metal colloids formed in silica by sequential ion implantation of In and Ag and In and Cu, *J. Electronic Mater.* 25, 27-35

Anderson, T.S.; Magruder III, R.H.; Kinser, D.L.; Zuhr, R.A. & Thomas, D.K. (1997). Formation and optical properties of metal nanoclusters formed by sequential implantation of Cd and Ag in silica, *Nucl. Inst. Meth. Phys. Res. B* 124, 40-46

Anderson, T.S.; Magruder III, R.H.; Kinser, D.L.; Wittig, J.E.; Zuhr, R.A. & Thomas, D.K. (1998). Ag:Cd and Cd:Ag implantations into high purity silica, *J. Non.-Cryst. Sol.* 224, 299-306

Anderson, T.S.; Magruder III, R.H.; Wittig, J.E.; Kinser, D.L. & Zuhr, R.A. (2000). Fabrication of Cu-coated Ag nanocrystals in silica by sequential ion implantation, *Nucl. Inst. Meth. Phys. Res. B* 171, 401-405

Antonello, M.; Arnold, G.W.; Battaglin, G.; Bertoncello, R.; Cattaruzza, E.; Colombo, P.; Mattei, G.; Mazzoldi, P. & Trivillin, F. (1998). Fluence and current density dependence of silver nanocluster dimensions in ion-implanted fused silica, *J. Mater. Chem.* 8, 457-461

Armelao, L.; Bertoncello, R.; Cattaruzza, E.; Gialanella, S.; Gross, S.; Mattei, G.; Mazzoldi, P. & Tondello, E. (2002). Chemical and physical routes for composite materials synthesis: Ag and Ag2S nanoparticles in silica glass by sol–gel and ion implantation techniques, *J. Mater. Chem.* 12, 2401-2407

Arai, N.; Tsuji, H.; Motono, M.; Goto, Y.; Adachi, K.; Kotaki, H. & Ishikawa, J. (2003). Formation of silver nanoparticles in thin oxide layer on Si by negative-ion implantation, *Nucl. Inst. Meth. Phys. Res. B* 206, 629-633

Arai, N.; Tsuji, H.; Ueno, K.; Matsumoto, T.; Gotoh, Y.; Abachi, K.; Kotaki, H. & Ishikawa, J. (2005). Evaluation of nanoparticles emvedded in thin silicon dioxide film by optical reflection property, *Surf. Coat. Technol.* 196: 44-49

Arai, N.; Tsuji, H.; Ueno, K.; Matsumoto, T.; Gotoh, Y.; Abachi, K.; Kotaki, H.; Gotoh, Y. & Ishikawa, J. (2006). Formation of silver nanoparticles aligned near the bottom of SiO_2 film on silicone substrate by negative-ion implantation and post-annealing, *Nucl. Inst. Meth. Phys. Res. B* 242, 217-220

Arai, N.; Tsuji, H.; Abachi, K.; Kotaki, H.; Gotoh, Y. & Ishikawa, J. (2007a). Silver negative-ion implantation into thermally growth this SiO_2 film on Si substrate and heat treatment for formation of silver nanoparticles, *Jap. J. Appl. Phys.* 46, 6260-6266

Arai, N.; Tsuji, H.; Gotoh, N.; Matsumoto, T.; Ishibashi, T.; Adachi, K.; Kotaki, H.; Gotoh, Y. & Ishikawa, J. (2007b). Nanocrystal formation of metals in thermally grown thin silicon dioxide layer by ion implantation and thermal diffusion of implanted atoms in heat treatment, *J. Phys.: Conf. Ser.* 61, 41-45

Arnold, G.W. (1975). Near-surface nucleation and crystallization of an ion-implanted lithia-alumina-silica glass, J. Appl. Phys. 46, 4466-4473

Arnold, G.W. & Borders, J.A. (1977). Aggregatin and migration of ion-implanted silver in lithia-alumina-silica glass, J. Appl. Phys. 48, 1488-1496

Battaglin, G.; Cattaruzza, E.; D'Acapito, F.; Gonella, F.; Mazzoldi, P.; Mobilio, S. & Priolo, F. (1998). EXAFS study on metal cluster doped silica glass obtained by ion implantation procedures, *Nucl. Inst. Meth. Phys. Res. B* 141, 252-255

Battaglin, G.; Catalano, M.; Cattaruzza, E.; D'Acapito, F.; De Julian C., Fernandez, G.; De Marchi, G.; Gonella, F.; Mattei, G.; Maurizio, C.; Mazzoldi, P.; Miotello, A. & Sada, C. (2001). Influence of annealing atmosphere on metal and metal alloy nanoclusters produced by ion implantation in silica, *Nucl. Inst. Meth. Phys. Res. B* 178, 176-179

Bertoncello, R.; Gross, S.; Trivillin, F.; Cattaruzza, E.; Mattei, G.; Caccavale, F.; Mazzoldi, P. & Battaglin, G. (1999). Mutually reactive elements in a glass host matrix: Ag and S ion implantation, *J. Mater. Res.* 14, 2449-2457

Buerhop, C.; Blumenthal, B. & Weissmann, R. (1990) Glass surface treatment with excimer and CO_2 laser, *Appl. Surf. Sci.* 46, 430-434

Caccavale, F. (1998). Metal-ion implantation in glasses: Physical and chemical aspects, *Pramana - J. Phys.* 50, 653-668

Cai, G.X.; Ren, F.; Xiao, X.H.; Fan, L.X. & Jiang, C.Z. (2008). Fabrication of hallow Ag nanoclusters in silica by irradiation with Ar ions, *Nucl. Inst. Meth. Phys. Res. B* 266, 889-893

Carles, R.; Farcău, C.; Bonafos, C.; Benassayag, G.; Pécassou, B. & Zwick, A. (2009). The synthesis of single layers of Ag nanocrystals by ultra-low-energy in implantation for large-scale plasmonic structures, *Nanotechnol.* 20, 1-6

Castro, T.; Reifenberger, R.; Choi, E. & Andres, R.P. (1990). Size-dependent melting temperature of individual nanometer-sized metallic clusters, Phys. Rev. B 42, 8549-8556

Cattaruzza, E.; Battaglin, G.; Polloni, R.; Cesca, T.; Gonella, F.; Mattei, G.; Maurizio, C.; Mazzoldi, P.; D'Acapito, F.; Zontone, F. & Bertoncello R. (1999). Nanocluster formation in silicate glasses by sequential ion implantation procedures, *Nucl. Inst. Meth. Phys. Res. B* 148, 1007-1011

Chandrasekharan, N.; Kamat, P.V.; Hu, J. & Jones II, G. (2000). Day-capped gold nanoparticles: Photoinduced morphological changes in gold/rhodamine 6G nanoassemblies, *J. Phys. Chem. B* 104, 11103-11109

Cheang-Wong, J.C.; Oliver, A.; Rodríguez-Fernandez, L.; Arenas-Alatorre, J.; Peña, O. & Crespo-Sosa, A. (2007). Optical absorption and HRTEM characterization of metallic nanoparticles produced by MeV ion implantation, *Revista Mexicana de Física S* 53, 49-54

D'Acapito, F. & Zontone, F. (1999). Grazing-incidence X-ray diffrectin in the study of metallic clusters buried in glass obtained by ion implantation, *J. Appl. Cryst.* 32, 234-240

Davenas, J.; Perez, A.; Thevenard, P. & Dupuy, C.H.S. (1973). Correlation between absorption bands and implanted alkali ions in LiF, *Phys. Stat. Sol. A* 19, 679-686

Deying, S.; Saito, & Y. Suganomata, S. (1994). Optical properties of $LiNbO_3$ implanted with Ag^+ ions, *Jpn. J. Appl. Phys.* 33, L966-L969

Dubiel, M.; Hofmeister, H. & Schurig, E. (1997). Compressive stresses in Ag Nanoparticle Doped Glasses by Ion Implantation, *Phys. Stat. Sol. B* 203, R5-R6

Dubiel, M.; Hofmeister, H.; Schurig, E.; Wendler, E. & Wesch, W. (2000). On the stress state of silver nanoparticles in ion-implanted silicate glasses, *Nucl. Inst. Meth. Phys. Res. B* 166-167, 871-876

Dubiel, M.; Hofmeister, H.; Tan, G.L.; Schicke, K.-D. & Wendler, E. (2003). Silver diffusion and precipitation of nanoparticles in glass by ion implantation, *Eur. Phys. J. D* 24, 361-364

Dubiel, M.; Hofmeister, H. & Wendler, E. (2008). Formation of nanoparticles in soda-lime glasses by single and double ion implantation, *J. Non.-Cryst. Solids* 354, 607-611

Flytzanis, C.; Hache, F.; Klein, M.C.; Ricard, D. & Rousignol P. (1991). *Nonlinear Optics in composite materials*, Elsevier Science, Amsterdam

Ganeev, R.A.; Ryasnyanskii, A.I.; Stepanov, A.L.; Kondirov, M.K. & Usmanov, T. (2003a). Nonlinear properties of composites based on dielectric layers containing copper and silver nanoparticles, *Opt. and Spectroscopy* 95, 967-975

Ganeev, R.A.; Ryasnyanskii, A.I.; Stepanov, A.L. & Usmanov, T. (2003b). Nonlinear optical susceptibilities of copper- and silver-doped silicate glasses in the ultraviolet range, *Phys. Stat. Sol. B* 238, R5-R7

Ganeev, R.A.; Ryasnyanskii, A.I.; Stepanov, A.L. & Usmanov, T. (2004a). Nonlinear optical response of silver and copper nanoparticles in the near-ultraviolet spectral range, *Phys. Solid. State.* 46, 341-346

Ganeev, R.A.; Ryasnyanskii, A.I.; Stepanov, A.L. & Usmanov, T. (2004b). Saturated absorption and nonlinear refraction of silicate glasses doped with silver nanoparticles at 532 nm, *Opt. Quant. Electr.* 36, 949-960

Ganeev, R.A.; Ryasnyanskii, A.I.; Stepanov, A.L. & Usmanov, T. (2004c). Characterization of nonlinear optical parameters of copper- and silver-doped silica glasses at λ=1064 nm, *Phys. Stat. Sol. B* 241, 935-944

Ganeev, R.A.; Ryasnyanskii, A.I.; Stepanov, A.L.; Marques, C.; da Silva, R.C. & Usmanov, T. (2005). Application of RZ-scan technique for investigation of nonlinear refraction of sapphire doped with Ag, Cu and Au nanoparticles, *Opt. Comm.* 253, 205-213

Ganeev, R.A.; Ryasnyanskii, A.I.; Stepanov, A.L.; Usmanov, T.; Marques, C.; da Silva, R.C. & Alves, E. (2006). Investigation of the nonlinear optical characteristics of composite materials based on sapphire with silver, copper, and gold nanoparticles by the reflection Z-scan method, *Opt. and Spectr.* 101, 615-622

Gonella, F.; Mattei, G.; Mazzoldi, P.; Sada, C.; Battaglin, G. & Cattaruzza E. (1999). Au-Cu nanoclusters in silica formed by ion implantation and annealing in reducing or oxidizing atmosphere, *Appl. Phys. Lett.* 75, 55-57

Hamanala, Y.; Hayashi, N.; Nakamura, A. & Omni, S. (2000). Dispersion of thired-order nonlinear optical susceptibility of silver nanocrystal-glass composites, J. Luminescence 87-89, 859-861

Hole, D.E.; Stepanov, A.L. & Townsend, P.D. (1999). Dependence of optical properties of implanted silver nanoparticles in float glass on substrate temperature, *Nucl. Instrum. Meth. Phys. Res.* 148, 1054-1058

van Huis, M.A.; Fedorov, A.V.; van Veen, A.; Falub, C.V., Eijt, S.W.H., Kooi, B.J.; De Hosson, J.Th.M., Hibma, T. & Zimmerman, R.L. (2002). Structural properties of Au and Ag nanoclusters embedded in MgO, *Nucl. Inst. Meth. Phys. Res. B* 191, 442-446

Ila, D.; Williams, E.K.; Sarkisov, S.; Smith, C.C.; Poker, D.B. & Hensley, D.K. (1998). Formation of metallic nanoclusters in silica by ion implantation, *Nucl. Inst. Meth. Phys. Res. B* 141, 289-293

Ishikawa, J.; Tsuji, H.; Motono, M.; Gotoh, Y.; Arai, N.; Adachi, K. & Kotaki, H. (2002). Negative-Ion Implantation into Thin SiO_2 Film on Si, and Formation of Silver Nanoparticles in the Film, *IEEE* 12, 690-693

Joseph, B.; Lenka, H.P.; Kuiri, P.K.; Mahapatra, D.P. & Kesavamoorthy, R. (2007a). Synthesis of alloy metal nanoclusters on silica glass by sequential ion implantation, *Intern. J. Nanoscience* 6, 423-430

Joseph, B.; Suchan Sandeep, C.S.; Sekhar, B.R.; Mahapatra, D.P. & Philip, R. (2007b). Nonlinear optical properties of MeV and keV ion beam synthesized Ag nanoclusters, *Nucl. Inst. Meth. Phys. Res. B* 265, 631-636

Jiang, C.Z. & Fan, X.J. (2000). In-situ TEM observation of silver nanocrystals in an Ag-implanted SiO_2 film, *Surf. Coat. Technol.* 131, 330-333

Khaibullin, R.I.; Osin, Y.N.; Stepanov, A.L. & Khaibullin, I.B. (1999). Synthesis of metal/polymer composite films by implantation of Fe and Ag ions in viscous and solid state silicone substrates, *Nucl. Inst. Meth. Phys. Res. B* 148, 1023-1028

Kurata, H.; Takami, A. & Koda, S. (1998). Size reduction of gold particles in aqueous solution by pulsed laser irradiation, *Appl. Phys. Lett.* 72, 789-791

Link, S.; Burda, C.; Mohamed, M.B.; Nikoobakht, B. & El-Sayed, M.A. (1999). Laser photothermal melting and fragmentation of gold nanorods: Energy and laser pulse-width dependence, *J. Phys. Chem A* 103, 1165-1170

Liu, Z.; Wang, H.; Li, H. & Wang, X. (1998a). Red shift of plasmon resonance frequency due to the interacting Ag nanoparticles embedded in single crystal SiO_2 by implantation, *Appl. Phys. Lett.* 72, 1823-1825

Liu, Z.; Li, H.; Feng, X.; Ren, S.; Liu, Z. & Lu, B. (1998b). Formation effects and optical absorption of Ag nanocrystals embedded in single crystal SiO2 by implantation, *J. Appl. Phys.* 84, 1913-1917

Liu, Z.; Li, H.; Wang, H.; Shen, D.; Wang, X. & Alkemade, P.F.A. (2000). Favored structure of Ag nanoparticles embedded in SiO_2 by implantation: Single crystal with contracted (111) lattice, *J. Mater. Res.* 15, 1245-1247

Liu, X.F.; Jiang, C.Z.; Feng, R. & Fu, Q. (2005). Optical absorption, Raman spectra and TEM study of Ag nanoparticles formed by ion implantation into α-SiO_2, *Acta Phys. Sinca* 54, 4633-4637

Mafune, F.; Kohno, J.; Takeda, Y. & Kondow, T. (2002). Growth of gold clusters into nanoparticles in a solution following laser-induced fragmentation, *J. Phys. Chem. B* 106, 8555-8561

Marques, C.; da Silva, R.C.; Wemans, A.; Maneira, M.J.P.; Kozanecki, A.; & Alves E. (2006). Optical properties tailoring by high fluence implantation of Ag ions on sapphire, *Nucl. Inst. Meth. Phys. Res. B* 242, 104-108

Matsunami, N. & Hosono H. (1993a). Anomalous fringe pattern of Ag colloid in phosphate-glasses by implantation, *Nucl. Inst. Meth. Phys. Res. B* 80, 1233-1236

Matsunami, N. & Hosono H. (1993b). Colloid formation effects on depth profile of implanted Ag in SiO_2 glass, *Appl. Phys. Lett.* 63, 2050-2053

Mazzoldi, P.; Tramontin, I.; Boscolo-Boscoletto, A.; Battaglin, G. & Arnold, G.W. (1993). Substrate effects in silver-implanted glasses, *Nucl. Inst. Meth. Phys. Res. B* 80-81, 1192-1196

Mazzoldi, P. & Mattei, G. (2007). Some aspects of ion implantation technique in nanostructured materials, *Phys. Stat. Sol. A* 204, 621-630

Magruder III, R.H.; Osborne Jr., & D.H. Zuhr, R.A. (1994). Non-linear optical properties of nanometer dimension Ag-Cu particles in silica formed by sequential ion implantation, *J. Non.-Cryst. Solids* 176, 299-303

Magruder III, R.H.; Zuhr, R.A. & Osborne Jr., D.H. (1995). Modification of the optical properties of glass by sequential ion implantation, *Nucl. Inst. Meth. Phys. Res. B* 99, 590-593

Magruder III, R.H.; Anderson, T.S.; Zuhr, R.A. & Thomas, D.K. (1996). Formation and optical characterization of multi-component Ag-Sb nanometer dimension colloids formed by sequential ion implantation in silica, *Nucl. Inst. Meth. Phys. Res. B* 108, 305-312

Magruder III, R.H. & Meldrum, A. (2007) The effect of Ti and O on the optical properties and microstructure of Ag nanocrystals formed in silica by sequential ion implantation, *J. Non.-Cryst. Solids* 353, 4813-4818

Magruder III, R.H.; Robinson, S.J.; Smith, C.; Meldrum, A.; Halabica, A.; & Haglund Jr., R.H. (2009). Dichroism in Ag nanoparticles composites with bimodal size distribution, *J. Appl. Phys.* 105, 24303-1 - 24303-1

Nistor, L.C.; von Landuyt, J., Barton, J.B., Hole, D.E.; Skelland, N.D. & Townsend P.D. (1993) Colloid size distribution in ion implanted glass, *J. Non.-Cryst. Solids* 162, 217-224

Osborne Jr., D.H.; Haglund Jr., R.F.; Gonella, F. & Garrido, F. (1998). Laser-induced sign reversal of the nonlinear refractive index of Ag nanoclusters in soda-lime glass, *Appl.Phys. B* 66, 517-521

Palpant, B (2006). Third-order nonlinear optical response of metal nanoparticles, In: *Non-Linear Optical Properties of Matter*, Papadopoulos, M.G. (Ed.) Springer, Berlin, 461-508

Peña, O.; Cheang-Wong, J.C.; Rodríguez-Fernandez, L.; Arenas-Alatorre, J.; Crespo-Sosa, A.; Rodríguez-Iglesias, V. & Oliver, A. (2007). Metal and metal oxide nanoparticles produced by ion implantation in silica: A microstructural study unsing HRTEM, *Nucl. Inst. Meth. Phys. Res. B 257, 99-103*

Pham, M.T.; Matz, W. & Seifarth H. (1997). Surface roughness with nanometer-scale Ag particles generated by ion implantation, *Anal. Chim. Acta* 350, 209-220

Qian, Y.; Ila, D.; Zimmerman, R.L.; Poker, D.B.; Boatner, L.A. & Hensley, D.K. (1997). MeV silver ion implantation induced changes in optical properties of MgO (100), *Nucl. Inst. Meth. Phys. Res. B* 127, 524-527

Rahmani, M.; Abu-Hassan, L.H.; Townsend, P.D.; Wilson, I.H.; and Destefanis, G.L. (1988). Silver colloid formation in Ag^+ implanted $LiNbO_3$, *Nucl. Inst. Meth. Phys. Res. B* 32, 56-60

Rahmani, M. and Townsend, P.D. (1989). Ag^+ implantation in Al_2O_3, $LiNbO_3$ and quartz, *Vacuum* 39, 1157-1162

Kreibig U. & Vollmer M. (1995). *Optical Properties of Metal Clusters*, Springer, Berlin

Ren, F.; Jiang, C.Z.; Chen, H.B.; Shi, Y.; Liu, C. & Wang, J.B. (2004a). Metal alloy and monoelemental nanoclusters in silica formed by sequential ion implantation and annealing in selected atmosphere, *Physics B* 353, 92-97

Ren, F.; Jiang, C.Z.; Zhang, L.; Shi, Y.; Wang, J.B. & Wang, R.H. (2004b). Formation and microstructural investigation of Ag-Cu alloy nanoclusters embedded in SiO_2 formed by sequential ion implantation, *Micron* 35, 489-493

Ren, F.; Jiang, C.Z.; Liu, C. & Shi, Y. (2005a). Alloy and core-shell nanoclusters formed by sequential ion implantation and thermal annealing, *J. Kor. Phys. Soc.* 46, S43-S47

Ren, F.; Jiang, C.Z.; Fu, D.J. & Ru, Q. (2005b). Raman scattering studies of Ag nanocluster composites formed by ion implantation into silica, *Jap. J. Appl. Phys.* 44, 8512-8514

Ren, F.; Jiang, C.Z.; Liu, C.; Fu, D. & Shi, Y. (2005c). Interface influence on the surface plasmon resonance of Ag nanocluster composite, *Solid State Comm.* 135, 268-272

Ren, F.; Jiang, C.Z.; Liu, C.; Wang, J. & Oku, T. (2006). Controlling the morphology of Ag nanoclusters by ion implantation to different doses and subsequent annealing, *Phys. Rev. Lett.* 97, 165501-1 - 165501- 4

Ren, F.; Jiang, C.Z.; Cai, G.X.; Fu, Q. & Shi, Y. (2007). Fabrication of hollow nanoclusters by ion implantation, *Nucl. Inst. Meth. Phys. Res. B* 262, 201-204

Ren, F.; Cai, G.X.; Xiao, X.H.; Fan, L.X.; Liu, C.; Fu, D.J.; Wang, J.B. & Jiang, C.Z. (2008). Ion irradiation induced hollow and sandwiched nanoparticles, *J. Appl. Phys.* 103, 843308-1 - 843308-4

Ren, F.; Xiao, X.H.; Cai, G.X.; Wang, J.B. & Jiang, C.Z. (2009). Engineering embedded metal nanoparticles with ion beam technology, *Appl. Phys. A* 96, 317-325

Rodríguez -Iglesias, V.; Silva-Pereyra, H.G.; Cheang-Wong, J.C.; Reyes-Esqueda, J.A.; Rodríguez-Fernandez, L.; Crespo-Sosa, A.; Kellerman, G. & Oliver, A. (2008). MeV Si ion irradiation effects on the optical absorption properties of metallic nanoparticles embedded in silica. Nucl. Inst. Meth. Phys. Res. B 266, 3138-3142

Roiz, J.; Oliver, A.; Munoz, E.; Rodrıguez-Fernandez, L.; Hernandez, J.M. & Cheang-Wong, J.C. (2004). Modification of the optical properties of Ag-implanted silica by annealing in two different atmospheres, *J. App. Phys.* 95, 1783-1791

Romanyuk, A.; Spassov, V. & Melnik, V. (2006). Influence of in situ ultrasound treatment during ion implantation on formation of silver nanoparticles in silica, *J. Appl. Phys.* 99, 034314-1 - 034314-4

Sahu, G.; Rath, S.K.; Joseph, B.; Roy, G.S. & Mahapatra, D.P. (2009). Saturation effects observed in high fluence heavy ion implantation at few tens of keV, *Vacuum* 83, 836-840

Saito, Y. & Kitahara, A. (2000). Absorption in the visible region of $LiNbO_3$ sequentially implanted with Ag and Cu, *J. Appl. Phys.* 87, 1276-1279

Sarkisov, S.S.; Williams, E.K.; Curley, M.J.; Smith, C.C.; Ila, D.; Venkateswarlu, P.; Poker, D.B. & Hensley, D.K. (1998a). Mechanisms of formation of nonlinear optical guide structures in metal cluster composites produced by ion beam implantation, *Mat. Res. Soc. Symp. Proc.* 504, 357- 362

Sarkisov, S.S.; Williams, E.K.; Curley, M.J.; Ila, D.; Venkateswarlu, P.; Poker, D.B. & Hensley, D.K. (1998b). Third order optical nonlinearity of colloidal metal nanoclusters formed by MeV ion implantation, *Nucl. Inst. Meth. Phys. Res. B* 141, 294-298

Sarkisov, S.S.; Williams, E.K.; Curley, M.J.; Smith, C.C.; Ila, D.; Poker, D.B.; Hensley, D.K.; Banks, C. & Penn, B. (1998c). Nonlinear optical waveguides based on metal nanocluster composites produced by ion beam implantation, *Proc. SPIE.* 3283, 942-948

Sarkisov, S.S.; Williams, E.K.; Curley, M.J.; Ila, D.; Svetchnikov, V.L.; Pan, V.M.; Poker, D.B.; Hensley, D.K.; Banks, C.; Penn, B. & Wang, J.W. (1998d). Nonlinear optical waveguides produced by Mev metal ion beam implantation of lithium niobate, *Proc. SPIE.* 3413, 98-110

Sarkisov, S.S.; Curley, M.J.; Williams, E.K.; Ila, D.; Svetchnikov, V.L.; Zandbergen, H.W.; Zykov, G.A.; Poker, D.B. & Hensley, D.K. (1999). Fabrication of nonlinear light-guiding nanocomposute structures by metal ion implantation. Proc. SPIE. 3790, 43-55

Sarkisov, S.S.; Curley, M.J.; Williams, E.K.; Ila, D.; Svetchnikov, V.L.; Zandbergen, H.W.; Zykov, G.A.; Banks, C.; Wang, J.-C.; Poker, D.B. & Hensley, D.K. (2000). Nonlinear optical waveguides produced by MeV ion implantation in $LiNbO_3$, *Nucl. Inst. Meth. Phys. Res. B* 166-167, 750-757

Sarychev, A. & Shalaev V. (2007). *Electrodynamics of metamaterials*, World Sci. Publ., New York

Shang, D.Y., Saito, Y., Kittaka, R., Taniguchi, S. & Kitahara, A. (1996). Optical properties of $LiNbO_3$ implanted with Ag ions, *J. Appl. Phys.* 80, 6651-6654

Sheik-Bahae, M.; Said A.A.; Wei, T.-Z.; Hagan, D.J. & van Stryland, E.W. (1990). Sensitive measurement of optical nonlinearities using a single beam, IEEE J. Quant. Electr. 29, 760-769

Stepanov, A.L.; Hole, D.E., Bukharaev, A.A.; Townsend, P.D. & Nurgazizov, N.I. (1998). Reduction of the size of the implanted silver nanoparticles in float glass during excimer laser annealing, *Appl. Surf. Sci.* 136, 298-305

Stepanov, A.L.; Hole, D.E. & Townsend, P.D. (1999a). Reflectance of the dielectrics layers containing metal nanoparticles formed by ion implantation, *J. Non.-Cryst. Solids* 224, 275-279

Stepanov, A.L.; Hole, D.E. & Townsend, P.D. (1999b). Formation of silver nanoparticles in soda-lime silicate glass by ion implantation near room temperature, *J. Non.-Cryst. Solids* 260, 67-74

Stepanov, A.L.; Hole, D.E. & Townsend, P.D. (1999c). Modification of size distribution of ion implanted silver nanoparticles in sodium silicate glass using laser and thermal annelaing, *Nucl. Instrum. Meth. Phys. Res.* 149, 89-98

Stepanov, A.L.; Hole, D.E. & Townsend, P.D. (2000a). Optical reflectance of insulators containing implanted metal nanoparticles, *Nucl. Instrum. Meth. Phys. Res.* 161-163, 913-916

Stepanov, A.L.; Hole, D.E. & Townsend, P.D. (2000b). Excimer laser annealing of glasses containing implanted metal nanoparticles, *Nucl. Instrum. Meth. Phys. Res.* 166-167, 26-30

Stepanov, A.L.; Abdullin, S.N.; Khaibullin, R.I., Valeev, V.F.; Osin, Yu.N.; Bazarov, V.V. & Khaibullin, I.B. (1995). Ion synthesis of colloidal silver nanoclusters in the organic substrate, Mat. Res. Soc. Proc. 392, 267-272

Stepanov, A.L. (2000c). Optical reflection from dielectric layers containing metal particles formed by ion implantation, *Optics and Spectroscopy* 89, 408-412

Stepanov, A.L.; Zhikharev, V.A.; Hole, D.E., Townsend, P.D.; Khaibullin, I.B. (2000d). Depth distribution of Cu, Ag and Au implanted at low energy into insulators, Nucl. *Instrum. Meth. Phys. Res.* 166-167, 882-886

Stepanov, A.L.; Popok, V.N.; Hole, D.E. & Bukharaev, A.A. (2001a). Interaction of high-power laser pulses with glasses containing implanted metallic nanoparticles, *Physics of the Solid State* 43, 2192-2198

Stepanov, A.L. & Popok, V.N. (2001b). Laser and thermal modification of silver-ion implanted glasses, *J. Appl. Spectr.* 68, 164-169

Stepanov, A.L.; Hole, D.E. & Popok, V.N. (2001c). Effect of the target surface temperature on the distribution of nanoparticles formed by ion implantation, *Tech. Phys. Lett.* 27, 554-556

Stepanov, A.L.; Hole, D.E. & Bukharaev, A.A. (2002a). Interaction of high-power excimer-laser pulses with soda-lime silicate glass containing ion-implanted metal nanoparticles, *Vacuum* 64, 169-177

Stepanov, A.L.; Popok, V.N. and Hole, D.E. (2002d). Formation of metallic nanoparticles in silicate glass througth ion implantation, *Glass Phys. Chem.* 28, 90-95

Stepanov, A.L. and Hole, D.E. (2003c). Laser annealing of metal-dielectric nanocomposites formed by ion implantation, *Phil. Mag. Lett.* 82, 149-155

Stepanov, A.L. (2002b). Formation of metal nanoparticles in dielectrics by low energy ion implantation, In: *Recent Res. Devel. Appl. Phys.* 5, 1-25, Transworld Res. Network, Kerala

Stepanov, A.L. & Popok, V.N. (2003a). Effect of the ion beam current density on the formation of implanted metal nanoparticles in a dielectric matrix, *Tech. Phys. Lett.* 29, 977-979

Stepanov, A.L. (2003b). Modification of implanted metal nanoparticles in the dielectrics by high-power laser pulses, *Rev. Adv. Mater. Sci.* 4, 45-60

Stepanov, A.L. (2002c). Laser annealing of glasses with implanted metal nanoparticles, In: *Recent Res. Devel. Non.-Cryst. Solids* 3, 177-198, Transworld Res. Network, Kerala

Stepanov, A.L. & Popok, V.N. (2004a). Nanosecond pulse laser and furnace annealing of silver nanoparticles formed by implantation is silicate glass, *Surf. Coat. Thechnol.* 185, 30-37

Stepanov, A.L. & Popok, V.N. (2004b). Nanostructuring of silicate glass under low-energy Ag-ion implantation, *Surf. Sci.* 566-568, 1250-1254

Stepanov, A.L. (2005a). Optical extinction of metal nanoparticles synthesised in polymers by ion implantation, In: *Metal-Polymer Nanocomposites*, edited by L. Nicolais, G. Carotenuto (Eds.), 241-263, John Wiley & Sons Publ, London

Stepanov, A.L. & Popok, V.N. (2005b). Synthesis of silver nanoparticles by the ion implantation method and investigation of their optical properties, *J. Appl. Spectr.* 72, 229-234

Stepanov, A.L.; Chichkov, B.N.; Valeev, V.F.; Nuzhdin, V.I. & Faizrakhmanov, I.A. (2008a). Modification of ion-synthesized silver nanoparticles in glass by high-power eximer laser pulses, *Tech. Phys. Lett.* 34, 184-186

Stepanov, A.L.; Valeev, V.F.; Nuzhdin, V.I.; Faizrakhmanov, I.A. & Chichkov, B.N. (2008b). laser annealing induced melting of silver nanoparticles in a glass matrix, *Tech. Phys. Lett.* 34, 1014-1017

Stepanov, A.L.; Valeev, V.F.; Nuzhdin, V.I.; Bazarov, V.V.; & Faizrakhmanov, I.A. (2009). Excimer laser-assisted annealing of silicate glass with ion-synthesised silver nanoparticles, *Tech. Phys.* 54, 1504-1510

Steiner, G.; Pham, M.T.; · Kuhne, Ch. & Salzer, R. (1998) Surface plasmon resonance within ion implanted silver clusters, *Fresenius J. Anal Chem.* 362, 9-14

Takeda, Y.; Plaksin, O.A.; Lu, J. & Kishimoto N. (2006). Optical nonlinearity of metal composites fabricated by negative ion implantation, *Vacuum* 80, 776-779

Townsend, P.T.; Chandler, P.J. & Zhang L. (1994). *Optical Effects of Ion Implantation*, Cambridge Univ. Press, Cambridge

Townsend, P.T. & Olivares, J. (1997) Laser processing of insulator surface, *Appl. Surf. Sci.* 109-110, 275-282

Treilleux, M.; Thevenard, P.; Ghassagne, G. & Hobbs L.H. (1978). Observation of implanted potassium aggregates in MgO single crystal, *Phys. Stat. Sol. A* 48, 425-430

Tsuji, H.; Kurita, K.; Gotoh, Y.; Kishimoto, N. & Ishikawa, J. (2002a). Optical absorption properties of Cu and Ag double negative-ion implanted silica glass, *Nucl. Inst. Meth. Phys. Res. B* 195, 315-319

Tsuji, H.; Kurita, K. & Motono, M. (2002b).Control of optical absorption band due to Cu/Ag nanoparticles in SiO_2 glass by dual ion implantation of Cu^- and Ag, *J. Vac. Soc. Jap.* 45, 528-532

Tsuji, H.; Sugahara, H.; Gotoh, Y. & Ishikawa, J. (2002c). Surface modification of TiO_2 (rutile) by metal negative ion implantation for improving catalytic properties. *Surf. Coat. Thechnol.* 158-159, 208-213

Tsuji, H.; Arai, N.; Motono, M.; Gotoh, Y.; Abachi, K.; Kotaki, H. & Ishikawa, J. (2003a). Study on optical reflection property from multilayer on Si substrate including nanoparticles in SiO_2 layer, *Nucl. Inst. Meth. Phys. Res. B* 206, 615-619

Tsuji, H.; Sugahara, H.; Gotoh, Y. & Ishikawa, J. (2003b). Improvement of photocatalytic efficiency of rutile titania, *Nucl. Inst. Meth. Phys. Res. B* 206, 249-253

Tsuji, H.; Arai, N.; Matsumoto, T.; Ueno, K.; Gotoh, Y.; Abachi, K.; Kotaki, H. & Ishikawa J. (2004). Silver nanoparticle formation in thin oxide layer on silicon by silver-negative-ion implantation for Coulomb blockade at room temperature, *Appl. Surf. Sci.* 238, 132-137

Tsuji, H.; Arai, N.; Matsumoto, T.; Ueno, K.; Abachi, K.; Kotaki, H.; Gotoh, Y. & Ishikawa, J. (2005a).Delta layer formation of silver nanoparticles in thin silicon dioxide film by negative ion implantation, *Surf. Coat. Technol.* 196: 39-43

Tsuji, H.; Arai, N.; Matsumoto, T.; Ueno, K.; Abachi, K.; Kotaki, H.; Gotoh, Y. & Ishikawa, J. (2005b). Delta layer formation of silver nanoparticles in thin silicon dioxide film by negative ion implantation, *Surf. Coat. Technol.* 196: 39-43

Tsuji, H.; Sakai, N.; Sugahara, H.; Gotoh, Y. & Ishikawa, J. (2005c). Silver negative-ion implantation to sol-gel TiO_2 film for improving photocatalytic property under fluorescent light, *Nucl. Inst. Meth. Phys. Res. B* 237, 433-437

Tsuji, H.; Sakai, N.; Gotoh, Y. & Ishikawa, J. (2006). Photocatalytic properties of sol-gel titania film under fluorescent-ligth irradiation improved by silver negative-ion implantation, *Nucl. Inst. Meth. Phys. Res. B* 242, 129-132

Wang Y.H.; Jiang C.Z.; Ren, F.; Wang, Q.Q.; Chen, D.J. & Fu, D.J. (2007). Effect of ingredient on optical properties of Ag/Cu metal alloy nanoclusters in silica glass, *J. Mater. Sci.* 42, 7294-7298

Wang, Y.H.; Jiang, C.Z.; Xiao, X.H. & Chen, D.J. (2008). Third-order nonlinear optical response of Cu/Ag nanoclusters by ion implantation under 1064 nm laser exitation, *Physica B* 403, 2143-2147

Wang, Y.H.; Peng, S.J.; Lu, J.D.; Wang, R.W.; Mao, Y.I. & Chen, Y.G. (2009a). Optical properties of Cu and Ag nanoparticles synthesized in glass by ion implantation, *Vacuum* 83, 408-411

Wang, Y.H.; Peng, S.J.; Lu, J.D.; Wang, R.W.; Chen, Y.G. & Mao Y.I. (2009b). Nonlinear optical properties of Ag nanocluster composite fabricated by 200 keV negative ion implantation, *Vacuum* 83, 412-415

White, C.W.; Thomas, D.K.; Hensley, D.K.; McCallum, J.C., Pogany, A.; Haglund Jr., R.F.; Magruder, R.H.; & Yang, L. (1993). Colloidal Au and Ag precipitation formed in Al_2O_3 by ion implantation and annealing, *Nanostruc. Mat.* 3, 447-457

Williams, E.K.; Ila, D.; Sarkisov, S.S.; Curley, M.J.; Poker, D.B.; Hensley, D.K. & Borel C. (1998a). Study of the effects of VeV Ag, Cu, Au and Sn implantation on the optical properties of $LiNbO_3$, *Mat. Res. Soc. Symp. Proc.* 504, 363-369

Williams, E.K.; Ila, D.; Sarkisov, S.S.; Curley, M.J.; Cochrane, J.C.; Poker, D.B.; Hensley, D.K. & Borel, C. (1998b). Study of the effects of MeV Ag and Au implantation on the optical properties of $LiNbO_3$, *Nucl. Inst. Meth. Phys. Res. B* 141, 268-273

Williams, E.K.; Ila, D.; Darwish, A.; Poker, D.B.; Sarkisov, S.S.; Curley, M.J.; Wang, J.-C.; Svetchnikov, V.L. & Zandbergen H.W. (1999). Characterisation of silver coloids formed in $LiNbO_3$ by Ag and O implantation at room and elevated temperatures, *Nucl. Inst. Meth. Phys. Res. B* 148, 1074-1078

Wood, R.A.; Townsend, P.D.; Skelland, N.D.; Hole, D.E.; Barton, J. & Afonso, C.N. (1993). Annealing of ion implanted silver colloids in glass, *J. Appl. Phys.* 74, 5754-5756

Xiao, X.H.; Jiang, C.Z.; Ren, F.; Wang, J. & Shi, Y. (2006). Ion implantation inducing nanovoids characterization by TEM and STEM, *Solid State Comm.* 137, 362-365

Xiao, X.H.; Ren, F.; Wang, J.B.; Liu, C. & Jiang, C.Z. (2007a). Formation of aligned silver nanoparticles by ion implantation, *Mater. Lett.* 61, 4435-4437

Xiao, X.H.; Guo, L.P.; Ren, F.; Wang, J.B.; Fu, D.J.; Chen, D.L.; Wu, Z.Y.; Jia, Q.J.; Liu, C. & Jang, C.Z. (2007b). Formation of metal nanopartilcles in silica by the sequential implantation of Ag and Cu, *Appl. Phys. A* 89, 681-684

Xiao, X.H.; Xu, J.X.; Ren, F.; Liu, C. & Jiang, C.Z. (2008). Fabrication of Ag nanoclusters in single-crystal MgO by high-energy ion implantation, *Physica E* 40, 705-708

Zhang, L.; Jiang, C.Z.; Ren, F.; Chen, H.-B.; Shi, Y. & Fu, Q. (2004). Optical absorption of nanoclusters by sequentially implanting into SiO_2 glass and subsequently annealing in a selected atmosphere, *Acta Phys. Sinica* 53, 2910-2914

Zimmerman, R.L., Muntele, C.I. & Ila, D. (2005). MeV ion beam induced change in the linear optical properties of MgO, *Surf. Coat. Technol.* 196, 85-88

Zuhr, R.A.; Magruder III, R.H. & Anderson, T.S. (1998). Formation and optical properties of intermetallic nanoclusters formed by sequential ion implantation, *Surf. Coat. Technol.* **101**, 401-408

6

Synthesis of silver nanoparticles with laser assistance

A. Pyatenko
National Institute of Advanced Industrial Science and Technology (AIST).
Nanotechnology research Institute.
Tsukuba, Japan

Silver nanoparticles of different sizes and shapes have been attracting much attention due to their unusual size and shape-depending optical, electrical, and magnetic properties [1 - 4]. Up to day the majority of silver colloids are still preparing by using different chemical methods. The general idea of all these methods is chemical reduction of silver ions, Ag^+, in aqueous or non aqueous solutions by different reduction agents, like citrate [5] or borohydride [5,6]. Many promising results were obtained with this method, but like any other experimental technique, the chemical reduction has its own disadvantages and restrictions. And it was quite natural that when many researchers were working under development of chemical reduction technique, some scientists had tried to create some new, alternative methods for nanoparticle production.
One of alternative methods to chemical reduction is the method of laser ablation in liquid phase. Laser ablation in gas phase was very well known method which used successfully for many years [7 - 10]. The idea of applying this technique for liquid phase was proposed by two groups of researchers, Cotton-Chumanov group [11] and Henglein group [12] in 1993. Let's consider one of these works in more details. For their new method, authors of [11] proposed the experimental scheme which remains practically unchangeable (with small variations) up to day. This scheme is shown in Figure 1a, while in Figure 1b the other experimental scheme from the paper of Mafune et al., published in 2000 [13] is depicted for comparison. For production of silver nanoparticles in both these schemes, surface of the metal plate immobilized in water solution is irradiated by pulsed laser beam with different parameters (wavelength, pulse duration and pulse energy, pulse repetition rate). Lens is used for the beam focusing on the metal surface. Naonosecond Nd:YAG laser were used in both these experiments as well as in most of experiments accomplished up to day. Because the motivation of the work [11] was the subsequent using of nanoparticles produced in surface- enhanced Raman scattering measurements, the synthesis itself was done in pure water, without any surfactants or other stabilizers. For comparison, the same measurements were made in pure methanol and acetone. Four different metals, silver gold, platinum and cupper were studied in [11]. In case of silver plate ablated in pure water, stable colloids with average size of the particle about 20 nm (with asymmetrical distribution of sizes ranging from approximately 10 to 50 nm) were synthesized. Electron microscopy (data were not

shown in paper) indicates the presence of some amount of large particles and particle agglomerates which can be removed from colloid by centrifugation. When liquid was changed from water to methanol or acetone, the amount of large particle and agglomerates into a colloid were increased. Such colloids became unstable and precipitated completely overnight at room temperature.

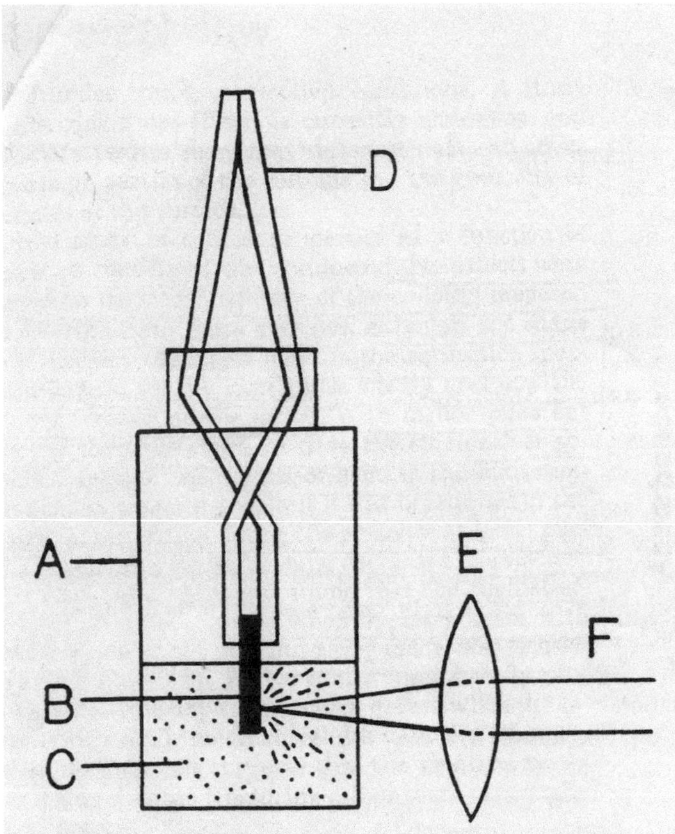

Fig. 1a. Experimental scheme proposed by Cotton-Chumanov group [11]

Further this method was developed in details by Mafune with coworkers [13 -16]. Typical experimental scheme is depicted in Figure 1b. As can be seen from Figure 1, this scheme is very similar to that one proposed by Cotton-Chumanov group [11], and is using by most of researchers in their experiments on laser ablation in liquid phase up to day. To make the process more homogeneous, some researchers rotate the metal plate or vessel or apply a stirring during an ablation [17-19].

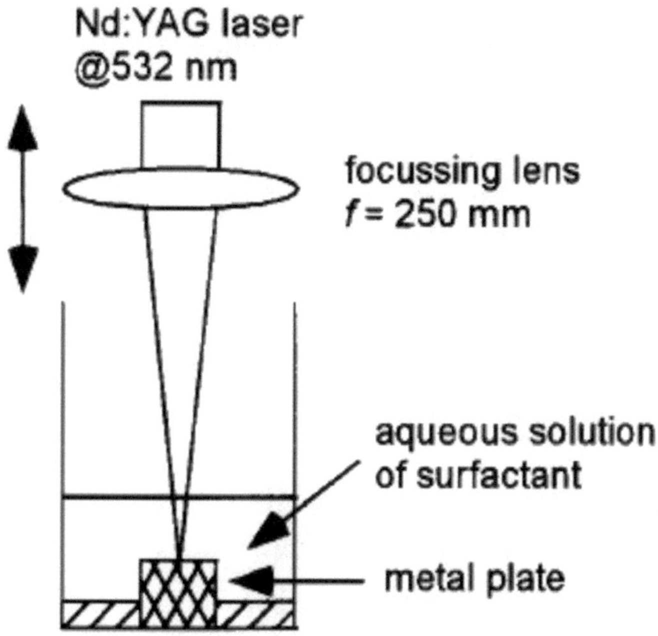

Fig. 1b. Experimental setup used in work of Mafune et al. [13]

To control the particle size, Mafune and co-authors used a surfactant. The different concentrations of surfactant, sodium dodecyl sulfate, SDS [13], or different chemical composition of surfactant [14] were used for particle size control. The following mechanism of nanoparticle's formation was proposed by Mafune et al. Absorption of laser pulse energy causes the plume formation, small cloud of hot dense plasma over metal surface contained high concentration of metal atoms and ions. Metal atoms in the plume aggregate rapidly into small embryonic particles (or nuclears) as fast as metal atoms collide mutually. After that, two concurrent processes take place: relatively slow particle growth and particle stabilization. Final average size of the particle depends on chemical composition and concentration of stabilizer. The results of experiments with different surfactant concentration are shown in Figure 2, where TEM photographs and correspondent particle size distributions are presented for three different concentrations of SDS [13].

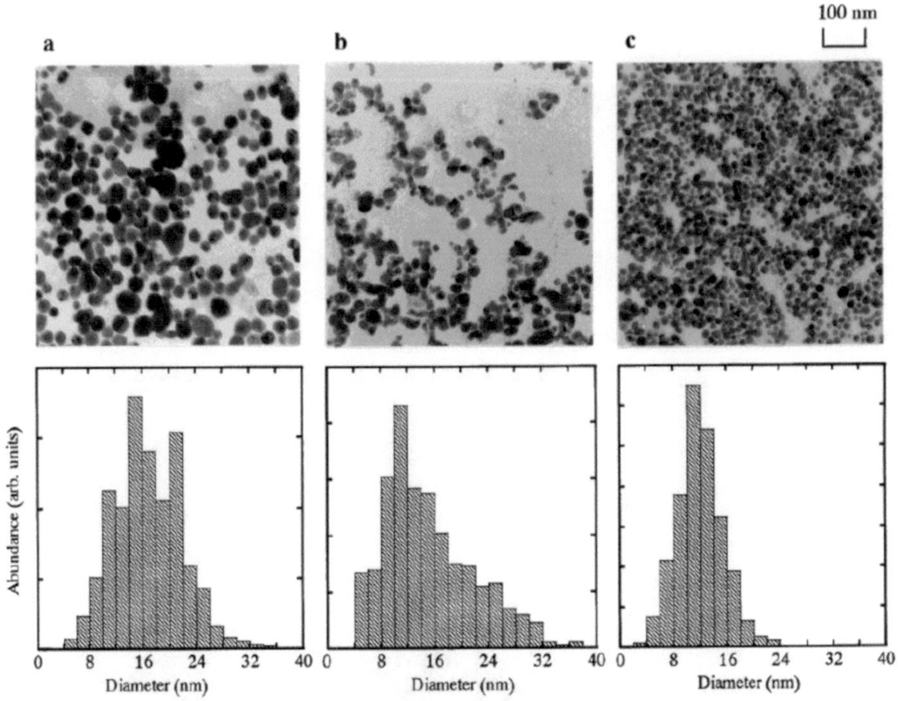

Fig. 2. Electron micrographs and size distributions of the silver nanoparticles produced by laser ablation at 90 mJ/pulse in a SDS aqueous solution at the various SDS concentrations. The concentrations of the solution in panels a-c are 0.003, 0.01, and 0.05 M, respectively. The average size decreases with an increase in the SDS concentration.From paper of Mafune [13].

Silver nanoparticles thus produced were practically spherical and had the average diameters of 16.2 ± 14.0, 14.9 ± 8.4, and 11.7 ± 5.3 nm in the 0.003, 0.01, and 0.05 M solutions, respectively. The results show that the average diameter decreases with an increase in the SDS concentration. But the relative changes in average diameter and standard deviation were very small. Moreover, further increase in SDS concentration had no effect on particle sizes [13]. Similar measurements were performed in a 0.01 M solution with changing the laser power. The results are shown in Figure 3. The average diameters of nanoparticles were found to be 7.9 ± 3.3, 10.7 ± 5.8, and 12.8 ± 4.1 nm for runs with the laser powers of 40, 55, and 70 mJ/pulse, respectively. The results show that the average diameter increases practically linearly with an increase in the laser power. Therefore, smallest and most homogeneous silver nanoparticles were produced for lowest laser pulse energy. And even at that experimental condition, the extent of particle monodispersity (42%) is not so good. Moreover, the efficiency of ablation process decreases with decrease in laser pulse energy, which puts a restriction on using of very low laser power.

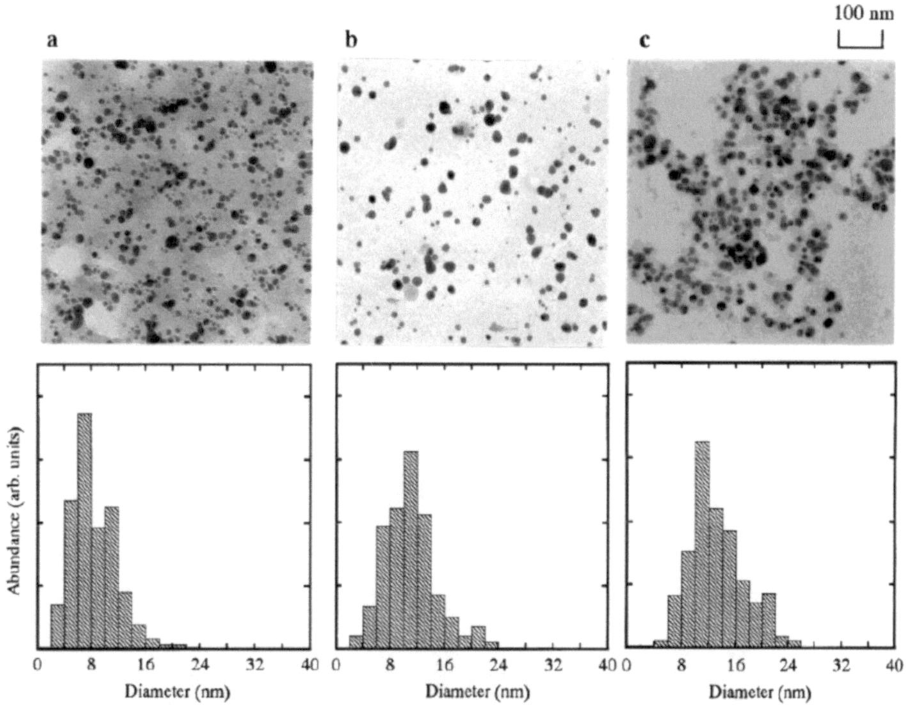

Fig. 3. Electron micrographs and size distributions of the silver nanoparticles produced by laser ablation in a 0.01 M SDS aqueous solution at the various laser powers. The laser powers in panels (a)-(c) are 40, 55, and 70 mJ/pulse, respectively. The average size increases and the distribution broadens with an increase in the laser power. From paper of Mafune [13].

Apart from the energy of laser pulse, the results of laser ablation in liquid phase depend on many other experimental parameters: laser wavelength, pulse duration, time of ablation experiment, mixing conditions and focusing conditions. Precise influence of many of these factors is not clear up to day. Let's consider the last factor, the laser focusing, in more details. Laser beam focusing was applied already in first work of Cotton - Chumanov group [11]. The authors of [11] observed that "tight focusing decreased the efficiency of ablation. It was found that the highest efficiency occurred when the laser beam with an initial diameter of 6 mm was focused into a spot about 1.5 mm." On the contrary, Prochazka et al. in their work published in 1997 [20] found that "the laser beam focusing substantially increases the efficiency of the ablation process". They found also that the colloid revealed the same optical characteristics regardless of the focusing of the beam." Let's note that the focusing conditions or the value of spot size on the metal surface did not report in [20]. Mafune et al. [13] varied the spot size of the laser beam on the surface of the metal plate in the range of 1-3 mm in diameter. Because they never mentioned how the results depend on the different focusing, it is possible to assume that in their experimental conditions the focusing does not play essential role. Pyatenko et al. in their paper published in 1994 [17] tried to vary the focusing conditions keeping the high laser beam power of 0.3 J/pulse for all experiments. In their experiments the spot size of the laser beam on the surface of a silver plate was changed from 1.2 to 0.7 nm when the initial laser beam had the diameter of 7 mm. All the results of ablation including the ablation efficiency, colloid optical characteristics, and particle size distribution, depended strongly on focusing conditions. The efficiency increased near linearly and average particle size decreased with higher focusing. The authors reported that "for small spot sizes, 0.6–0.7mm, most of the observed particles were very small, spherical, and their sizes were rather uniform, d_p = 2–5 nm. A typical TEM image is shown in Fig. 4. In some TEM pictures mid-sized particles with d_p = 5–15 nm were observed, surrounded by smaller particles. The number of such mid-sized particles was much lower than that of the smaller ones, and had practically no influence on the particle-size distribution, which is also shown in Fig. 4.When the spot size was further increased, the number of mid-sized and large (d_p more than 20 nm) particles increased. In the case of a spot size of about 1.5mm, a typical image contains mainly large and mid-sized particles, and only a few small particles could be observed. Also, a large number of particle agglomerates was observed for such poor focusing conditions." All these results show the necessity for additional experimental study with more precise control of the focusing conditions and by varying the values of spot size of the laser beam on a metal surface in wide range.

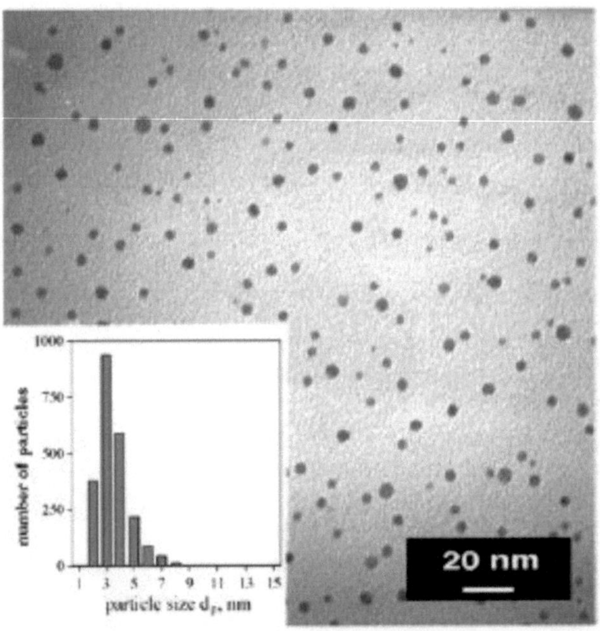

Fig. 4. A typical TEM image of small particles synthesized in [17] with strong laser focusing (spot size on the metal surface was 0.7 mm for the original beam diameter 7 mm). Insert shows the particle size distribution

As we already mentioned before, in works of Mafune et al. [13 - 16] the control under ablation process was accomplished by using the surfactant, and in particular SDS. There are two methods of particle stabilization, electrostatic (Coulombic repulsion) and steric (polymeric or other organic "overcoat") [21 - 23]. In first method, the particles prevent from agglomeration by electrical double layer formed by negative ions absorbed on the particle surface and positive charges induced on the metal particle surface by mirror effect. In second method, the particle agglomeration prevents by absorption of polymer, surfactant or ligand molecules at the surface of the particle. Both techniques of particle stabilization are actively using in chemical reduction method. Both these techniques were employed for the ablation method in liquid phase as well. Different researchers have tried to use the different surfactants, ionic species or other molecules for stabilization of nanoparticles and more precise particle size control.

Chen and Yeh [19] compared the results of ablation made in two different surfactants, the anionic SDS and cationic CTAB. They succeeded to synthesize the colloidal silver particles with average diameter of 4.2 ± 1.9 nm and 7.8 ± 4.5 nm in SDS and CTAB solutions correspondently. Prochazka et al. [20] showed that the addition of a proper amount of $NaCl$ "increased the efficiency of the ablation process, reduces the sizes of colloidal particles, and prevents formation of large particles. However the colloids prepared in the presence of $NaCl$ were less stable than those prepared in pure water". Similar results were obtained by Bae et al. [18]. The authors [18] showed that the addition of $NaCl$ improved the results of ablation only when the concentration of $NaCl$ is below 5 mM. At higher concentrations aggregation turned out to be dominant. They also observed that silver colloids produced in pure water were much more stable than those in $NaCl$ solution. Prochazka et al. [20] have also found that some other anions such as NO_3^-, did not produce the similar effect. They attributed such differences to a substantially higher affinity of the chlorine anion to the silver surface in comparison to other ions, such as nitrates [24]. Tsuji et al. [25] performed the ablation of a silver plate in polyvinylpyrrolidone (PVP) aqueous solutions. By using the different concentration of PVP with molecular weight of 10,000, they produce the silver colloids with different average sizes of the particles. When PVP concentration was increased from 0 to 2, and then to 6 mM, the average particle size decreased from 18 ± 9 to 13 ± 10, and then to 10 ± 5 nm. Further increase in PVP concentration had no effect on average particle size (d_p was equal to 11 ± 7 and 11 ± 8 nm when the PVP concentration increased to 12 and 18 mM). Moreover, because the standard deviation in d_p was found to increase in two last experiments, it is possible to expect that the high concentration of PVP can promote particle agglomeration. This result is looked similar to result obtained by Bae et al.[18] for $NaCl$ (see above).

Analysis of all available dada shows that in spite of the different experimental conditions and different techniques for particle stabilization, the values of average particle diameter varied not so dramatically, from several nanometers in experiments of Pyatenko et al. [17], and some experiments of Mafune [13] and Chen and Yeh [19], to 10 - 15 nm in other experiments. At the same time the average value of standard deviation in average diameter was usually more than 40 %. Very often agglomeration takes place in spite of special efforts to prevent it. Also, the production of large particles additionally to the main relatively small particles was reported in most of papers. The UV-Vis absorption curves reported in almost all papers revealed the red tails, that unambiguously evident to agglomeration process and large particle formation. Therefore, it is possible to conclude that the particle size control and the particle size homogeneity are the main difficulties of laser ablation method.

If the majority of the works on laser ablation in liquid was performed in pure water (the main motivation of these works is SERS application [11, 20]) or in aqueous solutions, some researchers made their experiments in different organic solvents [26 - 33]. In such case, inside of the boundary region between plume and liquid different high temperature processes, like chemical reactions between atoms and ions of ablated material and organic molecules can take place. The molecules of organic solvent themselves can be fragmented, and the reactions between such fragments with atoms or ions of ablated material can be observed. Furthermore, some metastable phase compounds can be formed at such high temperature-high pressure conditions. Therefore, the formation of new materials such as carbides, nitrides, hydrates and others can be observed. This could be new and very interesting direction in the field of ablation in liquid phase [34, 35].

Because the colloidal particles produced in result of ablation of metal target immersed in liquid can be irradiated by subsequent laser pulses, the secondary process of particle interaction with laser beam will occur simultaneously with primary ablation process. It makes the total process more complicated. This phenomenon was recognized already in the first paper published by Cotton-Chumanov group [11]. The authors call this phenomenon "self - absorption" and tried to minimize it effect on absorption process: "In order to reduce self-absorption of the laser radiation when the pulse penetrates through the optical dense colloid, the metal was placed close to the wall". To use this phenomenon for improving the results of primary ablation was proposed by Prochazka et al. [20]. They showed that "after additional irradiation of colloid when silver foil was removed from the cell, the particle size distribution was narrowed, the average particle size was reduced, and particles larger than about 40 nm were completely absent". Further this phenomenon was studied in more details in woks of Tsuji with co workers [25, 36 -39], Smejkal et al. [40] as well as by other research groups. They showed in particular that the results of self-absorption strongly depend on the wavelength of laser irradiation. This fact can be explained by the strong dependence of particle adsorption cross section on the wavelength [41] (see bellow). The values of particle cross section for absorption of light of different wavelengths, σ_{abs}^{λ}, are completely different for fundamental wavelength, λ = 1064 nm, second, λ = 532 nm, and third, λ = 532 nm, harmonics of Nd:YAG laser, which used in these researches. Usually, the σ_{abs}^{1064} value is about two orders smaller than σ_{abs}^{532} and σ_{abs}^{355} values. This fact can be used (and have been used by some researchers, for example by Mafune [13-16] even without any explanations) to prevent secondary absorption or to divide primary and secondary processes temporary by made the primary ablation with fundamental wavelength and studying the secondary self-absorption process with second or third harmonics. More fundamental study of this phenomenon was made by Pyatenko et al. [42]. They showed that by using the process of secondary laser treatment of silver colloid at appropriate conditions, it is possible not only reduce the average size of the particle dramatically, but to made the particles near monodisperse. The authors [42] explained their results by applying the mechanism proposed by Mafune with co-workers [13] and considering the dependence of σ_{abs}^{λ} on particle size.

As it was mentioned before, the absolute majority of the experiments on laser ablation in liquid phase was accomplished with nanosecond Nd:YAG laser. Recently some researchers tried to use the femtosecond laser to study the ablation process [43-44]. According to Tsuji et al. [43] the average size of the particles decreased from 27 nm to 41 nm when femtosecond laser was used instead of nanosecond one. At the same time, particle size distribution for colloid prepared with femtosecond laser was more narrow, and standard deviation in average particle diameter increased from 20 nm (about 70%) in case of ns to 16 nm (40%) for the case of fs laser. More detail study of this problem is needed with more detail analysis of the possible mechanisms of ablation process.

Most of experimental results obtained up to day in study of the process of laser ablation in liquid can be explained by using the mechanism proposed by Mafune et al. [13]. With this mechanism it is possible also to understand the difficulties with particle size control that arise in laser ablation method (see above). In dead, if embryonic particles were born and subsequently grown in a plume, hot and dense region just above the ablated surface, where the strong non homogeneity existed, the conditions for particle formation and growth are very non homogeneous. And because the process of particle formation and growth strongly depends on the temperature, pressure and concentrations, it is quite natural to expect that the particle sizes will vary in wide range even though the macroscopic parameters will control very precisely. However, nucleation and growth do not represent the only mechanism of nanoparticle formation. Some experimental data show that another formation process, consisting in the ejection of metal drops or solid fragments from the target, so called "explosive boiling", are also take place at particular experimental conditions [45-48]. This second mechanism can be responsible for the formation of large (with sizes larger than about 100 nm) particles, that in turn, leads to the bimodal particle size distribution. Let's emphasize again, that the particle size control, decreasing the width in particle size distribution, and the preventing of the particle agglomeration are still the main challenges in laser ablation method.

In our previous papers [41, 42, 49], we proposed a new approach to the synthesis of noble metal spherical particles. In our new method we did not ablate a metal plate immersed into a liquid, but undergo by laser treatment a colloid consisted from metal particles, synthesized by chemical reduction method. By chousing appropriate parameters of laser irradiation, we can transform the initial colloid, prepared by standard citric reduction method, to the colloid contained only small monodisperse spherical particles. We used these particles further as seeds for their subsequent growth by adding to the colloid appropriate amount of silver nitrate (as a metal precursor) and citrate (as a reduction agent). For final precise control of the particle size and sphericity, the secondary laser treatment of colloid with appropriate parameters was applied. Let's consider this new method in details now.

At the beginning the initial silver colloid was synthesized with standard citric reduction method [5, 6]. For this purpose 450 ml of pure water was mixed with 50 ml of 0.01M $AgNO_3$. The mixture was heated up to boiling with intensive stirring. After boiling started, 10 ml of trisodium citrate water solution (1g of trisodium citrate dissolved in 100 ml of pure water) was added to boiling solution. The boiling with stirring then continued for one hour. To prevent large evaporation of solution, a glass was closed with alumina foil during this 1 hour boiling process. After about 3 min of boiling, the solution turned yellow, and after about 5 min it turned gray-yellow and became opaque. The extinction spectrum of this initial colloid diluted 10 times with water is shown in Figure 5. The peak shape and position clearly shows that the colloid was composed of large, nonspherical particles. The results of TEM observation, shown in Figure 2, proved that the main component of the colloid was

large spheroid particles with typical size of about 120 nm and an aspect ratio of about 1.5. Apart from these particles, a relatively large amount of nanorods with different lengths and thicknesses were observed.
Seed particles were prepared then by "hard" laser treatment of initial colloid.
Here we have to stop talking about experimental procedure and explain about "hard" and "soft" laser irradiation regimes, which we used in our method for primary and secondary laser treatment of colloids synthesized in different stages.

Both these regimes were calculated in frame of particle heating-melting-evaporation model, proposed by Takami et al. [50] and further developed in our works [41, 42, 49] for interaction of nanosecond laser pulse with colloidal particles. As it was shown in [41] , when the laser intensity is smaller than the critical value of 10^{12} W/cm^2 (this condition is valid practically every time when we use a nanosecond Nd:YAG laser), namely this model adequately describes the process of colloidal nanoparticle interaction with laser beam. Thus, if we work with laser intensity lower than the limit value, we can use thermodynamics to calculate the parameters of laser beam for which the particle evaporation process will start and completed. The conditions corresponded to evaporation threshold or evaporation start we call "soft", when the conditions corresponded to evaporation completed or full evaporation call "hard". As it was shown in [41], the laser fluence (not laser power) is the value responsible for these processes, therefore, both "soft" and "hard" processes can be accomplished with different laser beam diameter and different laser power. Laser fluences for "soft" and hard regimes can be calculated by equations (1a) and (1b):

$$J^{soft} \cdot \sigma_{abs}^{\lambda} = \rho \cdot \frac{\pi d_p^3}{6} \cdot \left\{ c_p^s (T_m - T_0) + c_p^l (T_b - T_m) + \Delta H_m \right\} \tag{1a}$$

$$J^{hard} \cdot \sigma_{abs}^{\lambda} = \rho \cdot \frac{\pi d_p^3}{6} \cdot \left\{ c_p^s (T_m - T_0) + c_p^l (T_b - T_m) + \Delta H_m + \Delta H_{ev} \right\} \tag{1b}$$

All physical and thermodynamic constants used in equation (1), i.e., silver density, ρ, heat capacities, c_p^s for solid, and c_p^l for liquid silver, heat of melting, ΔH_m, and heat of evaporation of silver, ΔH_{ev} can be found in reference books. The values of particle cross section for absorption of light of different wavelengths, σ_{abs}^{λ}, can be calculated for spherical silver particles by classic Mie theory [1, 51, 52]. The results of such calculations are shown in Figure 7 for fundamental wavelength, λ = 1064 nm, second, λ = 532 nm, and third, λ = 355 nm, harmonics of Nd:YAG laser. It can be seen from this Figure that the absorption cross sections are strongly depends on particle size, and their size dependences are completely different for different wavelengths. When particle cross sections for light absorption are known, the "soft" and "hard" laser fluences can be calculated for particles of different sizes for different laser wavelengths. The results of these calculations are shown in Figure 8 for three main wavelengths of Nd:YAG laser. Let's note that the strong dependences of

J^{soft} and J^{hard} on particle size and laser wavelength are the direct consequence of these dependences for the absorption cross section, σ_{abs}^{λ}.

Now we can return to the experimental procedure of nanoparticle synthesis. As we already said, seed particles were prepared from initial colloid irradiating it with "hard" conditions. Because the initial colloid was very non homogeneous in particle sizes and shapes, the maximum "hard" laser fluence of 0.8 J/pulse·cm^2 for second harmonic of Nd:YAG laser (λ = 532 nm), was used. According to Figure 4, at such conditions all spherical particles will be evaporated completely independently on their sizes. As concern to non spherical particles, we assumed that at first these particles will be melted becoming spherical, and after that will be evaporated completely as other spherical particles. After 30 min of such "hard" treatment of initial colloid we produce the colloid contained only very small spherical nanoparticles. The average size of these particles was about 10 nm with average standard deviation ± 2 nm. Typical TEM photograph for this colloid is shown in Figure 13, and typical UV-vis absorption spectrum is shown in Figure 10. Because the measurement of particle size distribution for such small particles needs a time (sometimes totally more than 2 weeks), preliminary check has to be done by measuring the UV-Vis absorption spectra. Absorption peak for seed particle colloid has to be very narrow (FWHM is about 35-37 nm) without any shoulder and tail. Peak maximum is about 387-390 nm. All other results will show that some other particles existed into a colloid or agglomeration of seed particles takes place.

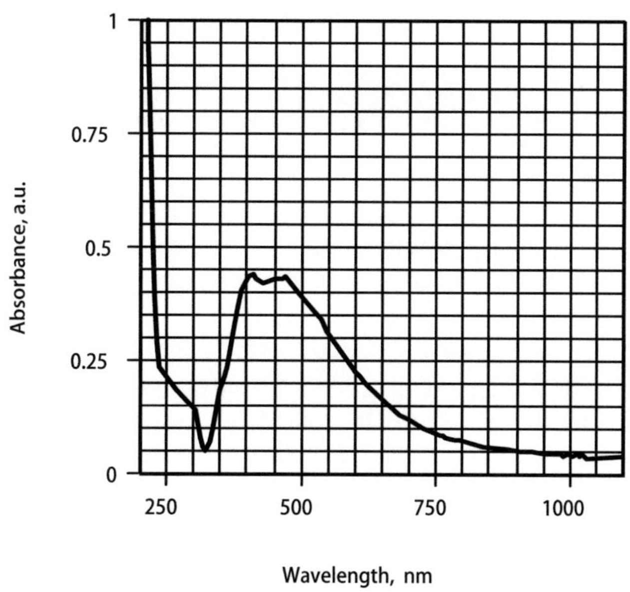

Fig. 5. UV-Vis absorption spectrum of initial colloid

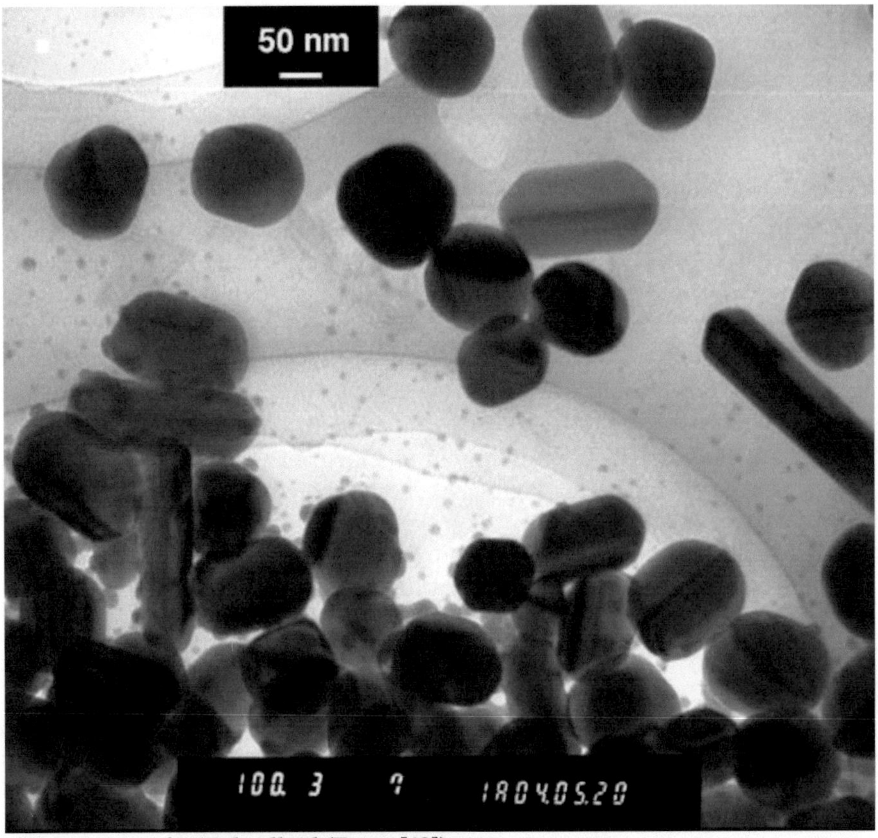

Fig. 6. TEM picture of initial colloid (From [42])

Further growth of seed particles was accomplished by applying seed method. Seed method has a long history [5, 53-55, 22]. Recently it was further developed by the groups of Marphy [56-58] and El-Sayed [59-60] mainly for the synthesis of noble method nanorods. The base of this method can easily be explained for a colloid consisted of monodisperse spherical metal particles (with initial diameter d_0). For subsequent growth of these particles, the appropriate amount of metal precursor (in our case it will be silver nitrate, $AgNO_3$) and reduction agent (in our case - citrate) are added to the colloid. The experimental conditions must be appropriate for reduction process (in our case – colloid boiling). Assuming there is no bulk nucleation and the reduction of silver ions occurs only on the surface of seed particles, that is total number of silver particles in colloid remains the same and only the homogeneous growth of seed particles takes place, we can write:

$$\frac{\Delta N_{Ag}}{N_{Ag}^{0}} = \frac{\Delta m_{Ag}}{m_{Ag}^{0}} = \frac{\Delta V}{V_0} = \frac{V - V_0}{V_0} = \frac{d^3 - d_0^3}{d_0^3} = \left(\frac{d}{d_0}\right)^3 - 1 \tag{2}$$

where d_0 and V_0 are the diameter and volume of the seed particle; d and V the diameter and volume of the particle after growth; N_{Ag}^{0} and m_{Ag}^{0} are total amount of silver atoms in seed colloid and their total mass; ΔN_{Ag} and Δm_{Ag} are the total amount of silver atoms in precursor (silver nitrate) added to seed colloid and the total mass of these silver atoms. Therefore, the value of relative particle growth, $\left(\frac{d}{d_0}\right)$, can easily be calculated by knowing the mass of seed particles and the total mass of silver in silver nitrate added to seed colloid:

$$\left(\frac{d}{d_0}\right) = \left(1 + \frac{\Delta m_{Ag}}{m^0_{Ag}}\right)^{1/3} \tag{2a}$$

Writing the equation (2) we assume also that the particle density, ρ, does not depend on the particle size. This assumption is quite natural for large particles, but could be not so obvious when the size order became less or about few nanometers.

Thus, we tried to increase the size of seed particles by using such seed technique. The experimental procedure was similar to that one used or preparation of initial colloid: some amount of silver nitrate water solution was heated up to boiling with intensive stirring. After that some amount of seed colloid and appropriate amount of citrate solution were added to boiled nitrate. The boiling with stirring then continued for one hour. The appropriate amount of seed colloid was calculated by using equation (2a), relative amount of citrate added was the same as for classic citric reduction method [5,6] (1 ml of citrate solution for 5 ml of 0.01M silver nitrate solution).

Several syntheses were achieved by changing the $\frac{\Delta m_{Ag}}{m^0_{Ag}}$ ratio in equation (2a). The results of TEM observation for four different syntheses are presented in Fig. 11. For small $\Delta m_{Ag} / m^0_{Ag}$ ratios, most of the particles were spherical, and the average size of a spherical particle increased with the $\frac{\Delta m_{Ag}}{m^0_{Ag}}$ ratio. When this ratio exceeded 20, however, the synthesis of spherical particles was accompanied by the synthesis of nanorods. The relative concentration and aspect ratio of these nanorods increased with the $\frac{\Delta m_{Ag}}{m^0_{Ag}}$ ratio.

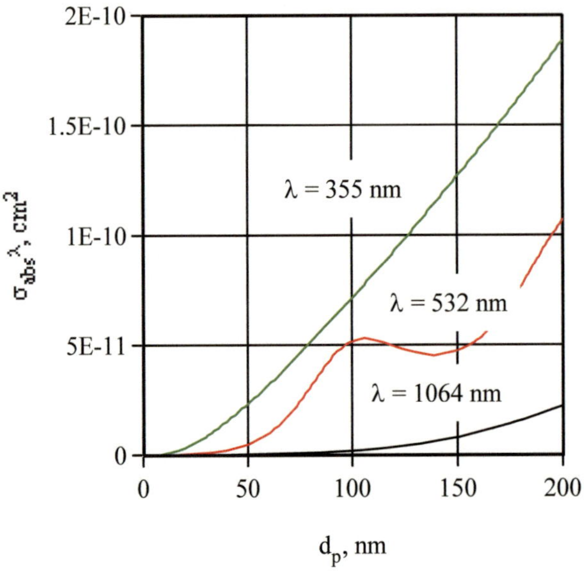

Fig. 7. Absorption cross sections, σ_{abs}^{λ}, for spherical silver nanoparticles calculated as a function of particle diameter, d_p, with Mie theory [51,52] for three characteristic wavelengths of Nd:YAG laser

Table 1 compares the average sizes of spherical particles measured from TEM photographs with the sizes calculated by equation (2a). The good coincidence between experimental and calculated values of d_p proves that the main process was the reduction of silver ions on the surface of seed particles and that the main products were the spherical particles, at least while the $\frac{\Delta m_{Ag}}{m_{Ag}^0}$ ratio did not exceed 100. Very small particles with sizes of few nanometers can be seen in Fig. 11d. These particles can be identified as the result of bulk nucleation, which can already start at $\frac{\Delta m_{Ag}}{m_{Ag}^0}$ ratio about 50. However, the total amount of these very small particles is still rather small, and those particle were not observed in most of other TEM photographs made for $\frac{\Delta m_{Ag}}{m_{Ag}^0} = 45$.

$\frac{\Delta m_{Ag}}{m^0_{Ag}}$	d_p, nm		
	measured		calculated by equation (2a)*
	before soft laser treatment	after soft laser treatment	
3.6	14.2	14.0	13.3
18.9	20.2	19.8	21.7
32.4	27.1	28.7	25.8
45	29.6	28.8	28.7
90	34.0	33.5	36.0

Table 1. Average size of the spherical particles synthesized in one-step experiments

UV-Vis extinction spectra of different colloids are presented in Fig. 8a. The positions of all plasmon peaks correspond well to Mie's theory for spherical particles [1]: the maximums were close to 400 nm and shifted to red as the particles became larger. However, a red tail appeared when the $\frac{\Delta m_{Ag}}{m^0_{Ag}}$ ratio exceeded 20. Small peaks can be distinguished in the tail region for curves 2 to 5 in the insert of Fig. 8a. The maximums of these peaks shifted from 600 to 700nm as the n^+/n_s ratio became larger. These peaks can be identified as the longitudinal peaks of the plasmon absorption of nanorods [61]. The low intensity of these peaks also proves the small relative amount of nanorods.

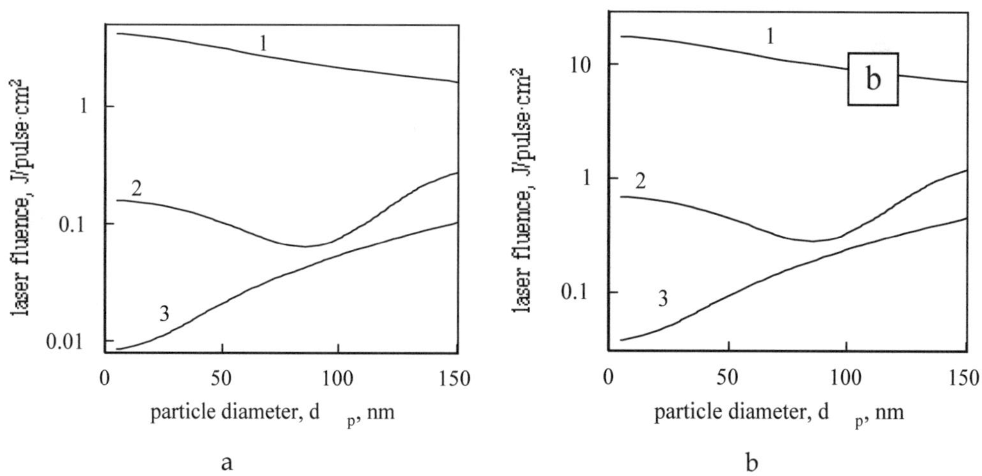

Fig. 8. "Soft" (a) and "hard" (b) laser fluences calculated as the functions of particle diameter for 3 characteristic wavelengths of Nd:YAG laser. (1) λ = 1064 nm, (2) λ = 532 nm, and (3) λ = 355 nm.

To eliminate these nanorods from colloid, we need to irradiate a colloid with "soft" laser irradiation. Second harmonic of Nd:YAG laser (λ = 532 nm) is used again for this irradiation, but the value of laser fluence will depend on the particle size as shown in Figure 8b. For particle size of 20, 30 and 40 nm this "soft" laser fluence will be equal to 0.14, 0.13 and 0.12 $J/puls \cdot cm^2$ correspondently. Normal duration time for such "soft" irradiation is 40 min. During this irradiation all nanorods were melted and became spherical, thus only spherical particles with the same (or practically the same) particle size distribution will remain into the colloid after such "soft" laser treatment. It can be checked by UV-Vis absorption spectra (see Figure 12b). All the red tails, as well as the small red peaks, have disappeared. At the same time, small red shifts of the primary peaks and their widening corresponding to increase in particle size can still be observed. Direct TEM observations shown in Figures 9 for synthesizes 4 from Table 1, also demonstrate that only spherical particles with practically the same particle size distribution remain into a colloid after "soft" irradiation.
Thus, by applying one-step synthesis and subsequent treatment of the synthesized colloid with soft laser irradiation, we obtained stable colloids containing only spherical and almost mono-disperse particles. The average particle size can be well controlled by varying the $\frac{\Delta m_{Ag}}{m^0_{Ag}}$ ratio and can easily be changed from 8 to 40nm. These results are summarized in Table 1.

A further increase in the $\frac{\Delta m_{Ag}}{m^0_{Ag}}$ ratio leads to a further increase in nanorod concentration, diameter and aspect ratio. The treatment of such colloids with soft laser irradiation cannot melt those nanorods completely. Increase in laser power starts the spherical particle evaporation and produces many very small particles. The following multi-step procedure was designed to produce larger particles.
In first step seed particles prepared as described before were used. For each further step, the particles synthesized in previous step used as the seed particles for the next step. Excluding first 3 steps, where relatively small particles were synthesized, all next synthesis includes two procedures: chemical synthesis and laser treatment of colloid synthesized. "Soft" conditions for laser treatment of different colloids were estimated by using the particle sizes calculated from equation (2b) and using the dependence of "soft" laser fluence on particle size depicted in Figure 8a.

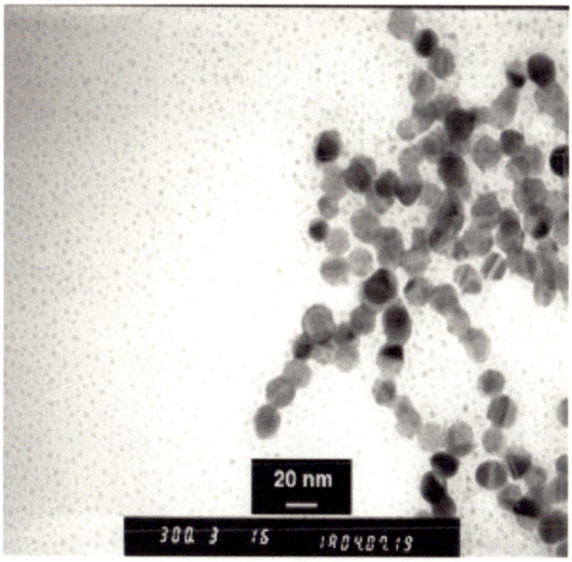

Fig. 9. TEM image of seed colloid (initial colloid treated by "hard" irradiation at λ =532 nm for 30 min).(From [42])

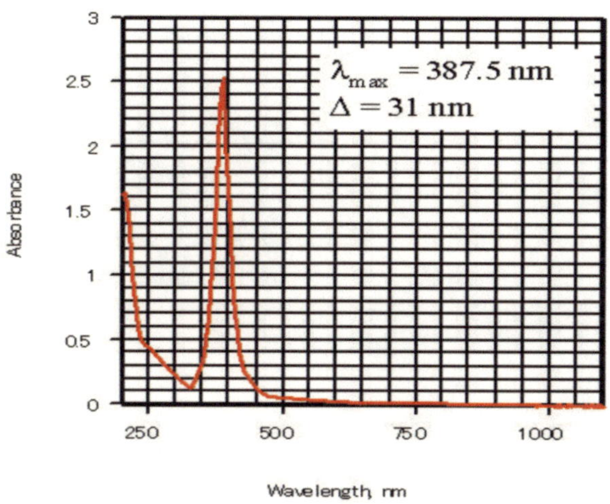

Fig. 10. Typical UV-Viis absorption spectrum of seed colloid (initial colloid treated by "hard" irradiation at λ =532 nm for 30 min).

Because of direct measurements of PSD rather time-spending process, the results of synthesizes can be checked preliminary by measuring the UV-Vis absorption spectra. The results of these measurements are shown in Figure 14a. It can be seen that for each next step excluding first two ones, the absorption peak became more and more wide and shifted to red compare with the previous one. The peaks for synthesis 1 and 2 are very similar to each other and very similar to the peak obtained for seed particles (not shown here). It shows that only spherical relatively small particles were synthesized in first two steps. It can be seen also that peak for synthesis 4 has small red shoulder indicating that some small amount of nanorods with small aspect ratio were formed along with spherical particles. Figure 14b demonstrates how the absorption peaks changed after the "soft" laser treatment of colloids. These results indicate that all nanorods were removed from a colloid as well as all other non spherical particles. Very small shifts of peak maximum show also that the average sizes of the particles remain unchanged after such "soft" treatment. Direct TEM observation also proves that only spherical particles with nearly the same sizes remain into a colloid after "soft" irradiation. The fact that average size of the particle remains unchanged after "soft" treatment can be seen in Table 2 where the results of such multi- step procedure are summarized. Finally the TEM pictures of colloids obtained in Synthesis 4,5,6, and 7 are shown in Figure 15.

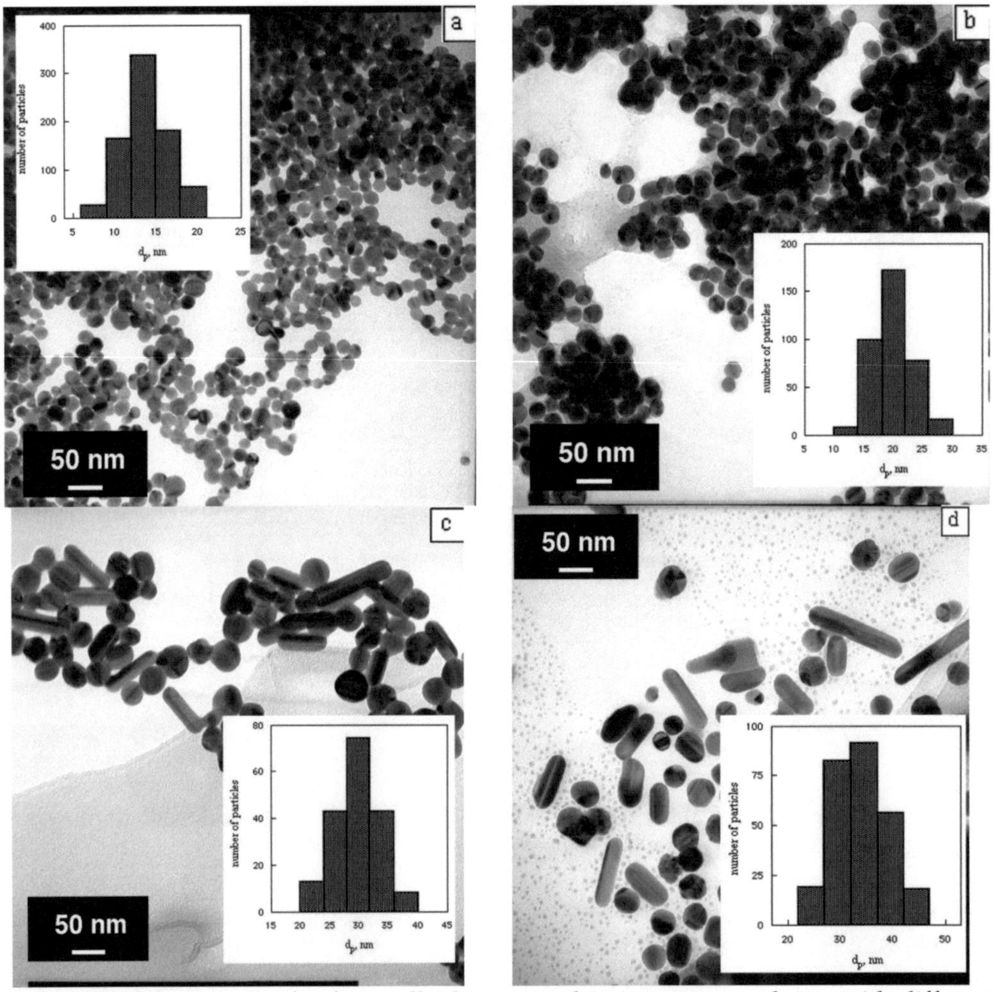

Fig. 11. TEM photographs for four colloids prepared in one-step syntheses with different $\Delta m_{Ag} / m^{0}_{Ag}$ ratios: a) 3.6; b) 18.9; c) 32.4; d) 45. All photos have the same magnification. The inserts are the histograms for particle size distribution.(From [49])

Step number	d_p, nm		
	measured		calculated by equation (2a)*
	before soft laser treatment	after soft laser treatment	
Syn1	23.3 +/- 3.0	-	21.5
Syn2	27.1 +/- 3.8	-	27.1
Syn3	35.9 +/- 4.3	-	34.1
Syn4	40.3 +/- 4.0	39.5 +/-3.5	43.0
Syn5	58.5 +/-5.4	53.6 +/- 5.0	54.2
Syn6	73.1 +/- 6.5	74.5 +/- 7.0	68.3
Syn7	84.1 +/-6.9	-	86.1

* d_{p0} = 10.0nm

Table 2. Average size of the spherical particles synthesized in multi-step experiments

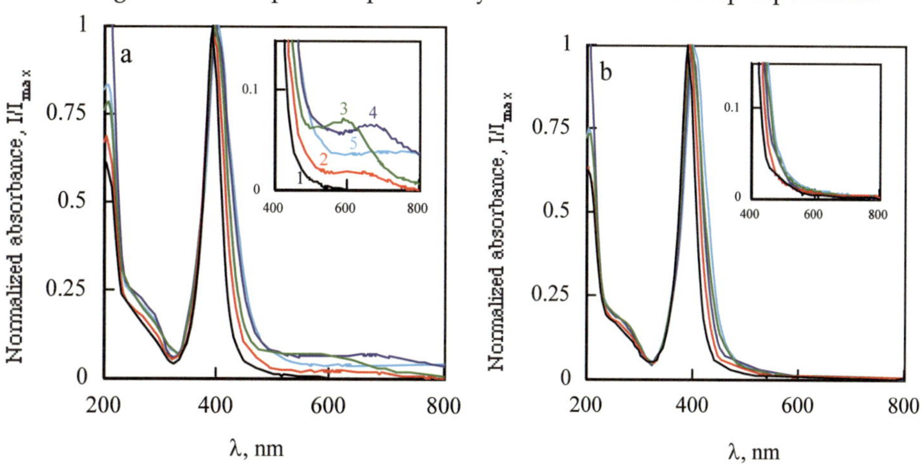

Fig. 12. UV-Vis absorption spectra of colloids synthesized in one-step synthesis, (a) before, and (b) after soft laser treatment. The inserts display details in the 400 to 800nm region. Curves 1 to 5 correspond to Synthesizes 1-5 in Table 1. (From [49])

Thus, we can conclude that our new developed procedure permits to synthesize spherical silver nanoparticles with different desirable size from 10 to 100 nm. All particles are monodisperse, with standard deviation in particle size of about 10%. For synthesis of relatively small particles with size from 10 to 40 nm one-step synthesis with "soft" laser treatment of final colloid can be used, while for synthesis of larger particles with sizes more than 40 nm, multi-step procedure with "soft" laser treatment of intermediate products has to be applied. The new method that we described here is more complicated and more time-spending compare with laser ablation technique, but it has one great advantage compare with laser ablation - the possibility of particle size control, and production of colloids with different desirable size of monodisperse nanoparticles.

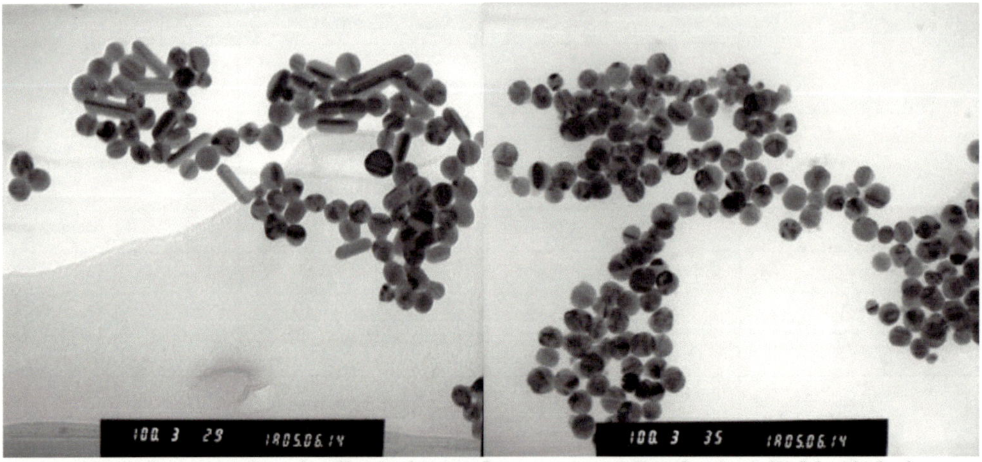

Fig. 13. TEM photographs of colloid obtained in one-step Synthesis 4 (Table 1), before and after "soft" laser treatment.

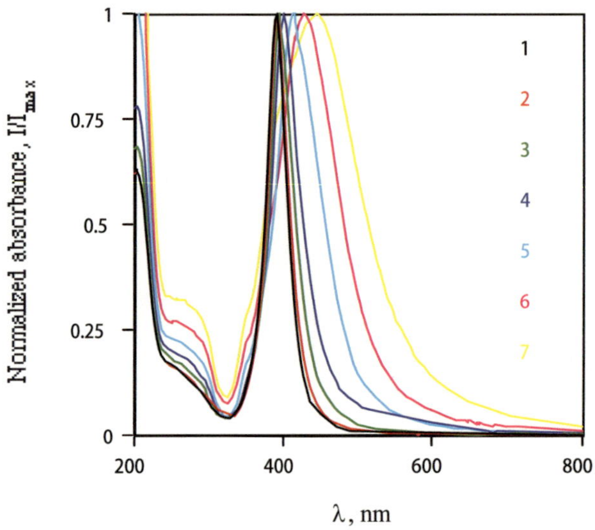

Fig. 14a. UV-Vis absorption spectra of colloids synthesized in multi-step synthesis, Curves 1 to 7 correspond to Synthesizes 1-7 in Table 2.

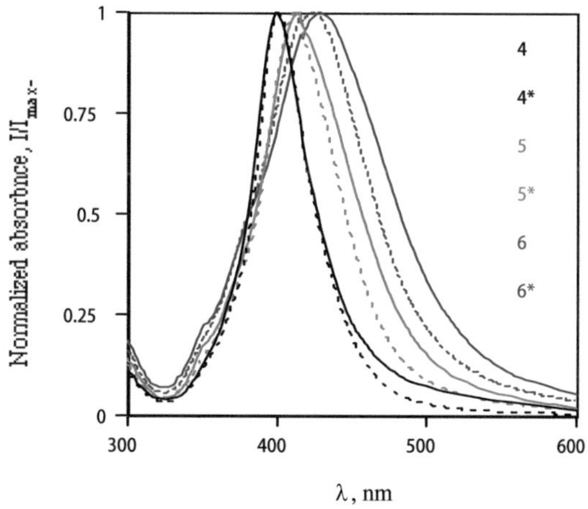

Fig. 14b. UV-Vis absorption spectra of colloids synthesized in multi-step synthesis before (4,5,6) and after (4*,5*,6*) "soft" laser treatment. Curves 4,5,6 correspond to Synthesizes 4,5,6 in Table 2.

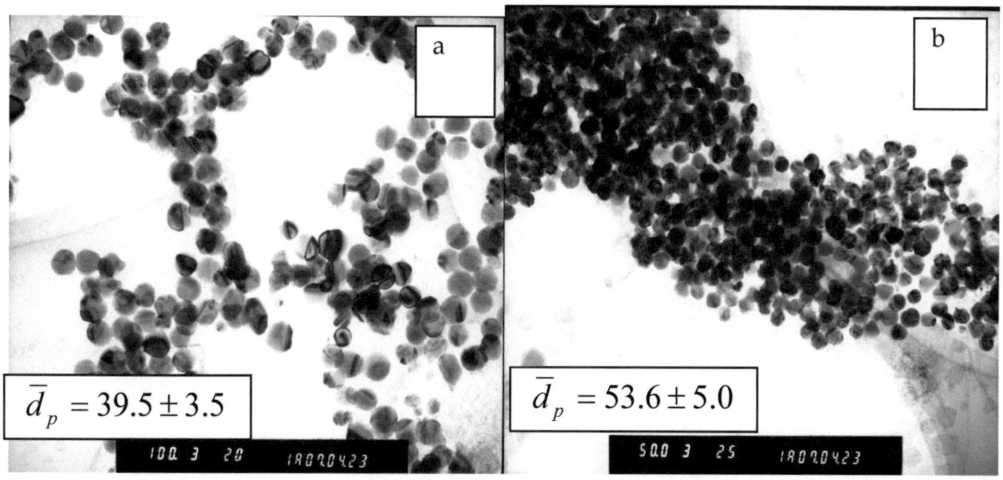

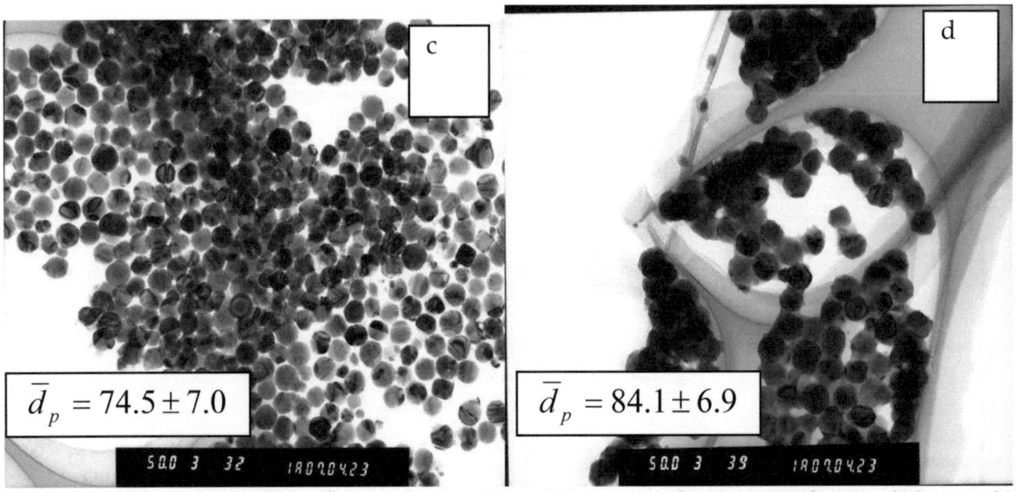

Fig. 15. TEM pictures for colloids obtained in different multi-step synthesis: a) Syn4*; b) Syn5*, c) Syn6* and d) Syn7. Magnification in picture a) is $100x10^3$, in other pictures $50x10^3$.

References

(1) Kreibig, U.; Vollmer, M. *Optical Properties of Metal Clusters;* Springer Series in Material Science 25; Springer: Berlin, 1995.

(2) Kamat, P. V. *J. Phys. Chem. B* 2002, *106,* 7729.

(3) Kelly, L.; Coronado, E.; Zhao, L; Schatz, G. C. *J. Phys Chem. B* 2003, *107,* 668.

(4) Noguez, C. *J. Phys. Chem. C* 2007, *111,* 3806.

(5) Turkevich, J.; Stevenson, P. C.; Hillier, *J. Discuss. Faraday Soc.,*1951, *11,* 55.

(6) Lee, P. C.; Meisel, D. *J Phys. Chem.* 1982, *86,* 3391.

(7) *Laser-induced Chemical Processes.* Steinfeld, J. (ed.), Plenum Press, New York, 1981.

(8) *Laser Ablation in Material Processing: Fundamental and Applications.* MRS Symposium Proceedings. Braren, B.; Dubowski, J. J.; Norton, D.P. (eds.), Pitsburg, MRS, 1993.

(9) Bauerle, D. Laser *Processing and Chemistry.* Second ed. Springer-Verlag Berlin Heidelberg New York, 1996.

(10) *Laser Ablation and Desorption.* In: Experimental Methods in the Physical Sciences. Miller, J. C.; Haglund, R. F. (eds.). Academic press, San Diego, 1998.

(11) Neddersen, J.; Chumanov, G.; Cotton, T. M. *Appl. Spectrosc.* 1993, *47,* 1959.

(12) Fojtic, A. Henglein, A. *Ber. Bunsenges. Phys. Chem.* 1993, *97,* 252.

(13) Mafune, F.; Kohno, J.; Takeda, Y.; Kondow, T.; Sawabe, H. *J. Phys. Chem.* B 2000, *104,* 9111.

(14) Mafune, F.; Kohno, J.; Takeda, Y.; Kondow, T.; Sawabe, H. *J. Phys. Chem. B* 2000, *104,* 8333

(15) Mafune, F.; Kohno, J.; Takeda, Y.; Kondow, T.; Sawabe, H. *J. Phys. Chem. B* 2001, *105,* 5114.

(16) Mafune, F.; Kohno, J.; Takeda, Y.; Kondow, T. *J. Phys. Chem. B* 2001, *105,* 9050.

(17) Pyatenko, A.; Shimokawa, K.; Yamaguchi, M.; Nishimura, O.; Suzuki, M. *Appl. Phys. A,* 2004, *79,* 803.

(18) Bae, C. H.; Nam, S. H.; Park, S. M. *Appl. Surface Sci.* 2002, *197,* 628.

(19) Chen, Y. H.; Yeh, C. S. *Colloid and Surf. A* 2001, *197,* 133.

(20) Prochazka, M.; Mojzes, P.; Stepanek, J.; Vickova, B.; Turpin, P. *Anal. Chem.* 1997, *69*, 5103.
(21) Overbeek J. T. G. *in Colloidal Dispersions;* J. W. Goodwin (ed.), Royal Society of Chemistry, London, 1981, p1.
(22) *Nanoparticles;* G. Schmid (ed.), WILEY-VCH Verlag GmbH & Co. KGaA, Weinheim, 2004.
(23) *Metal Nanoparticles. Synthesis, Characterization, and Applications;* D. L. Feldheim, and Foss C. A. Jr. (eds.), Dekker: New York, 2002.
(24) Aubard, J., Bagnasco, E., Pntigny, J., Ruasse, M., Levi, J., Wentrup-Byrne, E. *J Phys. Chem.* 1995, *99*, 7075.
(25) Tsuji, T., Thang, D., Okazaki, Y., Nakanishi, M., Tsuboi, Y., Tsuji, M. *Appl. Surface Sci.* 2008, *254*, 5224.
(26) Tilaki, R., Zad, A., Mahdavi, S. *Appl. Phys. A* 2006, *84*, 215.
(27) Tilaki, R., Zad, A., Mahdavi, S. *J. Nanopart. Res.* 2007, *9*, 853.
(28) Tilaki, R., Zad, A., Mahdavi, S. *Appl. Phys. A* 2007, *88*, 415.
(29) Compagnini, G., Scalisi, A., Puglisi, O. *Phys. Chem. Chem. Phys.* 2002, *4*, 2787.
(30) Compagnini, G., Scalisi, A., Puglisi, O. *J. Appl. Phys.* 2003, *94*, 7874.
(31) Amendola, V., Rizzi, G., Polizzi, S. Meneghetti, M. *J. Phys. Chem. B* 2005, *109*, 23125.
(32) Amendola, V., Polizzi, S. Meneghetti, M. *J. Phys. Chem. B* 2006, *110*, 7232.
(33) Amendola, V., Polizzi, S. Meneghetti, M. *Langmuir* 2007, *23*, 6766.
(34) Wang, C., Yang, Y., Yang G. *J. Appl. Phys.* 2005, 97, 066104.
(35) Liu, Q., Wang, C., Yang, G. *Europ. Phys. J. B* 2004, 41, 479.
(36) Tsuji, T., Iryo, K., Nishimura, Y., Tsuji, M. *J. Photochem. Photobio. A* 2001, *145*, 201.
(37) Tsuji, T., Watanabe, N., Tsuji, M. *Appl. Surface Sci.* 2003, *211*, 189.
(38) Tsuji, T., Higuchi, T., Tsuji, M. *Chem. Letters* 2005, *34*, 476.
(39) Tsuji, T., Okazaki, Y., Higuchi, T., Tsuji, M. *J. Photochem. Photobio. A* 2006, *183*, 297.
(40) Smejkal, P.; Peleger, J.; Siskova, K.; Viskova, B.; Dammer, O.; Slouf, M. *Appl. Phys. A*, 2004, *79*, 1307.
(41) Pyatenko, A., Yamaguchi, M., Suzuki, M. *J. Phys. Chem.* C 2009, *113*, 9078.
(42) Pyatenko, A., Yamaguchi, M., Suzuki, M. *J. Phys. Chem.* B 2005, *109*, 21608.
(43) Tsuji, T., Kakita, T., Tsuji, M. *Appl. Surface Sci.* 2003, *206*, 314.
(44) Barcikowski, S., Hahn, A., Kabashin, A., Chichkov, B. *Appl. Phys. A* 2007, *87*, 47.
(45) Nichols, W., Sasaki, T., Koshizaki, N. *J. Appl. Phys.* 2006, *100*, 114911.
(46) Nichols, W., Sasaki, T., Koshizaki, N. *J. Appl. Phys.* 2006, *100*, 114912.
(47) Kabashin, A., Meunier, M. *J. Appl. Phys.* 2003, *94*, 7941.
(48) Zhigilei, L., Kodali, P., Garrison, B. *J. Phys. Chem. B* 1998, *102*, 2845.
(49) Pyatenko, A., Yamaguchi, M., Suzuki, M. *J. Phys. Chem. C* 2007, *111*, 7910.
(50) Takami, A., Kurita, H., Koda, S. *J. Phys. Chem. B* 1999, *103*, 1226.
(51) Mie, G. *Ann. Phys.* 1908, *330*, 377.
(52) Kelly, L.; Jensen, T.; Lazarides, A.; Schatz, G. C. In *Metal Nanoparticles. Synthesis, Characterization and Application;* Feldheim, D. L., Foss, C. A., Jr., Eds.; Dekker: New York, 2002.
(53) Zsigmondy, R.; Thiessen, P. A. *Das Kolloide Gold;* Veragsges.:Leipzig, 1925.
(54) Schmid, G. *Chem. ReV.* 1992, *92*, 1709.
(55) Henglein, A.; Meisel, D. *Langmuir* 1998, *14*, 7392.
(56) Jana, N. R.; Gearheart, L.; Murphy, C. J. *J. Phys. Chem. B* 2001,*105*, 4065.
(57) Jana, N. R.; Gearheart, L.; Murphy, C. *J. AdV. Mater.* 2001, *13*, 1389.

(58) Murphy, C. J.; Jana, N. R. *AdV. Mater.* 2002, *14*, 80.
(59) Nikoobakht, B.; El-Sayed, M. A. *Langmuir* 2001, *17*, 6368.
(60) Nikoobakht, B.; El-Sayed, M. A. *Chem. Mater.* 2003, *15*, 1957.
(61) Jana, N. R.; Gearheart, L.; Murphy, C. *Chem. Commun.* 2001, 617.

7

Synthesis of Ag Nanoparticles by Through Thin Film Ablation

P. Terrence Murray and Eunsung Shin
University of Dayton
USA

1. Introduction

There are several applications that require a thin layer of nanoparticles that are spread uniformly and without agglomeration over a surface. Among these applications are carbon nanotube synthesis and the development of fuel cells, advanced sensors, and optoelectronic devices. There is a need for improved processes of forming these nanoparticles. Among the numerous methods that have been developed, pulsed laser ablation has proved (ElShall et al., 1995; Burr et al., 1997, Seraphin et al., 1997, Geohegan et al., 1998; Lowndes et al, 1998, Makimura et al., 1998; Becker et al., 1998; Lowndes et al., 1999; Makino et al., 1999; Link and El-Sayed, 2000; Mafune et al., 2000; Mafune et al., 2001; Ogawa et al., 2000; Tang et al., 2001; Ozawa et al., 2001; Hata et al., 2001; Barnes et al., 2002) to be especially effective because of the potential for congruent ablation, the ability to produce nanoparticles of high purity, the ability to deposit nanoparticles on room temperature substrates, and the relative simplicity of the process. One problem with conventional pulsed laser ablation is the observation that the deposited material, in addition to containing nanoparticles, also contains large, μm-sized particles. These are formed through a process denoted splashing (Ready, 1963), which occurs when a transient liquid layer is formed in the irradiated volume of the target; liquid droplets can be ejected from the target by the recoil pressure of the expanding gas on the transient liquid layer. Traditionally, splashing has been minimized (but not eliminated) by ablating very smooth targets or by using low laser energy densities. Target splashing is an issue in nanoparticle synthesis since the deposition of large particles within a field of nanoparticles makes it considerably more difficult to exploit the unique properties of isolated, non-agglomerated nanoparticles. Target splashing is also an issue because it represents material waste. There is, therefore, a need for a laser ablation process that further reduces or, ideally, eliminates splashing in nanoparticle synthesis. In addition, there is the further need for a process that minimizes nanoparticle agglomeration

Presented here is a process, denoted Through Thin Film Ablation (TTFA), which enables the efficient synthesis of nanoparticles. TTFA, in addition to substantially reducing the probability of large particle splashing, also reduces significantly the problem of nanoparticle agglomeration. The TTFA process is based on irradiating a very thin layer of target material from the backside. The use of such a thin target allows one to reduce the thickness of the liquid layer to such an extent that splashing of large particles is eliminated. Use of a very thin target has the added benefit of minimizing nanoparticle agglomeration. TTFA was originally developed as a method to synthesize transition metal nanoparticles for catalytic growth of single walled carbon nanotubes. The TTFA process has since been used to synthesize nanoparticles of several other materials including Fe (Murray and Shin, 2009), Pt (Petry, 2009; Tschopp et al., 2009), Ni, Ti, and Si. TTFA is illustrated here by presenting the results for Ag (Murray and Shin, 2008) nanoparticle formation.

2. Experimental

Nanoparticle synthesis by TTFA is carried out by ablating, from the backside, a very thin (10's of nm) film of the selected nanoparticle material. Ablation is carried out in the presence of a background gas (typically Ar), which serves to slow the species that are ejected from the target with initial kinetic energy of a few 10's of eV. The slowed atoms recombine in the gas phase to form nanoparticles, which are easily collected on an appropriately placed substrate. The experimental setup is presented in Figure 1(a), and a false color photograph of a TTFA plume is presented in Figure 1(b).

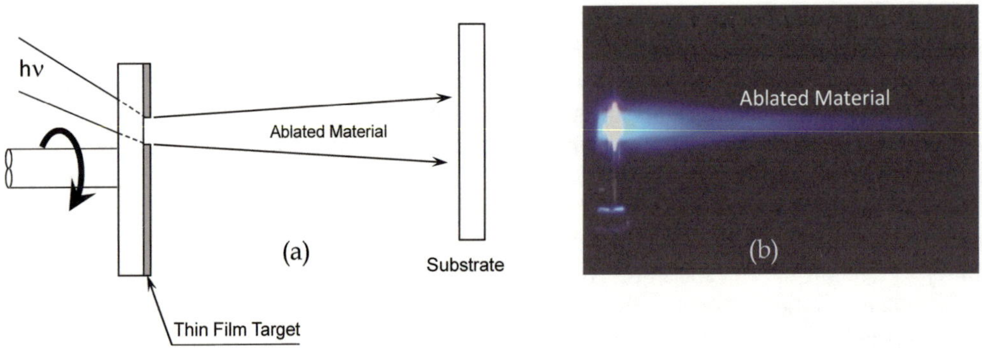

Fig. 1. (a) Schematic of the TTFA experiment. (b) photograph (false color) showing light emission from the ablated material.

For the work described here, the target consisted of a thin film (20 nm) of Ag that was sputter deposited onto a 5 cm diameter disk of UV-transparent fused silica. This structure was irradiated from the back side through the silica support with the output of a KrF excimer laser (λ= 248 nm). The laser beam struck the target at an angle of 45°, and the laser energy density at the target was 0.6 J/cm^2. The target was rotated after each laser pulse in order to expose a fresh area for subsequent ablation.

Ablation was carried out in vacuum (p=2x10^{-4} Pa) and with the ablation chamber filled with Ar at pressures of 133, 266, and 665 Pa. Ag nanoparticles were deposited onto TEM grids placed 5 cm from the target on a substrate of single crystal Si. Each deposition was carried out by using a single laser pulse, and the substrates were nominally at room temperature during deposition. All nanoparticle samples described here were prepared by ablating different area of the same target.

3. Results and Discussion

Shown in Figures 2(a)-2(d) are TEM micrographs of Ag nanoparticles deposited by TTFA; each sample was prepared by using a single laser pulse.

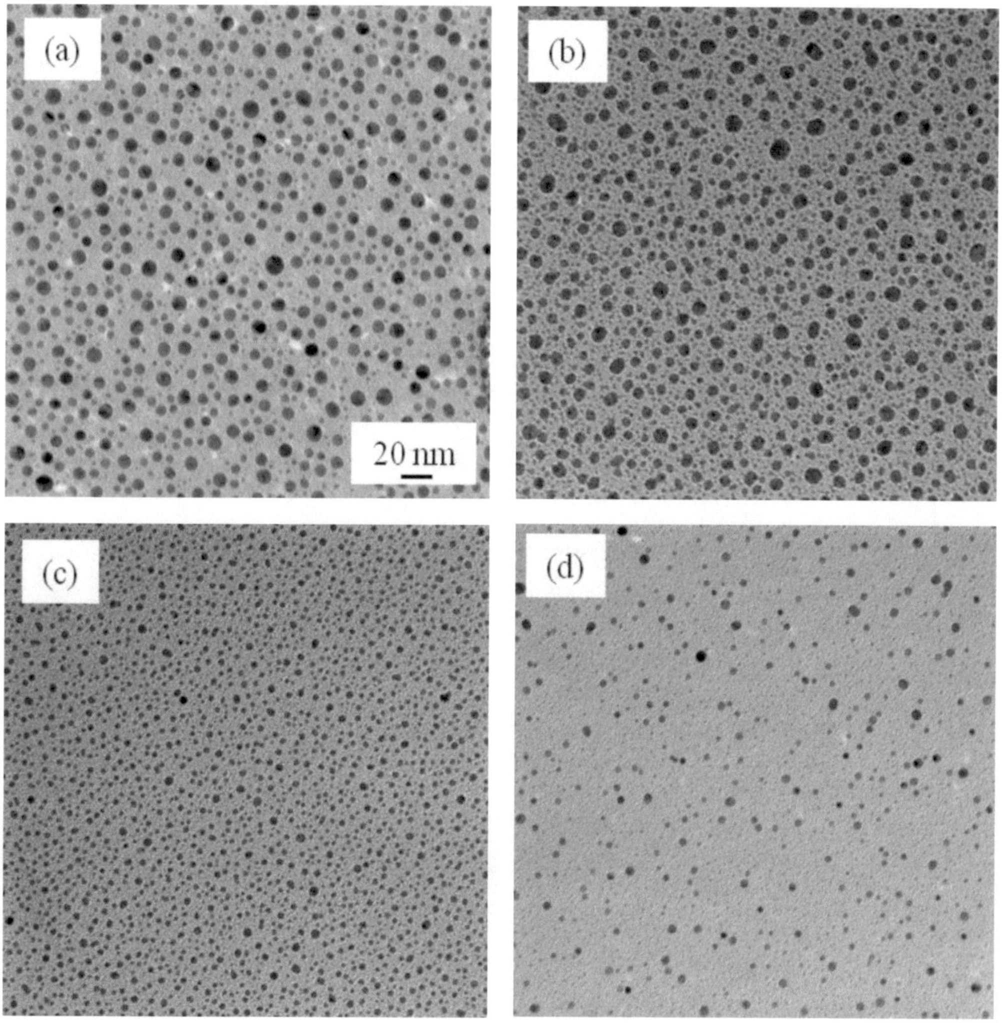

Fig. 2. TEM micrographs of Ag nanoparticles synthesized by TTFA in: (a) Vacuum, (b) 133 Pa, (c) 266 Pa, and (d) 665 Pa of Ar.

All micrographs were acquired with the same magnification, and the scale bar seen in Figure 2(a) represents a distance of 20 nm. The micrograph presented in Figure 2(a) was obtained from a sample formed by TTFA in vacuum. There is no discernible agglomeration of the nanoparticles, and the largest particles have diameters less than 20 nm. Presented in

Figure 2(b) is a micrograph of the Ag nanoparticles formed by TTFA in 133 Pa of Ar. There appear to be additional, even smaller particles formed in 133 Pa of Ar. Presented in figure 2(c) is a micrograph acquired from a sample prepared in 266 Pa of Ar. It is interesting to note that there are numerous smaller nanoparticles as well as fewer of the larger particles. Presented in Figure 2(d) is a micrograph of the sample synthesized in 665 Pa of Ar. There is no evidence of agglomeration in any of the TEM micrographs shown in Figure 2. This is in contrast to previous work (Murray et al., 2006; Koehler et al., 2006) that reported extensive agglomeration of nanoparticles made by conventional pulsed laser ablation. The absence of agglomeration in the TTFA process is attributed to the smaller amount of material ablated per laser shot compared to the conventional process. There are no particles larger than 20 nm seen in Figure 2 or in other, lower magnification images.

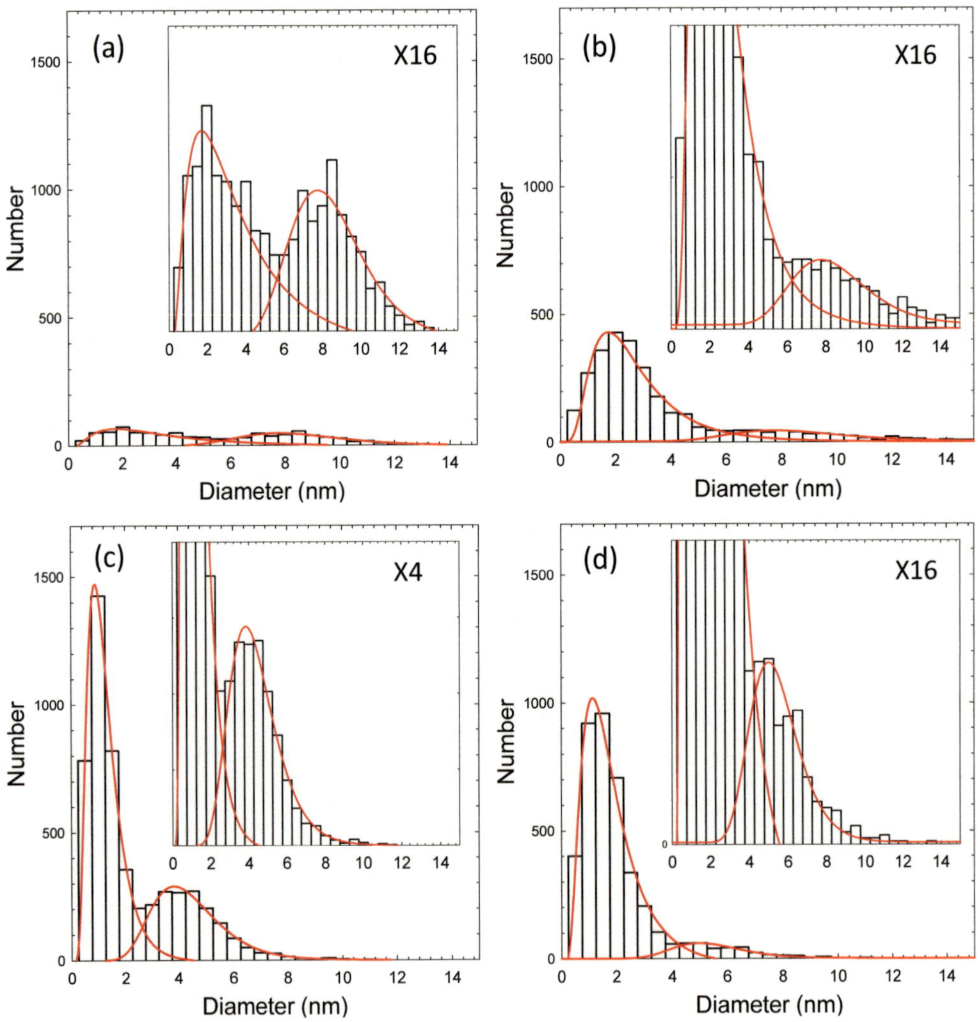

Fig. 3. Size distributions of Ag nanoparticles synthesized by TTFA in: (a) vacuum, (b) 133 Pa, (c) 266 Pa, and (d) 665 Pa of Ar.

Presented in Figures 3(a) through 3(d) are the Ag nanoparticle size distributions that were extracted from the TEM images. The distribution presented in Figure 3(a) was obtained from the sample made by TTFA in vacuum, and the inset shows an expanded region of the same distribution. The distribution of particles formed in vacuum is bimodal, with one component centered near 2 nm and another near 8 nm. We hypothesize that the 8 nm component was formed by splashing of a very thin, transient liquid layer on the irradiated target. Gated, high speed images recorded during ablation indicated the presence of a slow (~ 100m/s) component that is attributed to hot, Ag nanoparticles ejected directly from the target. The 2 nm component is most likely due to gas phase recombination of ablated atomic and molecular species. Indeed, analysis by optical emission spectroscopy showed a large (excited) atomic Ag component in the ablated flux. Furthermore, gated, high speed images recorded during ablation (in vacuum) indicated the presence of a fast (~ 3 km/s) component due to fast atomic Ag (which was confirmed by optical emission spectroscopy). The observation of a fast atomic component suggests that TTFA deposition in vacuum is a process in which a thin film of Ag is formed on the substrate by the atomic component, and the slower nanoparticles land on this nascent thin film some time later.

Shown in Figure 3(b) is the size distribution of Ag nanoparticles formed by TTFA in 133 Pa of Ar. Comparison of Figures 3(a) and 3(b) indicates that the number of small (~ 2 nm) Ag nanoparticles formed by TTFA in Ar is nearly an order of magnitude greater than that formed in vacuum. This is reasonable given the fact that these species were formed by recombination of ablated Ag atoms following collision with background Ar. The inset in Figure 3(b) presents an expanded portion of the size distribution. It can be seen that the most probable nanoparticle size is approximately 2 nm. It can also be seen that the larger (8 nm) component is still present, albeit with a smaller intensity.

Presented in Figure 3(c) is the size distribution of nanoparticles formed in 266 Pa of Ar. The smaller component is nearly a factor of 3 more intense than seen at 166 Pa, and the most probable size is near 1 nm. It is interesting to note that the most probable size of the larger component is near 4 nm, suggesting that the hot, molten nanoparticles splashed from the target undergo enough collisions with background Ar (and with slowed atomic Ag and with nanoparticles newly formed by gas phase recombination) to undergo fragmentation by the time they land on the substrate.

The size distribution of particles formed in 665 Pa of Ar is shown in Figure 3(d). The overall number of particles (large and small component) is smaller than that seen at lower pressure; this suggests that, at this pressure, there are a sufficient number of collisions with background gas (and other species) that prevent the nanoparticles from reaching the substrate. These data illustrate the importance of finding the proper combination of background pressure and substrate distance to collect the optimal number of nanoparticles.

TTFA of Ag in Ar under the current conditions results in the formation of non-agglomerated nanoparticles with a most probable size of between 1 and 2 nm. The most probable size can, most likely, be altered by varying the nature of the background gas (i.e. by using He instead of Ar), the background gas pressure, and the target-substrate distance. The results presented here were obtained with a laser energy density at the target of 0.6 J/cm^2.

Ablation at energy densities greater than ~ 1 J/cm^2 resulted in damage to the fused silica support structure (as indicated by the presence of excited Si lines in the optical emission spectra). Ablation at energy densities less than 1 J/cm^2 caused no detectable damage to the silica support and no detectable contamination of the nanoparticle deposit.

The TTFA process allows one to synthesize nanoparticles with minimal contribution from larger particles. This is in contrast to the conventional laser ablation process in which larger, splashed particles are formed in addition to nanoparticles. We hypothesize that this difference lies in the two distinct target geometries. Conventional laser ablation entails irradiating a bulk target from the front. With this configuration, the region of highest transient temperature rise lies at the surface of the target; the temperature rise will decrease as a function of depth into the target. At some critical distance from the surface, the temperature rise will be insufficient to cause explosive ejection of atomic species but will be sufficient to cause target melting. At this point and deeper, the transient liquid layer of molten target will be formed, and this will be the source of large particles that are splashed from the target. In contrast, TTFA entails irradiating the target film from the back side. In this geometry, the region of highest transient temperature rise will lie at the interface between the target support and target film. Target material at this interface will be predominantly atomized, and regions away from the interface (toward the target surface) will experience smaller temperature rises. At some critical distance from the interface, the temperature rise will be insufficient to cause explosive ejection but will be sufficient to melt the target. By choosing a target thickness that is less than this critical distance, one can ablate a target with minimal melting and therefore with a minimal amount of splashed particles.

4. Conclusion

Well dispersed Ag nanoparticles have been formed by TTFA. The nanoparticles were deposited on room temperature substrates, had a most probable size between 1 and 2 nm, and were not agglomerated. The nanoparticle deposit produced by TTFA showed no evidence of the larger particles μm-sized commonly observed from conventional pulsed laser ablation from a bulk target. Synthesis of nanoparticles by TTFA should be possible for any material that can be made as a thin film target and may enable the unique properties of isolated, non-agglomerated nanoparticles to be exploited more fully.

5. Acknowledgment

The excellent technical work of Scott Streiker in the Nanoscale Engineering Science and Technology Laboratory at the University of Dayton is greatly appreciated.

6. References

Barnes, P.N.; Murray, P.T.; Haugan, T.; R. Rogow, R.; and Perram, G. (2002). In situ creation of nanoparticles from YBCO by pulsed laser deposition. *Physica C- Superconductivity and Its Applications.* Vol. 377, Issue 4. (Sep 15, 2002) 578-584, ISSN: 0921-4534.

Becker, M.F.; Brock, J.R.; Cai, H.; Henneke, D.E.; Keto, J.W.; Lee, J.Y.; Nichols, W.T.; Glicksman, H.D. (1998). Metal Nanoparticles Generated by Laser Ablation. *Nanostructured Materials*. Vol. 10, Issue 5. (Jul 1998). 853-863. ISSN: 0965-9773.

Burr, T.A.; Seraphin, A.A.; Werwa, E.; and K. D. Kolenbrander, K.D. (1997). Carrier transport in thin films of silicon nanoparticles. *Physical Review B*. Vol. 56, Issue 8 (Aug 15, 1997). 4818-4824, ISSN: 0163-1829.

ElShall, M.S.; Li, S.; Turkki, T.; Graiver,D., Pernis, U.C.; and Baraton, M.I. (1995). Synthesis and Photoluminescence of Weblike Agglomeration of Silica Nanoparticles. *Journal of Physical Chemistry*, Vol.99, Issue 51 (Dec. 21, 1995) 17805-17809, ISSN: 0022-3654.

Geohegan, D.B.; Puretzky, A.A.; Duscher, G.; and Pennycook, S.J. (1998). Time-Resolved Imaging of Gas Phase Nanoparticle Synthesis by Laser Ablation. *Applied Physics Letters*. Vol. 72, Issue 23 (Jun 8, 1998). 2987-2989 ISSN: 0003-6951.

Hata, K.; Fujita, M.; Yoshida, S.; Yasuda, Makimura, T.; Murakami, K.; Shigekawa, H.; Mizutani, W.; and Tokumoto, H. (2001). Selective adsorption and patterning of Si nanoparticles fabricated by laser ablation on functionalized self-assembled monolayer. *Applied Physics Letters*. Vol. 79, Issue 5 (Jul 30, 2001). 692-694, ISSN: 0003-6951.

Koehler, B.; Murray, P.T.; Shin, E.; Lipfert, S.; and Schreiber, J. (2006). Nanomaterials produced by laser ablation techniques, Part II - High-spatially resolved nondestructive characterization of nanostructures, *Proceedings of SPIE*, pp. U144-U152, ISBN 0-8194-6228-4, San Diego, CA, Feb 28-Mar 02, 2006, SPIE-Int. Society for Optical Engineering, Bellingham, WA.

Link, S. and El-Sayed, M.A. (2000). Shape and size dependence of radiative, non-radiative and photothermal properties of gold nanocrystals. International Reviews in Physical Chemistry. Vol. 19, Issue 3. (Jul-Sep 2000). 409-453 Published: JUL-SEP 2000. ISSN: 0144235X.

Lowndes, D.H.; Rouleau, C.M.; Thundat, T.G., Duscher, G.; Kenik,E.A.; and Pennycook, S.J. (1999). Silicon and Zinc Telluride Nanoparticles Synthesized by Low Energy Density Pulsed Laser Ablation into Ambient Gases. *Journal of Materials Research*. Vol. 14, Issue 2. (Feb 1999). 359-370.

Lowndes, D.H.; Rouleau, C.M.; Thundat, T.; Duscher, G.; Kenik,E.A.; and Pennycook, S.J. (1997). Silicon and Zinc Telluride Nanoparticles Synthesized by Pulsed Laser Ablation: Size Distributions and Nanoscale Structure. 4th International Conference on Laser Ablation (COLA 97), *Applied Surface Science*. Vol. 127 (Jul, 1997). 355-361 ISSN: 0169-4332.

Mafune, F.; Kohno, J.; Takeda, Y.; Kondow, T.; and Sawabe,H. (2000). Structure and stability of silver nanoparticles in aqueous solution produced by laser ablation. *Journal of Physical Chemistry B*. Vol. 104, (Sep 2000). 8333-8337, ISSN: 1089-5647.

Mafune, F.; Kohno, J.Y.; Takeda, Y.; Kondow, T.; and Sawabe, H. (2001). Formation of gold nanoparticles by laser ablation in aqueous solution of surfactant, *Journal of Physical Chemistry B*. Vol. 105, Issue 22. (Jun 7, 2001). 5114-5120. ISSN: 1089-5647.

Makimura, T.; Kunii, Y.; Ono, N., and Murakami, K. Silicon Nanoparticles Embedded in SiO_2 Films with Visible Photoluminescence. (1997). 4th International Conference on Laser Ablation (COLA 97), *Applied Surface Science*, Vol. 127 (Jul 1997) 388-392, ISSN: 0169-4332.

Makino, T.; Suzuki, N.; Yamada, Y., Oshida, T.; Seto,T.; and Aya, N. (1999). Size Classification of Si Nanoparticles Formed by Pulsed Laser Ablation in Helium Background Gas. 5th International Conference on Laser Ablation (COLA'99). *Applied Physics-A. Materials Science and Processing*. Vol. 69 (1999). S243-S247, ISSN: 0947-8396.

Murray, P.T and Shin, E. (2008). Formation of silver nanoparticles by through thin film ablation. *Materials Letters*. Vol. 62, Issue 28 (Nov. 15, 2008) 4336-4338, ISSN: 0167-577X.

Murray, P.T.; Koehler, B; Lipfert, S.; and Schreiber, J. (2006). Nanomaterials produced by laser ablation techniques Part I-Synthesis and passivation of nanoparticles, *Proceedings of SPIE*, pp. U153-U160, ISBN 0-8194-6228-4, San Diego, CA, Feb 28-Mar 02, 2006, SPIE-Int. Society for Optical Engineering, Bellingham, WA.

Murray, P.T. and Shin, E. (2009). Thin film, nanoparticle, and nanocomposite fabrication by through thin film ablation. *Proceedings of SPIE*, ISSN 0277-786X, San Diego, CA, Aug 5, 2009, SPIE-Int. Society for Optical Engineering, Bellingham, WA.

Petry, L.. Ph.D. Dissertation, University of Dayton (2009).

Ogawa, K.; Vogt, T.; Ullmann, M., Johnson, S.; and Friedlander, S.K. (2000). Elastic properties of nanoparticle chain aggregates of TiO_2, Al_2O_3, and Fe_2O_3 generated by laser ablation. *Journal of Applied Physics*. Vol. 87, Issue 1 (Jan 2000). 63-73. ISSN: 0021-8979.

Ozawa, E; Kawakami, Y; Seto, T. (2001). Formation and size control of tungsten nano particles produced by Nd : YAG laser irradiation. 5th International Conference on Nanostructured Materials (NANO 2000), *Scripta Materialia*. Vol. 44, Issue: 8-9 (2001) 2279-2283, ISSN: 1359-6462.

Ready, J.F. (1963). Development of Plume of Material Vaporized by Giant-Pulse Laser, *Applied Physics Letters*. Vol. 3, 11 (1963).

Seraphin, A.A.; Werwa, E.; Kolenbrander, K.D. (1997). Influence of nanostructure size on the luminescence behavior of silicon nanoparticle thin films, *Journal of Materials Research*. Vol. 12, Issue 12 (Dec. 1997). 3386-3392 .

Tang, Y.H.; Sun, X.H.; Au, F.C.K.; Liao, L.S.; Peng, H.Y.; Lee, C.S.; Lee, S.T.; and Sham,T. (2001). Microstructure and field-emission characteristics of boron-doped Si nanoparticle chains. *Applied Physics Letters*. Vol. 79, Issue 11 (Sep 10, 2001). 1673-1675, ISSN: 0003-6951.

Tschopp, M. A.; Bartha, B. B.; Porter, W. J.; Murray, P. T.; Fairchild, S. B. (2009). Microstructure-Dependent Local Strain Behavior in Polycrystals through In-Situ Scanning Electron Microscope Tensile Experiments. *Metallurgical and Materials Transactions A- Physical Metallurgy and Materials Science*. Vol. 40A, Issue 10 (Oct 2009). 2363-2368.

8

o-Phenylenediamine encapsulated silver nanoparticles and their applications for organic light-emitting devices

Chang-Sik Ha[1]*, Jin-Woo Park[1] and Md. Habib Ullah[2]
[1]*Department of Polymer Science and Engineering, Pusan National University Busan 609-735, Korea*
[2]*Department of Physics, American International University-Bangladesh (AIUB), 94, Road 4, Kemal Ataturk Avenue, Banani, Dhaka, Bangladesh*

1. Introduction

Quantum-size (<10 nm) metal nanoparticles (NPs) have attracted a great deal of interest for a variety of applications (Gittins & Caruso, 2001; Brust & Kiely, 2002), many of which require that the particles be water-soluble(Ackerson, et al. 2005; Kim, et al., 2005; Bao, et al. 2003; Manna, et al. 2001) or remain suspended in water with no loss of their physical or chemical properties over extended periods of time. In addition, many applications require colloidal precipitation (Li, et al. 2005; Pan & Rothberg, 2005; Haes & Van Duyne, 2004) of these NPs to allow them to be film-cast using suitable solvents. The choice of synthetic methods and precursor elements are vitally important for determining the phase (water-soluble or colloidal precipitation) of the NPs. Water-based syntheses of NPs, however, are fraught with problems that result from ionic interactions, which are typically overcome by using very low reactant concentrations (ca. 1×10^{-4} M) (Ahmadi, et al. 1996). Moreover, diffusion, which is much faster in aqueous solutions, can lead to aggregated metal NPs in the absence of a stabilizer (Rele, et al. 2004). In contrast, particles synthesized in suitable organic solvents can be prepared at relatively high concentrations of reactants (Sun & Xia, 2002), with predefined sizes and shapes (Gree & O'Brien, 1999; Pileni, 1998), and with improved monodispersity relative to that of NPs prepared in aqueous solutions.

Here, we describe the preparation of stable, monodisperse, quantum-sized silver nanoparticles (NPs) in water soluble form through a simple one-pot procedure that uses glycerol for reducing Ag^+ in the presence of o-phenylenediamine (o-PDA) (Habib, et al., 2006).

On the other hand, organic light emitting devices (OLEDs) have attracted great attention for flat panel display (FPD) applications due to their sophisticated device structure that is simpler than liquid crystal displays (LCD) as well as their fast response time and wide viewing angle which are similar to the typical cathode ray tube (CRT) display (Hung & Chen, 2002).

OLEDs are commonly fabricated on glass substrates coated with a thin film of indium tin oxide (ITO). The ITO glass is generally employed as the anode in OLEDs because of its good electrical conductivity, high transparency over the visible region, ease of patterning, and relatively high work function. However, it has been suggested that the inefficiency of hole injection from bare ITO, poor adhesion at the ITO-organic film interface, and the diffusion of materials across the ITO-organic interface have led to the poor device performance of OLEDs (Kim, et al. 1998; Kugler, et al. 1998; Mason, et al. 1999). As a result, great effort has been made to modify the ITO anode, such as various treatments of the ITO surface and the deposition of very thin buffer layers between the ITO and the hole transport layer in order to enhance the OLED performances. In particular, poly(3,4-ethylene dioxythiophene)/polystyrene sulfonic acid (PEDOT/PSS) is in the limelight because of high conductivity, stability, and transparency (Groenedaal, et al. 2000). In this aspect, we also investigated the performance of OLEDs by systematic incorporation of these Ag NPs in PEDOT/PSS matrix, which is used as a hole-injection layer for OLEDs (Park, et al. 2007).

2. Experimental

All chemical used were of analytical grade and purchased form Sigma-Aldrich. In a typical synthesis, *o*-PDA flakes (0.40 g) were dissolved in glycerol (24 mL) by heating at 140–160 °C for 30 min under vigorous magnetic stirring. The formation of a transparent light-green solution (0.154 M) indicated that o-PDA had completely dissolved. Meanwhile, $AgNO_3$ (1.0 g) was dissolved in glycerol (8 mL) over 15 min at 90–100 °C under magnetic stirring; once a clear solution (0.735 M) had formed, the stirring was continued for another 15 min. The hot colloidal Ag(0) solution was then added dropwise continually into the solution of o-PDA. Heating of the reaction mixture was continued at 140–160 °C for 15 min and then the reaction mixture was quenched in cold (0–2 °C) water [1.6 L; 50 times the solvent volume (32 ml)]. We adopted a direct quenching technique (pouring the reaction mixture in cold water) to prevent aggregation because of highly concentrated particles being aggregated due to slow cooling. The colloidal mixture was maintained under static conditions for 48 h (room temperature) to effect complete precipitation. The precipitated product was separated by decantation and washed several times with water. The decanted silver solution was then subjected to centrifugation, but no further precipitation or phase separation occurred. We termed this phase as a 'water-soluble'. Details of the single-phase solution product are described elsewhere (Habib, et al., 2006).

PEDOT/PSS (2.8 wt% dispersed in H_2O) was purchased from Aldrich. It was filtered through a 0.45 µm membrane filter (Millipore, PVDF) before preparing a hybrid solution. Tris(8-hydroxyquinolinato)aluminum (Alq3) and *N,N'*-diphenyl-*N,N'*-bis(3-methylphenyl)-1,1-biphenyl-4,4'-diamine) (TPD) were purchased from Tokyo Kasei Kogyo Co., Ltd. and used without further purification.

Hybrid PEDOT/PSS–Ag solutions were prepared by dispersing a given mass ratios of *o*-PDA encapsulated Ag NPs in the aqueous PEDOT/PSS solution in an ultrasonic bath for 4 h. Three different weight ratios (5, 10, and 20 wt%) of these Ag NPs were employed to investigate the influence of composition. For the fabrication of OLEDs, ITO coated glass substrates were patterned in a strip of size (2 × 30 mm) and were cleaned ultrasonically in deionized water using a non-phosphorus detergent and then rinsed with ethanol before drying. A layer of PEDOT/PSS-Ag nanocomposite (thickness = 40 nm) was spin-coated onto the pre-cleaned ITO coated glass at 4000 rpm for 60 s and then soft-baked at 100 °C for 6 h in a vacuum oven. On top of the PEDOT/PSS–Ag nanocomposite layer, a 50 nm TPD layer and a 60 nm thick Alq_3 layer were deposited as a hole transporting layer and an emission layer, respectively, and finally cathode layer of aluminum (Al) was deposited by vacuum evaporation (5×10^{-6} Torr) technique (as shown in Fig. 1), defining the active emission area of 4 mm^2. For comparison, the device with the following structure was also prepared; PEDOT/PSS (thickness = 40 nm) / TPD (thickness = 50 nm)/ Alq_3 (thickness = 60 nm)/ Al (see Fig. 1).

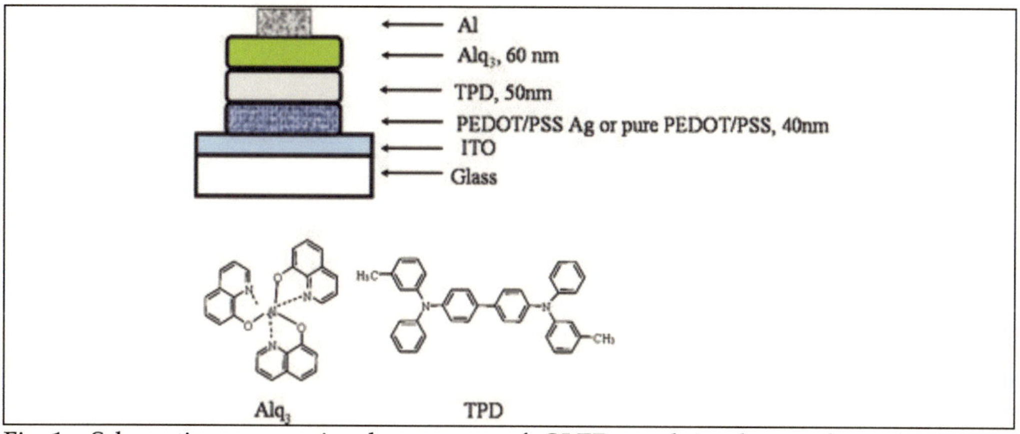

Fig. 1. Schematic cross-sectional structure of OLED used in this study and chemical structures of Alq_3 and TPD (Park, et al. 2007).

The current density-voltage-luminance (J-V-L) characteristics of OLEDs were measured using a customized device measurement system equipped with a photomultiplier tube (PMT, Hamamatsu Photonics Co.) and an electrometer (SMU 236, Keithley). All devices were mounted in a dark sample chamber for the J-V-L measurements in order to get rid of any influence of ambient light.

3. Results and Discussion

Figure 2 displays a typical TEM image of the water-soluble Ag NPs. The nanoparticles exhibit a bimodal distribution, with distinct size ranges of 2-3 and 5-6 nm (Habib,et al.) (Habib, et al., 2006). We verified the TEM results through an analysis of the sample's UV-Vis absorption spectra. The water-soluble Ag NPs displayed a strong plasmon absorbance (λmax = 448 nm) possessing a bimodal-type pattern (Taubert, et al. 2003); XRD data (Figure 3) of Ag NPs displayed strong Bragg reflections corresponding to the (111), (200), (220), and (311) reflection planes of fcc metallic silver.

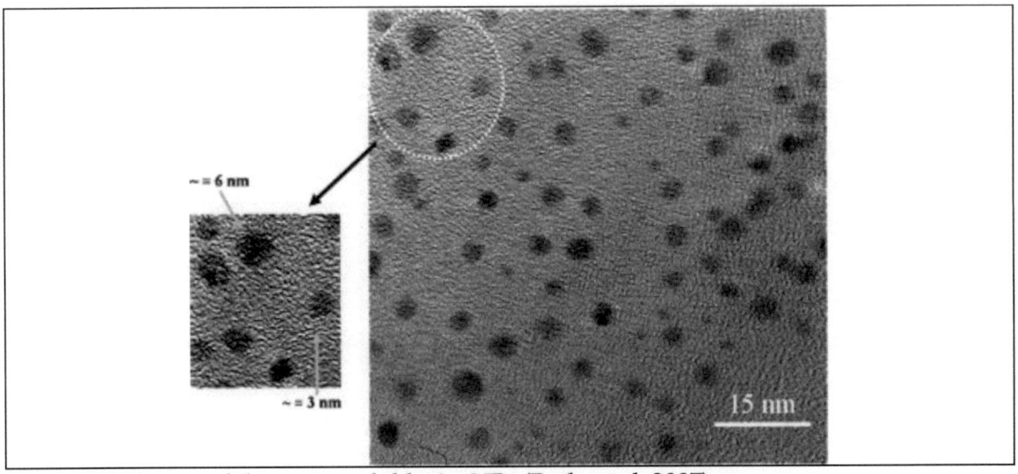

Fig. 2. TEM image of the water-soluble Ag NPs (Park, et al. 2007).

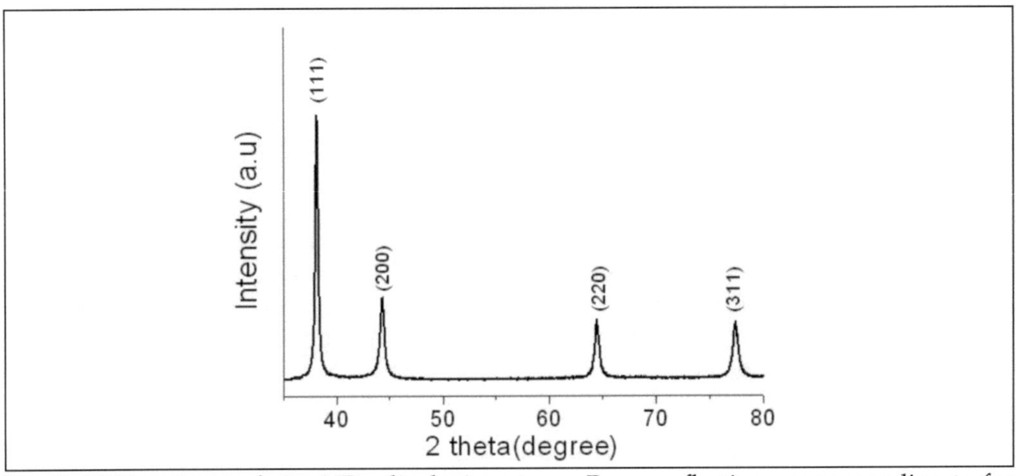

Fig. 3. XRD pattern of Ag NPs, displaying strong Bragg reflections corresponding to fcc metallic silver (Habib, et al. 2006).

The UV–Vis spectrum (Figure 4) of silver sol (prior to addition into the o-PDA solution) indicates that initially the particle sizes are <2 nm (Wilson, et al. 2005; Kim, et al. 2004), they may be stabilized temporarily through complexion with hydroxypyruvic acid (i.e., oxidized glycerol) and glycerol (Rele, et al. 2004). Relatively larger particles were formed after adding the mixture into the o-PDA solution. This finding may indicate that when the small (<2 nm) particles entered into the comparatively hot (140–160 °C) *o*-PDA solution, they became uncapped from their temporary complexes and underwent a melting-like process to form relatively larger particles (Pastoriza-Santos & Liz-Marzan, 2002).

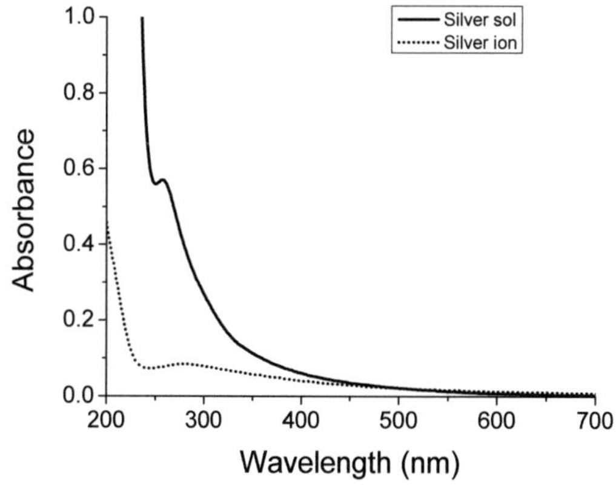

Fig. 4. The UV–Vis spectrum of silver sol prior to addition into *o*-PDA solution, which exhibits a monotonically increasing absorbance toward higher energies because, for very small particles, the distinctive plasmon band is replaced by a featureless absorbance, which has been compared with silver ion sol (Habib, et al. 2006).

Figure 5 presents the UV-vis absorption spectra of the PEDOT/PSS and the PEDOT/PSS–Ag nanocomposites with different concentrations of Ag NPs. *o*-PDA encapsulated water-soluble Ag NPs display a surface plasmon absorption band centered at 448 nm, which is large red shift compared to the uncapped metallic silver that has an intense plasmon absorption band in the visible region at ≈390 nm (He et al. 2001; Kumar et al. 2003). The UV-vis absorption coefficient (at wavelength ≈450 nm) of the PEODT/PSS–Ag nanocomposite increases with increasing the concentration of Ag NPs, where pure PEDOT/PSS does not show any plasmon band. This result indicates that *o*-PDA encapsulated silver NPs are well embedded in the composite layer.

Current density (J)–voltage (V) and luminance (L)–voltage (V) characteristics of various devices fabricated in this work are shown in Fig. 6 (Park et al. 2007). As expected, the TPD/Alq3 device (see (a) curves in Figs. 6A and 6B) without PEODT/PSS layer exhibits the poorest performance among the devices. It is noteworthy that the performance of the OLEDs is significantly improved (see (c), (d), and (e) curves in Figs. 6A and B when the PEDOT/PSS–Ag nanocomposite layer (with the loading concentrations of Ag NPs in the range of 5–20 wt%) is used compared to the OLED (see (b) curves in Figs. 6A and B) using PEDOT/PSS only as a hole injection layer. The devices with 10 or 20 wt% Ag NPs and PEDOT/PSS in the hole injection layer exhibited high brightness ((Fig. 6B). The turn-on voltages, 6.0, 7.8, 8.0, and 8.7 V, were obtained for the loading of 20, 10, 5, 0 wt% Ag, respectively; where the turn-on voltage, 14 V was measured for the OLED without PEDOT/PSS. These results demonstrate that the turn-on voltage decreases with increasing the Ag NPs' concentration. In this work, the maximum luminance of the device fabricated with 20 wt% Ag nanoparticles in the PEDOT/PSS matrix as a hole injection layer was ca. 9,000 cd/m^2 at around 16 V (see (e) curve in Fig. 6B).

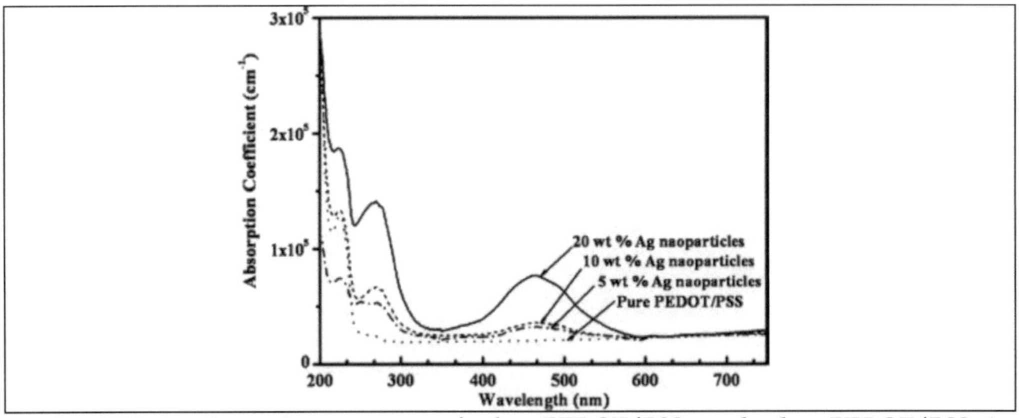

Fig. 5. UV-vis absorption spectra of the PEDOT/PSS and the PEDOT/PSS–Ag nanocomposites with different concentrations of Ag NPs in the PEDOT/PSS matrix (Park, et al. 2007).

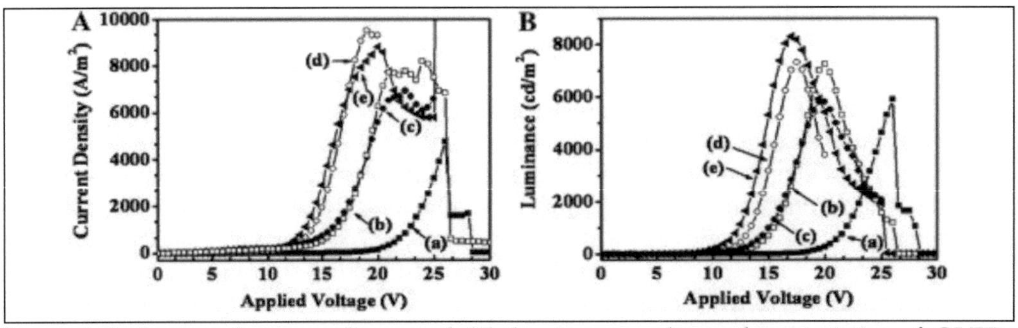

Fig. 6 (A) Current density-voltage and (B) luminance-voltage characteristics of OLEDs made with the PEDOT/PSS and the PEDOT/PSS–Ag nanocomposites with different concentrations of Ag NPs in the PEDOT/PSS matrix in the hole injection layer [(a) OLED without PEDOT/PSS, (b) OLED with PEDOT/PSS only, (c) PEDOT/PSS–Ag nanocomposite with 5 wt% Ag NPs, (d) PEDOT/PSS–Ag nanocomposite with 10 wt% Ag NPs, (e) PEDOT/PSS–Ag nanocomposite with 20 wt% Ag NPs] (Park, et al. 2007).

Further study is, however, needed in order to understand more exact relationship between the optimum concentration and size of the Ag nanoparticles and the device performances.

4. Conclusions

We have synthesized aggregation free quantum-size silver NPs using a high $AgNO_3$ concentration through a novel technique. The in-situ production of binary phases quantum-size Ag NPs having narrow size distributions will open the new avenue for the applied fields. The nanoparticles, especially, water-soluble Ag NPs are suitable for electronics applications because the π-conjugated aromatic amine molecules are well-coordinated with silver surface. In this sense, we investigated the performance of OLEDs consisting of PEDOT/PSS–Ag nanocomposites with different concentrations of quantum-size silver NPs as a hole injection layer. We found that the OLEDs with PEDOT/PSS–Ag nanocomposites exhibit improved performance. The turn-on voltage is significantly decreased when increasing the Ag NPs' concentration. In this work, the maximum luminance of the device consisting of 20 wt% Ag NPs in the PEDOT/PSS matrix as a hole injection layer was ca. 9,000 cd/m^2 at around 16 V.

5. Acknowledgements:

This work was supported by the National Research Foundation of Korea (NRF) Grant funded by the Ministry of Education, Science and Technology, Korea (MEST) (Acceleration Research Program (No. 2009-0078791)), and the Brain Korea 21 Project of the MEST.

6. References

Ackerson, C. J., Jadzinsky, P. D. & Kornberg, R. D. (2005). Thiolate ligands for synthesis of water-soluble gold clusters. *J. Am. Chem. Soc.*, 127, 18, 6550-6551, ISSN 0002-7863

Ahmadi, T. S., Wang, Z. W., Green, T. C., Henglein, A. & El-Sayeed, M. A. (1996). Shape-controlled synthesis of colloidal platinum nanoparticles. *Science,* 272, 5270, 1924-1925, ISSN 0036-8075

Bao, C., Jin, M., Lu, R., Xue, P., Zhang, T., Tan, C. & Zhao, Y. (2003). Synthesis of hyperbranched poly(amine-ester)-protected noble metal nanoparticles in aqueous solution. *J. Mater. Res.,* 18, 6, 1392-1398, ISSN 0884-2914

Brust, M. & Kiely, C. (2002). Some recent advances in nanostructure preparation from gold and silver particles: a short topical review. *J. Colloids Sur. A,* 202, 2-3, 175-186, ISSN 0927-7757

Gittins, D. I. & Caruso, F. (2001). Spontaneous Phase Transfer of nanoparticulate metals from organic to aqueous media, *Angew. Chem. Int. Ed.,* 40, 16, 3001-3004, ISSN 1433-7851

Green, M. & O'Brien, P. (1999). Recent advances in the preparation of semiconductors as isolated nanometric particles: new routes to quantum dots. *Chem. Commun.,* 22, 2235-2241, ISSN 1359-7345

Groenedaal, L., Jonas, F., Freitag, D., Pielartzik, H. & Reynolds, J. R. (2000). Poly(3,4-ethylenedioxythiophene) and its derivatives: past, present, and future. *Adv. Mater.,* 12, 7, 481-494, ISSN 0935-9548

Habib, M. H., Kim, I. & Ha, C. S. (2006). In-situ preparation of binary-phase silver nanoparticles at a high Ag+ concentration. *J. Nanosci. Nanotechnol.,* 6(3), 777 -782. ISSN 1533-4880.

Haes , A. J. & Van Duyne, R. P. (2004). A unified view of propagating and localized surface plasmon resonance biosensors. *Anal. Bioanal. Chem.,* 379, 7-8, 920-930, ISSN 1618-2642

He, S., Yao, J., Jiang, P., Shi, D., Zhang, H., Xie, S., Pang, S. & Gao, H. (2001). Formation of silver nanoparticles and self-assembled two-dimensional ordered superlattice. *Langmuir,* 17, 5, 1571-1575, ISSN 0743-7463

Hung, L. S. & Chen, C. H. (2002). Recent progress of molecular organic electroluminescent materials and devices. *Mater. Sci. Eng. Rep.,* 39, 5-6, 143-222, ISSN 0927-796X

Kim, J. S., Granstrom, M., Friend, R. H., Johansson, N., Salaneck, W. R., Daik, R., Feast, W. J. & Cacialli, F. (1998). Indium-tin oxide treatments for single- and double-layer polymeric light-emitting diodes: The relation between the anode physical, chemical, and morphological properties and the device performance. *J. Appl. Phys.,* 84, 12, 6859-6870, ISSN 0921-8979

Kim, S. -W., Kim, S., Tracy, J. B., Jasanoff, A. & Bawendi, M. G. (2005). Inter- and intramolecular [4 + 3] cycloadditions using epoxy enol silanes as functionalized oxyallyl cation precursors. *J. Am. Chem. Soc.*, 127, 13, 4556-4557, ISSN 0002-7863

Kim, Y. -G., Oh, S. -K. & Crooks, R. M. (2004). Preparation and characterization of 1−2 nm dendrimer-encapsulated gold nanoparticles having very narrow size distributions. *Chem. Mater.*, 16, 1, 167-172, ISSN 0897-4756

Kugler, T., Lögdlund, M. & Salaneck, W. R. (1998). Polymer surfaces and interfaces in light-emitting devices. *IEEE J. Sel. Top. Quantum Electron.*, 4, 1, 14-23, ISSN 1077-260X

Kumar, A., Joshi, H., Pasricha, R., Mandale, A. B. & Sastry, M., J. Phase transfer of silver nanoparticles from aqueous to organic solutions using fatty amine molecules. *Colloid Interface Sci.*, 264, 2, 396-401, ISSN 0021-9797

Li, Y., Wu, Y. & Ong, B. S. (2005). Facile synthesis of silver nanoparticles useful for fabrication of high-conductivity elements for printed electronics. *J. Am. Chem. Soc.*, 127, 10, 3266-3267, ISSN 0002-7863

Manna, A., Imae, T. Aoi, K., Okada, M. & Yogo, T. (2001). Synthesis of dendrimer-passivated noble metal nanoparticles in a polar medium: comparison of size between silver and gold particles. *Chem. Mater.*, 13, 5, 1674-1681, ISSN 0897-4756

Mason, M. G., Hung, L. S., Tang, C. W., Lee, S. T., Wong, K. W. & Wang, M. (1999). Characterization of treated indium–tin–oxide surfaces used in electroluminescent devices. *J. Appl. Phys.*, 86, 3, 1688-1692, ISSN 0921-8979

Pan, S. & Rothberg, L. J. (2005). Enhancement of platinum octaethyl porphyrin phosphorescence near nanotextured silver surfaces. *J. Am. Chem. Soc.*, 127, 16, 6087-6094. ISSN 0002-7863

Park, J. W., Hllah, M. H., Park, S. S. & Ha, C. S., (2007). Organic electroluminescent devices using quantum-size silver nanoparticles. *J. Mater. Sci.: Mater. Electron.*, 18, 393-397. ISSN: 0957-4522

Pastoriza-Santos, I. & Liz-Marzán, L. (2002). Synthesis of silver nanoprisms in DMF. *Nano Lett.*, *2*, 8, 903-905, ISSN 1530-6984

Pileni, M. P. (1998). Optical properties of nanosized particles dispersed in colloidal solutions or arranged in 2D or 3D superlattices. *New J. Chem.*, 22, 7, 693-702, ISSN 1141-0546

Rele, M., Kapoor, S., Sharma, G. & Mukherjee, T. (2004). Reduction and aggregation of silver and thallium ions in viscous media. *Phys. Chem. Chem. Phys.*, 6, 3, 590-595, ISSN 1463-9076

Sun, Y. & Xia, Y. (2002). Shape-controlled synthesis of gold and silver nanoparticles. *Science*, 298, 5601, 2176-2179, ISSN 0036-8075

Taubert, A., Wiesler, U. -M. & Müllen, K. (2003). Dendrimer-controlled one-pot synthesis of gold nanoparticles with a bimodal size distribution and their self-assembly in the solid state. *J. Mater. Chem.*, 13, 5, 1090-1093, ISSN 0959-9328

Wilson, O. M., Scott,R. W. J., Garcia-Martinez, J. C. & Crooks, R. M.(2005). Synthesis, characterization, and structureselective extraction of 1-3-nm diameter AuAg dendrimer-encapsulated bimetallic nanoparticles. *J. Am. Chem. Soc.*,

Wilson, O. M., Scott, R. W. J., Garcia-Martinez, J. C. & Crooks, R. M. (2005). *J. Am. Chem. Soc.*, 127, 3, 1015-1024, ISSN 0002-7863

9

High Surface Clay-Supported Silver Nanohybrids

Jiang-Jen Lin[1,2], Rui-Xuan Dong[1] and Wei-Cheng Tsai[2]

[1]*Institute of Polymer Science and Engineering, National Taiwan University, Taipei,*
[2]*Department of Materials Science and Engineering, National Chung Hsing University, Taichung, Taiwan*

Abstract

Conventionally, silver nanoparticles (AgNPs) have been prepared by using either physical methods such as electron beam- and photo-reductions or chemical methods with various reducing agents and organic stabilizers. Many researches have been conducted previously in the following areas: tailoring of particle size, polydispersity, geometric shape, and nucleation. Low-molecular-weight surfactants or functional polymers such as poly(vinylpyrrolidone) have also been commonly employed for stabilizing the generated AgNPs. The presence of organic stabilizers may provide soft templates for controlling the growth of the AgNPs with different shapes such as spherical, triangular, and fibrous. In this review, various syntheses involving the applications of inorganic supports such as alumina and aluminosilicate clays in place of organic stabilizers are discussed; in this manner, the synthesis of AgNPs supported on inorganic substrates is reviewed. The function of inorganic supports is primarily to stabilize the homogeneity of colloidal Ag^0. In the absence of contamination by organic components during the synthesis, the prepared AgNPs were found to exhibit unique properties such as catalytic performance, high stability for long-term storage, low-temperature Ag melting, and high efficacy for antimicrobial properties. This new class of AgNP nanohybrids on inorganic supports is expected to have considerable impact on biomedical fields and on several applications such as optoelectronic devices.

1. Introduction

Inorganic materials on the nanometer scale may have different physical and chemical properties from their bulk analogs. In recent years, tremendous efforts have been made to develop new nanomaterials for various applications such as antimicrobial [1], magnetic recording materials [2], DNA detection [3], and photodetection [4]. The synthesis and manipulation of nanoparticles with various sizes and shapes are important research issues. Silver nanoparticles (AgNPs) are of particular interest because they have unique properties that can be applied to antibacterial applications [5], cryogenic superconducting materials [6], biosensor materials [7], and as additives in composite fibers [8]. In particular, recent developments in electronic devices have led to reductions in the thickness of conductive films and the width of printed circuits. This has led to the development of new AgNP materials having different physical forms such as conductive silver pastes [9]. As catalysts, the performance of AgNPs is highly dependent on their size distribution [10], structure [11], shape [12], and support species [13].

In this review, conventional synthesis methods of AgNPs using various organic surfactants and reducing agents are first discussed. Different organic species used in the process may have an effect on the sizes and shapes of nanoparticles. Further, in place of organic reagents, various inorganic alumina and silicates are found to be effective for stabilizing the in-situ generated AgNPs. Thus, a new class of Ag nanohybrids on inorganic supports has been characterized to have different compositions and morphologies. The nanohybrids possessing unique physical properties such as low melting temperatures have been reported. Lastly, new trends for expanding the scope of applications are discussed.

2. Synthesis of silver nanoparticles (AgNPs)

2.1. Conventional syntheses and stabilization of AgNPs in solution

Several methods have been reported for synthesizing AgNPs. Physical methods include metal ablation using a laser [14–17] and metal vapor deposition [18,19]. Physical vapor deposition is an efficient process, and it has been extensively employed in combination with the evaporation and sputtering of metals with plasma as well as with magnetron sputtering. The advantages of physical over chemical processes are the uniformity of nanoparticle distribution and the relative absence of solvent contamination in the prepared thin films. However, in the case of polymeric substrates, the adhesion between the deposited metal and polymer matrix is generally poor. On the other hand, the following chemical methods have been well documented—reduction of silver ions in aqueous solutions [20–24] or non-aqueous solutions [25,26], electrochemical reduction [27–29], ultrasonic-assisted reduction [30,31], photoinduced or photocatalytic reduction [32,33], microwave-assisted synthesis [34–36], irradiation reduction [37–39], microemulsion [40–42], and biochemical reduction [43–45].

In general, by using a wet chemical process, AgNPs can be made into different shapes. The mechanism involves initial interactions of silver ions with organic stabilizers before their reduction into nanoparticles. Further aggregation leads to optimal sizes as well as the generation of a repelling layer. On the surface of small particles, the absorbed silver ions can be further reduced forming larger silver crystallites. However, the inherent problems of nanoparticle aggregation or coalescence might still be encountered. One of the key issues for synthesizing AgNPs is the stabilization and prevention of particle agglomeration. The presence of surfactants comprising functionalities such as amines, thiols, acids, and alcohols [46] for interactions with the particle surface can stabilize the particle growth. Polymeric compounds such as poly(vinylpyrrolidone) [23,24,29,30], poly(vinyl alcohol) [47], poly(ethylene glycol) [25], and various block copolymers [48] have been found to be effective stabilizers. These organic surfactants or functional polymers enable the protection of small particles from agglomeration or losing their surface properties. In addition, the presence of surfactants might also affect the growth process of nanocrystallites and allow control of AgNP shapes and sizes. By controlling the stability, size distribution, and shape, the surface activity and performance in the targeted applications can be influenced. Recent literature reports on various synthetic methods with different process parameters, solvents, stabilizers, and organic templates are summarized in Table 1. Various morphologies including spherical, triangular, wire, cubic, and dendritic shapes have been reported.

Synthetic methods	Reducing agents	Organic stabilizers	Silver morphologies	Ref.
chemical reduction (water-system)	sodium citrate	citrate	nanowire or spheroid	[20,21]
	$NaBH_4$	PVP and PVA	nanospheroids (7−20 nm)	[22]
	Polyol	PVP	nanowire or nanocubes	[23]
	ethylene glycol	PVP	nanocubes	[24]
chemical reduction (organic solvents)	dimethyl acetamide	PEG	nanospheroids	[25]
	acetonitrile	TTF	dendritic particles	[26]
electrochemical reduction	Cyclic voltammetry	polyphenylpyrrole	nanospheroids (3−20 nm)	[27]
	zeolite film modified electrodes	zeolites	nanospheroids (1−18 nm)	[28]
	rotating platinum cathode	PVP	nanospheroids (10−20 nm)	[29]
ultrasonic-assisted reduction	sonoelectrodeposition	PVA	nanosphere, nanowire and dendrite	[30]
	sonoelectrochemistry reduction	NTA	nanosphere	[31]
photoinduced or photocatalytic reduction	dual-beam illumination	citrate and PSS	triangular nanoprisms (3−120 nm)	[32]
	PSS, polychromatic irradiation	PSS	nanospheroids (~8 nm)	[33]

microwave-assisted synthesis	sodium citrate	citrate	nanorods or spheroids	[34]
	ethylene glycol	PVP	nanospheroids	[35]
	formaldehyde	citrate	nanospheroids	[36]
irradiation reduction	γ-irradiation	mesoporous silica	nanospheroids (1–4 nm)	[37]
	γ-irradiation	PVP	nanospheroids (5–20 nm)	[38]
	Fs and ns laser ablation	No	nanospheroids (20–50 nm)	[39]
microemulsion method	V_c	CTAB and SDS	nanowires and dendrites	[40]
	KBH_4	Ellipsoidal micelles	needle-shaped and wire-shaped particles	[41]
	$NaBH_4$	PFPE-NH_4	nanospheroids	[42]
biochemical reduction	Peptide	peptide	hexagonal, spherical and triangular	[43]
	Yeast strain	proteins	nanospheres (2–5 nm)	[44]
	Neem leaf extension	flavanones, terpenoids	polydisperse spheres (5–35 nm)	[45]

Table 1. Various methods for preparing silver structures with different morphologies

PVA: poly(vinylalcohol); PVP: poly(vinylpyrrolidone); PEG: polyethyleneglycol; TTF: tetrathiafulvalene; NTA: nitrilotriacetate; PSS: Poly(styrene sulfonate); CTAB: cetyltgrimethlammonium bromide; SDS: sodium dodecyl sulphate; PFPE: ammonium carboxylate perfluoropolyether

2.2. Chemical method—reduction of silver salts

In addition to the use of organic stabilizers for controlling the particle morphologies, the reduction process offers another possibility—tailoring of particle characteristics. Reducing agents such as sodium borohydride [22], sodium citrate [20,21], N, N-dimethylformamide (DMF) [49], polyols [50], ascorbate [51], Tollens reagent [52], and poly(ethylene glycol)-block copolymers [24,35] are popular. In general, strong reducing agents such as sodium borohydride can often afford fine particle sizes, as demonstrated by studies conducted with sodium citrate [21] and sodium borohydride [22]. Systematic studies were performed by varying the reductant/$AgNO_3$ ratios. The resultant colloids were characterized by UV-Vis spectroscopy immediately after the preparation as well as during its long-term stability studies. The AgNPs generated by the citrate method were large particles with diameters of around 40-60 nm; they were stable during storage. In contrast, the borohydride reduction afforded smaller AgNPs (3−10 nm). DMF is another example of a strong reducing agent for the silver ion reduction; it achieves fast reaction rate and nanoprism morphology [49]. AgNPs having diameters in the range of ~40 nm were prepared by using weak reducing agents such as polyols at elevated temperatures [50]. Large colloidal silver particles using ascorbic acid as the reducing agent have also been prepared in aqueous media [51]. The Tollens process has been recently recognized as a simple one-pot synthetic route for AgNPs with a narrow size distribution [52]. The basic reaction involves the reduction of silver solution from glucose. Stable aqueous dispersion of silver colloids having a size of 20−50 nm can be obtained. In this manner, the choice of reducing agents and synthetic conditions can significantly affect the size and stability of AgNPs. In Table 2, the commonly used reducing agents given in literature have been cited.

Reducing agents	References
Sodium borohydride ($NaBH_4$)	[22]
N, N-dimethylformamide (DMF)	[49]
Polyols	[50]
Ascorbate	[51]
Tollens reagent	[52]
Poly(ethylene glycol)-based	[24,35]

Table 2. Common reducing agents for converting silver salts to nanoparticles

2.3. Controlling factors for forming various shapes of silver nanoparticles

The activity and reaction selectivity of AgNPs are strongly dependent on their sizes, shapes, and surface composition [53]. Hence, a protective agent is essential for tailoring the shape formation of AgNPs; different shapes may be required for different targeted applications.

Organic surfactants can be provided as templates for altering the nanoparticle shapes to spherical [22,25], triangular [32,43], wire [20,23,40,41], cubical [24], and fibrous [54] morphologies. These strategies have proven to be useful for controlling the growth of anisotropic AgNPs with different shapes as a result of the effects of surfactants or polymers [34] that may induce and maintain directional growth. The molar ratio of the surfactant/$AgNO_3$ is also an important factor in determining the morphology of the final product [22]. Covalent linkage or surface capping with an organic molecule at an early stage of cluster growth usually yields smaller-sized clusters. Such controlling factors have been widely used to prepare specific shapes of AgNP nanocrystallites.

3. Inorganic supports for AgNPs without organic stabilizers

3.1. Preparation of AgNPs in the presence of inorganic supports

The synthesis of AgNPs using organic dispersants has several disadvantages. In addition to the possible contamination by toxic chemical agents, the presence of organic residues may reduce particle surface activity. In order to overcome the limitations of the chemical reduction method for manufacturing AgNPs with fine dispersion, inorganic dispersants are introduced. By using inorganic dispersants, nanoparticles with a specific diameter and morphology can be synthesized advantageously. The use of inorganic supports for AgNPs may lead to a new form of nanoparticles that are not contaminated by organic components [55–59]. In literature, there have been some reports on the synthesis of water-dispersible AgNPs in the absence of organic components. Some inorganic supports are effective for controlling the uniformity of particle size distribution and for stabilizing nanosized metal particles. For example, laponite clay was used for creating silver colloidal nanoparticles from silver ions by using the chemical reduction method with sodium borohydride as the reducing agent [57]. Controls in size growth and particle aggregation were achieved, and a high concentration of silver colloid with a narrow size distribution was obtained. The UV-vis absorption for the nanoparticles is shown in Figure 1. Different molar ratios of $NaBH_4$ to $AgNO_3$ were used for generating the AgNPs. When less $NaBH_4$ was used, two broad-absorption UV peaks at around 390 and 535 nm were observed. The latter peak disappeared with increasing $NaBH_4$ concentration. Hence, the formation of AgNPs is highly sensitive to and influenced by the reducing conditions and process control. Transmission electron microscopy (TEM) micrograms (Figure 2) have demonstrated that the particle size and aggregated morphologies varied with the experimental conditions using the same reducing agent.

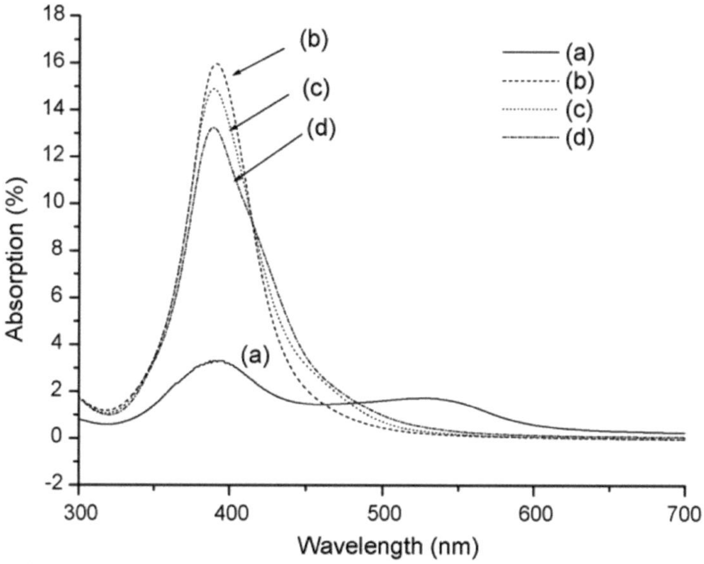

Fig. 1. UV-vis absorption of AgNPs as the function of $NaBH_4$ to $AgNO_3$ molar ratios at (a) 1/2, (b) 2, (c) 5, and (d) 10. Reproduced with permission from [57], J. Liu et al., Preparation of high concentration of silver colloidal nanoparticles in layered laponite solution. *Colloids Surf. A* 302, 276 (**2007**).

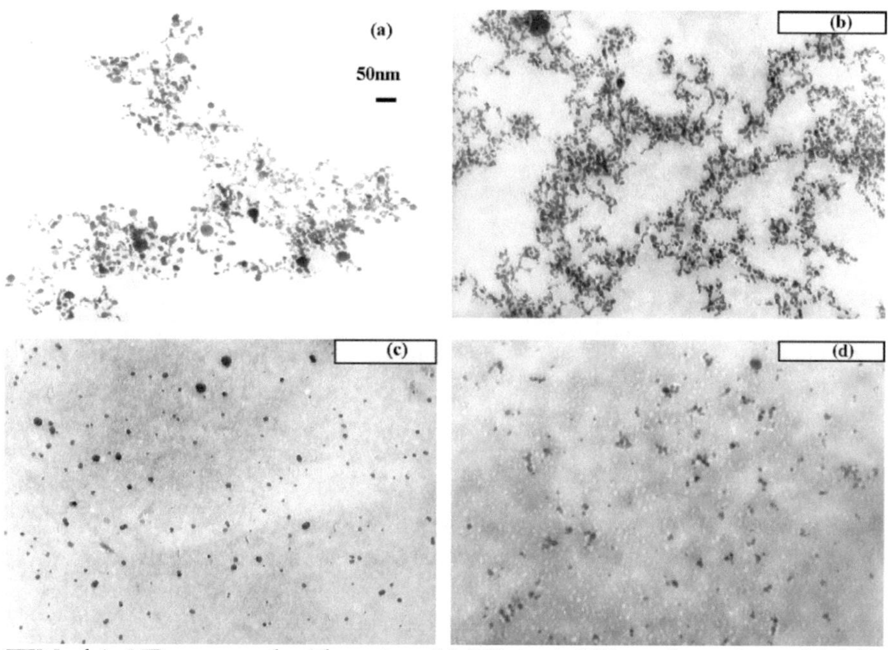

Fig. 2. TEM of AgNPs prepared with various $NaBH_4$ to $AgNO_3$ molar ratios at (a) 1/2, (b) 2, (c) 5, and (d) 10. Reproduced with permission from [57].

Another case of AgNP formation was examined by adding different silicate clays such as montmorillonite, bentonite, and mica [59]. The particle size was predominately affected by the $AgNO_3$/clay ratio and the species of the reducing agents (methanol, $NaBH_4$, etc.). Silicate clays with a lamellar geometric shape offered a high surface area for supporting AgNPs having a size of 17–88 nm, as shown in Figure 3. The particle size distribution was also influenced by the type of clay used. A narrow size distribution (polydispersity D_w/D_n = 1.2 when D_n = 26 nm; analysis by FE-SEM) was obtained.

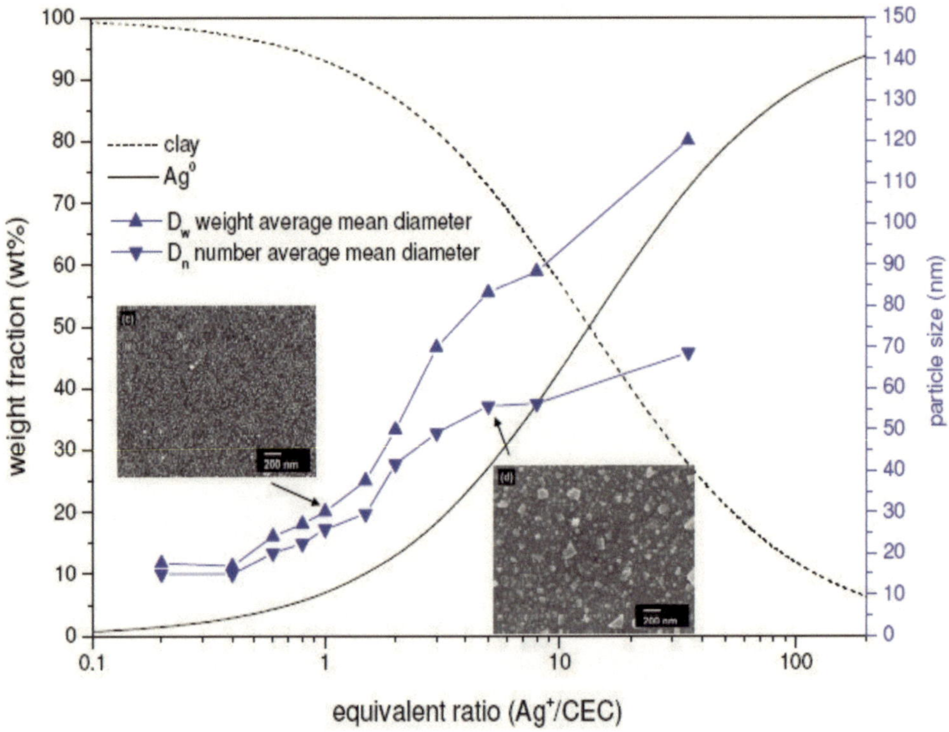

Fig. 3. Particle size and composition of AgNPs on clay using various Ag^+/CEC ionic ratios in the preparation. Reproduced with permission from [59], R. X. Dong et al., Synthesis of immobilized silver nanoparticles on ionic silicate clay and observed low-temperature melting. *J. Mater.Chem.* 19, 2184 **(2009)**.

3.2. Properties of AgNPs on clay supports

3.2.1 AgNPs with a low-temperature melting property

The development of metal composites for miniature interconnection materials is important for applications to microelectronic packaging. For meeting the requirements of low-processing temperatures and environmentally benign conditions, low-temperature-melting AgNPs can be one of the suitable candidates. It is generally believed that the melting temperature depends on the particle size. Recent studies have demonstrated that fine sizes of ~5 nm can be synthesized by using polymeric dispersants. Figure 4 shows different morphologies for the creation of AgNPs having an average size of 5 nm at ambient temperatures; when water-soluble hyperbranched polyglycidol (HPG) is employed as the template, the AgNPs are obtained in the form of a silver tube at 165 °C. Ostwald ripening of small Ag crystals with surface interactions with HPG is probably the key to the controlled formation of tubes, as schematically illustrated in Scheme 1. The low-melting-temperature properties of the AgNPs might also have contributed to the easy melting into tubes. By using differential scanning calorimetry (DSC), it was found that the melting process began at approximately 120 °C to form the tube morphology within 40 h. This study has demonstrated the hierarchical derivation of AgNPs having tubular shapes involving organics as templates and low-melting-temperature behavior.

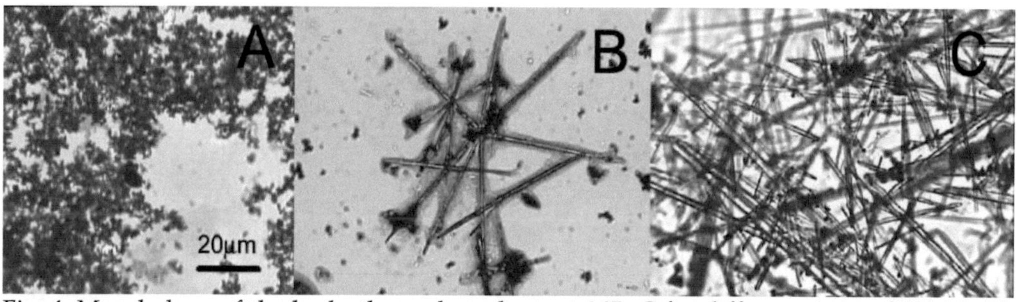

Fig. 4. Morphology of the hydrothermal products at 165 °C for different times of (A) 20, (B) 30, and (C) 40 h. Reproduced with permission from [60], X. Ding et al., Hyperbranched Polymer-Assisted Hydrothermal In situ Synthesis of Submicrometer Silver Tubes. *Crystal Growth & Design* 8, 2982 (**2008**).

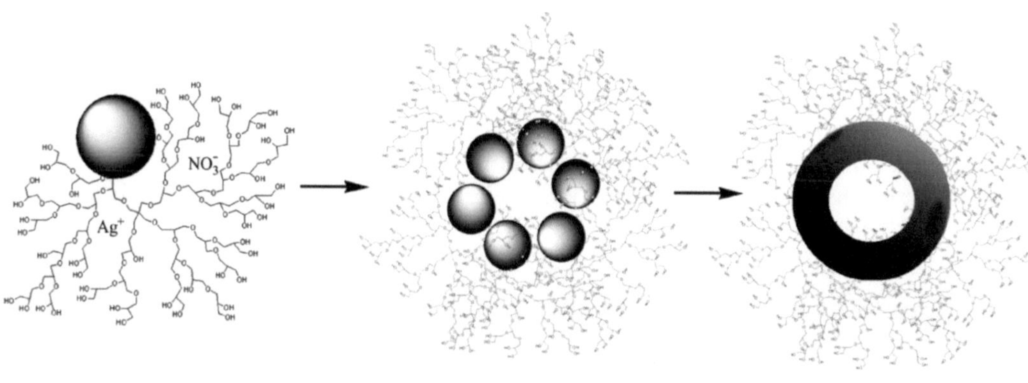

Scheme 1. Ostwald Ripening of Small Ag crystals with Surface Passivated by HPG to Form Tubes. Reproduced with permission from [60], X. Ding et al., Hyperbranched Polymer-Assisted Hydrothermal In situ Synthesis of Submicrometer Silver Tubes. *Crystal Growth & Design* 8, 2982 (**2008**).

In a recent study, AgNPs having a size of 26 nm on a clay surface were prepared at 80 °C and examined by using a field emission-scanning electron microscope (FE-SEM) (Figure 5a); they were then sample dried and treated at 110 °C for 2 h (Figure 5b). A comparison between Figures 5a and 5b reveals that AgNPs change into lumps, implying that they melted at 110 °C. Figure 6 illustrates the role of the clay supports in the initial formation of the $AgNO_3$/clay complex and reduction into AgNPs. When the clay supports were used, the AgNPs were stabilized and easily melted into an Ag matrix at low temperatures (110 °C).

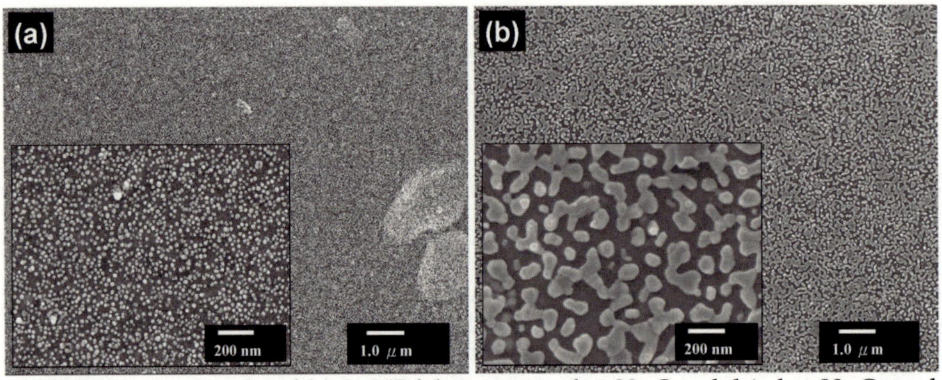

Fig. 5. FE-SEM micrographs of (a) AgNP/clay, prepared at 80 °C and dried at 80 °C on glass, (b) prepared at 80 °C and dried at 110 °C. Reproduced with permission from [59], R. X. Dong et al., Synthesis of immobilized silver nanoparticles on ionic silicate clay and observed low-temperature melting. *J. Mater.Chem.* 19, 2184 (**2009**).

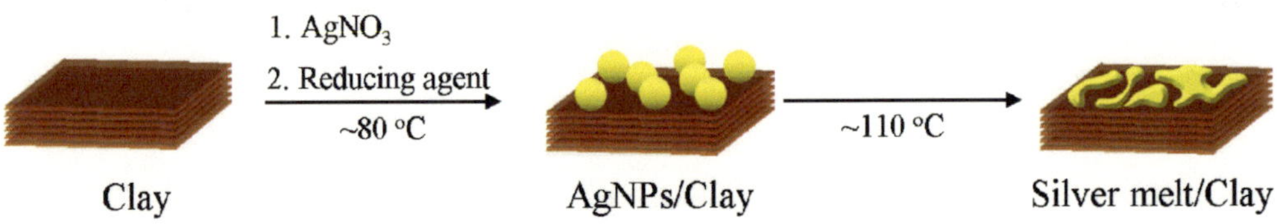

Fig. 6. Illustration of AgNP formation from $AgNO_3$/clay complex and the melting on the plate substrates. Reproduced with permission from [59], R. X. Dong et al., Synthesis of immobilized silver nanoparticles on ionic silicate clay and observed low-temperature melting. *J. Mater.Chem.* 19, 2184 (**2009**).

3.2.2 Antimicrobial properties of AgNPs/clay

AgNPs/clay has been shown to exhibit strong inhibition of bacterial growth. The plate-like clay of anionic surface charges provided a high surface support for AgNPs having a diameter of ~30 nm diameter. The AgNP nanohybrids have demonstrated the ability of inhibiting the growth of dermal pathogens such as *Staphylococcus aureus (S. aureus), Pseudomonas aeruginosa*, and *Streptococcus pyrogens,* as well as the methicillin- and oxacillin-resistant *S. aureus* (MRSA and ORSA). SEM has revealed the mechanism of nanohybrids adhering on the surface of individual bacteria (Figure 7). Cytotoxicity studies have indicated that surface contact with the reduced AgNPs on clay is the initial cause for cell death. Detailed mechanisms have been shown to involve a loss in membrane integrity due to reactive oxygen species (ROS) generation, as shown in Scheme 2. The AgNPs supported on clay surface are a new class of nanohybrids that can be applied to biomedical applications.

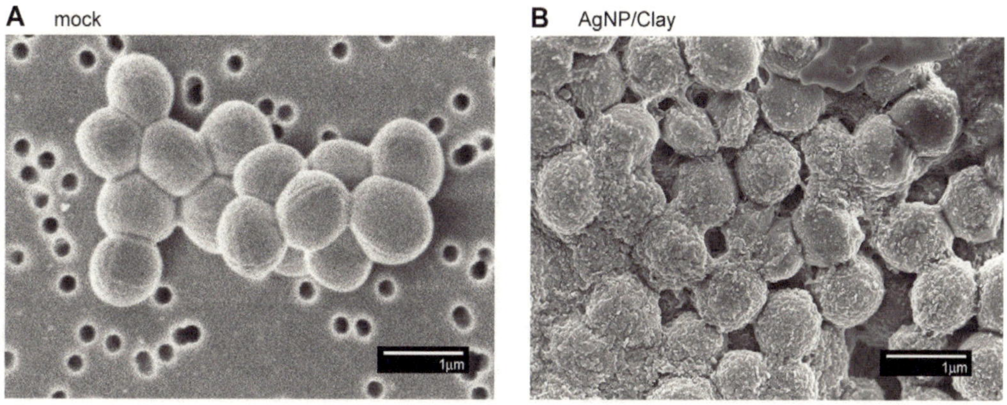

Fig. 7. AgNP/Clay inhibited bacterial proliferation. FE-SEM images showed the untreated (A) and the AgNP/Clay-encapsulated *S. aureus* (B). In contrast to the proliferative *E. coli* on clay. Reproduced with permission from [61], H. L. Su et al., The disruption of bacterial membrane integrity through ROS generation induced by nanohybrids of silver and clay. *Biomaterials* 30, 5979 (**2009**).

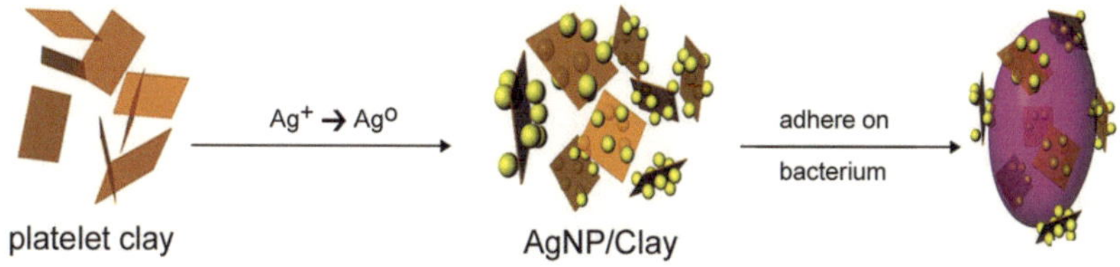

Scheme 2. Illustration of the AgNP/Clay synthesis and interaction with a bacterium: AgNPs are fabricated on the surface of platelet clay and electrostatic attraction to a bacterial cell wall. Reproduced with permission from [61], H. L. Su et al., The disruption of bacterial membrane integrity through ROS generation induced by nanohybrids of silver and clay. *Biomaterials* 30, 5979 (**2009**).

4. Conclusion

In the literature, various synthetic methods and stabilizers have been reported for AgNPs. By using different reducing agents and organic stabilizers, the fine particle diameter and changes in the morphology of AgNPs have been controlled. In addition to the influence by organic additives, the synthetic processes may alter the occurrence of chemical reduction from silver ions to silver metal with different morphologies. The particle sizes can be controlled by varying the species of reducing agent and its amount of addition. By using inorganic supports such as smectite clay, homogeneously dispersed AgNPs with a narrow size distribution on the clay surface can be synthesized. Furthermore, the presence of clay stabilized the AgNP stability for long storage durations and high surface "naked" activity. An unusual property is the low melting temperature at 110 °C; which might be useful in low-temperature fabrications of silver arrays, conductive wires, and biosensor devices. Furthermore, the AgNP nanohydrid also demonstrated its ability to inhibit bacterial growth by disruption of cell membranes and thus a high potency as an antimicrobial agent.

5. References

[1] Cho, K. H.; Park, J. E.; Osaka, T. & Park, S. G. (2005). The Study of Antimicrobial Activity and Preservative Effects of Nanosilver Ingredient. *Electrochimica Acta*, Vol. 51, pp. 956-960, 2005.

[2] Sun, S.; Murray, C. B.; Weller, D.; Folks, L. & Moser, A. (2000). Monodisperse FePt Nanoparticles and Ferromagnetic FePt Nanocrystal Superlattices. *Science*, Vol. 287, pp. 1989-1992, 2000.

[3] Park, S. J.; T. Taton, A. & Mirkin, C. A. (2002). Array-Based Electrical Detection of DNA with Nanoparticle Probes. *Science*, Vol. 295, pp. 1503-1506, 2002.

[4] Wang, J.; Gudiksen, M. S.; Duan, X.; Cui, Yi. & Lieber, C. M. (2001). Highly Polarized Photoluminescence and Photodetection from Single Indium Phosphide Nanowires. *Science*, Vol. 293, pp. 1455-1457, 2001.

[5] Jiang, H.; Manolache, S.; Wong, A. C. L. & Denes, F. S. (2004). Plasma-Enhanced Deposition of Silver Nanoparticles onto Polymer and Metal Surfaces for the Generation of Antimicrobial Characteristics. *Journal of Applied Polymer Science*, Vol. 93, pp. 1411-1422, 2004.

[6] Hirano, S.; Wakasa, Y.; Saka, A.; Yoshizawa, S.; Oya-Seimiya, Y.; Hishinuma, Y.; Nishimura, A.; Matsumoto, A. & Kumakura, H. (2003). Preparation of Bi-2223 Bulk Composed with Silver-Alloy Wire. *Physica C*, Vol. 392, pp. 458-462, 2003.

[7] Ren, X. L. & Tang, F. Q. (2002). Enhancement Effect of Ag-Au Nanoparticles on Glucose Biosensor Sensitivity, *Acta Chim. Sinica*, Vol. 60, pp. 393-397, 2002.

[8] Yeo, S. Y.; Lee, H. J. & Jeong, S. H. (2003). Preparation of Nanocomposite Fibers for Permanent Antibacterial Effect. *Journal of Materials Science*, Vol. 38, pp. 2143-2147, 2003.

[9] Wang, H.; Qiao, X.; Chen, J. & Ding, S. (2005). Preparation of Silver Nanoparticles by Chemical Reduction Method. *Colloids and Surfaces A: Physicochem. Eng. Aspects*, Vol. 256, pp. 111-115, 2005.

[10] Zhang, J.; Tanha, J.; Hirama, T.; To, N. H. K. R.; Tong-Sevinc, H.; Stone, E.; Brisson, J. & MacKenzie, C. R. (2004). Pentamerization of Single-domain Antibodies from Phage Libraries: A Novel Strategy for the Rapid Generation of High-avidity Antibody Reagents. *Journal of Molecular Biology*, Vol. 335, pp. 49-56, 2004.

[11] Chimentao, R. J.; Kirm, I.; Medina, F.; Rodriguez, X.; Cesteros, Y.; Salagre, P. & Sueiras, J. E. (2004). Different Morphologies of Silver Nanoparticles as Catalysts for the Selective Oxidation of Styrene in the Gas Phase. *Chemical Communications*, Vol. 7, pp. 846-847, 2004.

[12] He, B.; Tan, J. J.; Liew, K. Y. & Liu, H. (2004). Synthesis of Size Controlled Ag Nanoparticles. *Journal of Molecular Catalysis A: Chemical*, Vol. 221, pp. 121-126, 2004.

[13] Wang, H.; Qiao, X.; Chen, J. & Ding, S. (2005). Preparation of Silver Nanoparticles by Chemical Reduction Method. *Colloids and Surfaces A: Physicochem. Eng. Aspects*, Vol. 256, pp. 111-115, 2005.

[14] Yang, S.; Cai, W.; Liu, G.; Zeng, H. & Liu, P. (2009). Optical Study of Redox Behavior of Silicon Nanoparticles Induced by Laser Ablation in Liquid. *The Journal of Physical Chemistry C*, Vol. 113, pp. 6480-6484, 2009.

[15] Muto, H.; Yamada, K.; Miyajima, K. & Mafune, F. (2007). Estimation of Surface Oxide on Surfactant-Free Gold Nanoparticles Laser-Ablated in Water. *The Journal of Physical Chemistry C*, Vol. 111, pp. 17221-17226, 2007.

[16] Liu, P.; Cai, W. & Zeng, H. (2008). Fabrication and Size-Dependent Optical Properties of FeO Nanoparticles Induced by Laser Ablation in a Liquid Medium. *The Journal of Physical Chemistry C*, Vol. 112, pp. 3261-3266, 2008.

[17] Hashimoto, S.; Uwada, T.; Masuhara, H. & Asahi, T. (2008). Fabrication of Gold Nanoparticle-Doped Zeolite L Crystals and Characterization by Optical Microscopy: Laser Ablation- and Crystallization Inclusion-Based Approach. *The Journal of Physical Chemistry C*, Vol. 112, pp. 15089-15093, 2008.

[18] Wegner, K.; Walker, B.; Tsantilis, S. & Pratsinis, S. E. (2002). Design of Metal Nanoparticle Synthesis by Vapor Flow Condensation. *Chemical Engineering Science*, Vol. 57, pp. 1753-1762, 2002.

[19] Chen, C. C. & Yeh, C. C. (2000). Large-Scale Catalytic Synthesis of Crystalline Gallium Nitride Nanowires. *Advanced Materials*, Vol. 12, pp. 738-741, 2000.

[20] Caswell, K. K.; Bender, C. M. & Murphy, C. J. (2003). Seedless, Surfactantless Wet Chemical Synthesis of Silver Nanowires. *Nano Letters*, Vol. 3, pp. 667-669, 2003.

[21] Pillai, Z. S. & Kamat, P. V. (2004). What Factors Control the Size and Shape of Silver Nanoparticles in the Citrate Ion Reduction Method. *The Journal of Physical Chemistry B*, Vol. 108, pp. 945-95, 2004.

[22] Chou, K. S. & Ren, C. Y. (2000). Synthesis of Nanosized Silver Particles by Chemical Reduction Method. *Materials Chemistry and Physics*, Vol. 64, pp. 241-246, 2000.

[23] Sun, Y.; Mayers, B.; Herricks, T. & Xia, Y. (2003). Polyol Synthesis of Uniform Silver Nanowires: A Plausible Growth Mechanism and the Supporting Evidence. *Nano Letters*, Vol. 3, pp. 955-960, 2003.

[24] Sun, Y. & Xia, Y. (2002). Shape-Controlled Synthesis of Gold and Silver Nanoparticles. *Science*, Vol. 298, pp. 2176-2179, 2002.

[25] Chen, D. H. & Huang, Y. W. (2002). Spontaneous Formation of Ag Nanoparticles in Dimethylacetamide Solution of Poly(ethylene glycol). *Journal of Colloid and Interface Science*, Vol. 255, pp. 299-302, 2002.

[26] Wang, X.; Itoh, H.; Naka, K. & Chujo, Y. (2003). Tetrathiafulvalene-Assisted Formation of Silver Dendritic Nanostructures in Acetonitrile. *Langmuir*, Vol. 19, pp. 6242-6246, 2003.

[27] Johans, C.; Clohessy, J.; Fantini, S.; Kontturi, K. & Cunnane, V. J. (2002). Electrosynthesis of Polyphenylpyrrole Coated Silver Particles at a Liquid-Liquid Interface. *Electrochemistry Communications*, Vol. 4, pp. 227-230, 2002.

[28] Zhang, Y.; Chen, F.; Zhuang, J.; Tang, Y.; Wang, D.; Wang, Y.; Dong, A. & Ren, N. (2002). Synthesis of Silver Nanoparticles *via* Electrochemical Reduction on Compact Zeolite Film Modified Electrodes. *Chemical Communications*, Vol. 24, pp. 2814-2815, 2002.

[29] Ma, H.; Yin, B.; Wang, S.; Jiao, Y.; Pan, W.; Huang, S.; Chen, S. & Meng, F. (2004). Synthesis of Silver and Gold Nanoparticles by a Novel Electrochemical Method. *ChemPhysChem*, Vol. 24, pp. 68-75, 2004.

[30] Zhou, Y.; Yu, S. H.; Wang, C. Y.; Li, X. G.; Zhu, Y. R. & Chen, Z. Y. (1999). A Novel Ultraviolet Irradiation Photoreduction Technique for the Preparation of Single-Crystal Ag Nanorods and Ag Dendrites. *Advanced Materials*, Vol. 11, pp. 850-852, 1999.

[31] Socol, Y.; Abramson, O.; Gedanken, A.; Meshorer, Y.; Berenstein, L. & Zaban, A. (2002). Suspensive Electrode Formation in Pulsed Sonoelectrochemical Synthesis of Silver Nanoparticles. *Langmuir*, Vol. 18, pp. 4736–4740, 2002.

[32] Shchukin, D. G.; Radtchenko, I. L. & Sukhorukov, G. B. (2003). Photoinduced Reduction of Silver inside Microscale Polyelectrolyte Capsules. *ChemPhysChem*, Vol. 4, pp. 1101–1103, 2003.

[33] Jin, R.; Cao, Y. C.; Hao, E.; Metraux, G. S.; Schatz, G. C. & Mirkin, C. A. (2003). Controlling Anisotropic Nanoparticle Growth Through Plasmon Excitation. *Nature*, Vol. 425, pp. 487–490, 2003.

[34] Liu, F. K.; Huang, P. W.; Chang, Y. C.; Ko, F. H. & Chu, T. C. (2004). Microwave-Assisted Synthesis of Silver Nanorods. *Journal of Materials Research*, Vol. 19, pp. 469–473, 2004.

[35] Komarneni, S.; Li, D.; Newalkar, B.; Katsuki, H. & Bhalla, A. S. (2002). Microwave-Polyol Process for Pt and Ag Nanoparticles. *Langmuir*, Vol. 18, pp. 5959–5962, 2002.

[36] Yin, H.; Yamamoto, T.; Wada, Y. & Yanagida, S. (2004). Large-Scale and Size-Controlled Synthesis of Silver Nanoparticles under Microwave Irradiation. *Materials Chemistry and Physics*, Vol. 83, pp. 66–70, 2004.

[37] Hornebecq, V.; Antonietti, M.; Cardinal, T. & Treguer-Delapierre, M. (2003). Stable Silver Nanoparticles Immobilized in Mesoporous Silica. *Chemistry of Materials*, Vol. 15, pp. 1993–1999, 2003.

[38] Choi, S. H.; Lee, S. H.; Hwang, Y. M.; Lee, K. P. & Kang, H. D. (2003). Interaction Between the Surface of the Silver Nanoparticles Prepared by G-irradiation and Organic Molecules Containing Thiolgroup. *Radiation Physics and Chemistry*, Vol. 67, pp. 517–521, 2003.

[39] Tsuji, T.; Kakita, T. & Tsuji, M. (2003). Preparation of Nano-Size Particle of Silver with Femtosecond Laser Ablation in Water. *Applied Surface Science*, Vol. 206, pp. 314–320, 2003.

[40] Zheng, X.; Zhu, L.; Wang, X.; Yan, A. & Xie, Y. (2004). A Simple Mixed Surfactant Route for the Preparation of Noble Metal Dendrites. *Journal of Crystal Growth*, Vol. 260, pp. 255–262, 2004.

[41] Zhang, J.; Han, B.; Liu, M.; Liu, D.; Dong, Z.; Liu, J.; Li, D.; Wang, J.; Dong, B.; Zhao, H. & Rong, L. (2003). Ultrasonication-Induced Formation of Silver Nanofibers in Reverse Micelles and Small-Angle X-ray Scattering Studies. *The Journal of Physical Chemistry B*, Vol. 107, pp. 3679–3683, 2003.

[42] McLeod, M. C.; McHenry, R. S.; Beckman, E. J. & Roberts, C. B. (2003). Synthesis and Stabilization of Silver Metallic Nanoparticles and Premetallic Intermediates in Perfluoropolyether/CO2 Reverse Micelle Systems. *The Journal of Physical Chemistry B*, Vol. 107, pp. 2693–2700, 2003.

[43] Naik, R. R.; Stringer, S. J.; Agarwal, G.; Jones, S. E. & Stone, M. O. (2002). Biomimetic Synthesis and Patterning of Silver Nanoparticles. *Nature*, Vol. 1, pp. 169–172, 2002.

[44] Kowshik, M.; Ashtaputre, S.; Kharrazi, S.; Vogel, W.; Urban, J.; Kulkarni, S. K. & Paknikar, K. M. (2003). Extracellular Synthesis of Silver Nanoparticles by a Silver-Tolerant Yeast Strain MKY3. *Nanotechnology*, Vol. 14, pp. 95–100, 2003.

[45] Shankar, S. S.; Rai, A.; Ahmad, A. & Sastry, M. (2004). Rapid Synthesis of Au, Ag, and Bimetallic Au Core–Ag Shell Nanoparticles Using Neem (*Azadirachta indica*) Leaf Broth. *Journal of Colloid and Interface Science*, Vol. 275, pp. 496–502, 2004.

[46] Sardar, R.; Park, J. W. & Shumaker-Parry, J. S. (2007). Polymer-Induced Synthesis of Stable Gold and Silver Nanoparticles and Subsequent Ligand Exchange in Water. *Langmuir*, Vol. 23, pp. 11883-11889, 2007.

[47] Longenberger, L. & Mills, G. (1995). Formation of Metal Particles in Aqueous Solutions by Reactions of Metal Complexes with Polymers. *The Journal of Physical Chemistry*, Vol. 99, pp. 475–478, 1995.

[48] Sakai, T. & Alexandridis, P. (2006). Ag and Au Monometallic and Bimetallic Colloids: Morphogenesis in Amphiphilic Block Copolymer Solutions. *Chemistry of Materials*, Vol. 18, pp. 2577–2583, 2006.

[49] Pastoriza-Santos, I. & Liz-Marzan, L. M. (2002). Synthesis of Silver Nanoprisms in DMF. *Nano Letters*, Vol. 2, pp. 903–905, 2002.

[50] Kurihara, L.K.; Chow, G. M. & Schoen, P. E. (1995). Nanocrystalline Metallic Powders and Films Produced by the Polyol Method. *NanaShuchued Materials*, Vol. 5, pp. 607–613, 1995.

[51] Sondi, I.; Goia, D. V. & Matijevic, E. (2003). Preparation of Highly Concentrated Stable Dispersions of Uniform Silver Nanoparticles. *Journal of Colloid and Interface Science*, Vol. 260, pp. 75–81, 2003.

[52] Yin, Y.; Li, Z. Y.; Zhong, Z.; Gates, B.; Xia, Y. & Venkateswaran, S. (2002). Synthesis and Characterization of Stable Aqueous Dispersions of Silver Nanoparticles through the Tollens Process. *Journal of Materials Chemistry*, Vol. 12, pp. 522–527, 2002.

[53] Balogh, L.; Swanson, D. S.; Tomalia, D. A.; Hagnauer, G. L. & McManus A. T. (2001). Dendrimer-Silver Complexes and Nanocomposites as Antimicrobial Agents. Nano Letters, Vol. 1, pp. 18–21, 2001.

[54] Zhang, S.; Ni, W.; Kou, X.; Yeung, M. H.; Sun, L.; Wang, J. & Yan, C. (2007). Formation of Gold and Silver Nanoparticle Arrays and Thin Shells on Mesostructured Silica Nanofibers. *Advanced Functional Materials*, Vol. 17, pp. 3258–3266, 2007.

[55] Es-Souni, M.; Fischer-Brandies, H. & Es-Souni, M. (2008). Versatile Nanocomposite Coatings with Tunable Cell Adhesion and Bactericidity. *Advanced Functional Materials*, Vol. 18, pp. 3179–3188, 2008.

[56] Aihara, N.; Torigoe, K. & Esumi, K. (1998). Preparation and Characterization of Gold and Silver Nanoparticles in Layered Laponite Suspensions. *Langmuir*, Vol. 14, pp. 4945-4949, 1998.

[57] Liu, J.; Lee, J. B.; Kim, D. H. & Kim, Y. (2007). Preparation of High Concentration of Silver Colloidal Nanoparticles in Layered Laponite Sol. *Colloids and Surfaces A: Physicochem. Eng. Aspects*, Vol. 302, pp. 276–279, 2007.

[58] Papp, S.; Patakfalvi, R. & Dekany, I. (2008). Metal Nanoparticle Formation on Layer Silicate Lamellae. *Colloid and Polymer Science*, Vol. 286, pp. 3–14, 2008.

[59] Dong, R. X.; Chou, C. C. & Lin, J. J. (2009). Synthesis of Immobilized Silver Nanoparticles on Ionic Silicate Clay and Observed Low-Temperature Melting. *Journal of Materials Chemistry*, Vol. 19, pp. 2184–2188, 2009.

[60] Ding, X.; Xu, R.; Liu, H.; Shi, W.; Liu, S. & Li, Y. (2008). Hyperbranched Polymer-Assisted Hydrothermal In situ Synthesis of Submicrometer Silver Tubes. *Crystal Growth & Design*, Vol. 8, pp. 2982–2985, 2008.

[61] Su, H. L.; Chou, C. C.; Hung, D. J.; Lin, S. H.; Pao, I. C.; Lin, J. H.; Huang, F. L.; Dong, R. X.& Lin, J. J. (2009). The Disruption of Bacterial Membrane Integrity through ROS Generation Induced by Nanohybrids of Silver and Clay. Biomaterials, Vol. 30, pp. 5979-5987, 2009.

10

"Silver nanoparticles in oxide glasses: technologies and properties"

N.V. Nikonorov, Sidorov A.I. and Tsekhomskii V.A.
Saint-Petersburg State University Of Information Technologies, Mechanics, and Optics, Russia, 199034 St. Petersburg 4, Birzhevaya line

1. Introduction

Nano-size particles of noble metals (Ag, Au, Pt) are today a subject of thorough attention of scientific researchers and of technologists, because these particles reveal the wide range of physical features, absent in the case of bulk media.

Size-controlled optical properties of Ag-nanoparticles open good prospects for potential technological applications - such as diffraction elements, optical filters, nanoplasmonic devices, bi-sensors, and nonlinear media.

Silver nanoparticles can be produced in various media - in water solutions, in polymers, in glasses, and in crystalline media. For instance, the long period in photographic imaging technique was based on the photoinduced production of Ag nanoclusters in the crystals of AgHal (the latent image). At the same time one can say that the inorganic glasses are the unique matrix for silver nanoparticles formation. The wide temperature range of glass viscosity growth provides the possibility to control over the Ag nanoparticles size within the wide range by means of modifying the temperature and duration of thermal processing. In the fact, only such kind of matrix makes it possible to control and investigate all the stages of Ag nanoparticles formation, including the starting stage.

Glass coloring by introduction of the metallic Ag nanoclusters is one of the ancient methods, known by humanity [1]. The mechanisms, leading to nucleation and growth of such nanoclusters, are the scientific and technology problem, important both for colored glass production and for all other potential applications of glass with silver nanoparticles [2-5].

The researches on silver nanoclusters formation comprise numerous technical approaches. Especially interesting are the methods, based on the use of ionizing radiation, are especially interesting due to their ability provide either local irradiation or to irradiate the whole volume of the glass. In this case it is comparatively simple to provide the control over concentration and average size of metallic nanoclusters. For instance, the controlled photon irradiation is used for production of photosensitive [6], polychrome [7] and photo-thermo-refractive [8] glasses. Very often all these types of glasses are called the photosensitive one (PS). Further within this chapter we shall use this word.

2. Formation of silver nanoclusters in PS glasses by UV irradiation method

PS glasses contain Ag in the dispersed oversaturated solution. The raw PS glasses (non-irradiated and non-thermally processed), similarly to the silver-less silicate glasses, which are used for PS glass production, reveal the fundamental absorption edge nearby 3.75 eV. Their absorption spectra do not reveal significant difference, which could have been attributed to the silver presence.

The energies of the resonant doublet of free Ag atoms ($5sS_{1/2}$ – $5pP_{1/2}$ and $5sS_{1/2}$ – $5pP_{3/2}$) are equal correspondingly to 3.665 and 3.780 eV, while ions of Ag^+ do not absorb in this range. Hence one can conclude that in the non-irradiated and non-thermally processed PS glasses Ag is dissolved in the form of Ag^+-ions. Thermal processing under 500°C for up to 10^4 minutes does not result in change of absorption of these non-irradiated specimens.

The following components can be additionally inserted into PS glasses: (1) the ions of Ce^{3+}, which modify the spectral range of glass photosensitivity (the effective radiation energy is 3.88 eV, i.e. approximately in 1 eV less than in glasses without ions of Ce^{3+}); (2) the ions of Sb^{5+} and Sn^{4+}, which are electrons acceptors, playing the important role in production of silver nanoparticles under thermal processing and providing the necessary oxidation – reduction conditions during glass production; and (3) the ions of F^-, Br^- and Cl^-, which participate in formation of various crystalline phases under the thermal processing.

Let us consider the process of photo-thermo-induced nucleation in PS glasses. The action of the UV radiation, filling into the band of Ce^{3+} absorption, the electron-hole pair is produced in the glass. The hole is trapped by the three-valence cerium and the free electron is produced within the following reaction:

$$Ce^{3+} + \hbar\nu = Ce^{(3+)}h + e^- \quad (1)$$

In the absence of Ce^{3+} ions the action of the shorter wavelength UV radiation, X-rays or γ-quanta results in trapping the holes by the non-bridged oxygen (NBO) with production of the non-bridged oxygen hole center (NBOHC). The energy of radiation, which is efficiently producing such centers in PS-glasses, exceeds 4.5 eV. There is also some possibility of such centers production in glasses with Ce^{3+}, especially in the case of high temperature.

The fluorine ions provide significant influence onto the absorption band of Ce^{3+} ions (Fig.1).

Increase of fluorine concentration in glass results in this band maximum shifting towards shorter wavelength range, while the band intensity is slightly reduced. One can explain the spectral shift of absorption band to the shorter wavelength range with fluorine concentration by arrival of fluorine ions in the coordination environment of Ce^{3+} ions. The rate of this shift is some 10-15 nm. Hence one can say that the cerium ions reveal with the fluorine ions the so-called halide effect [9].

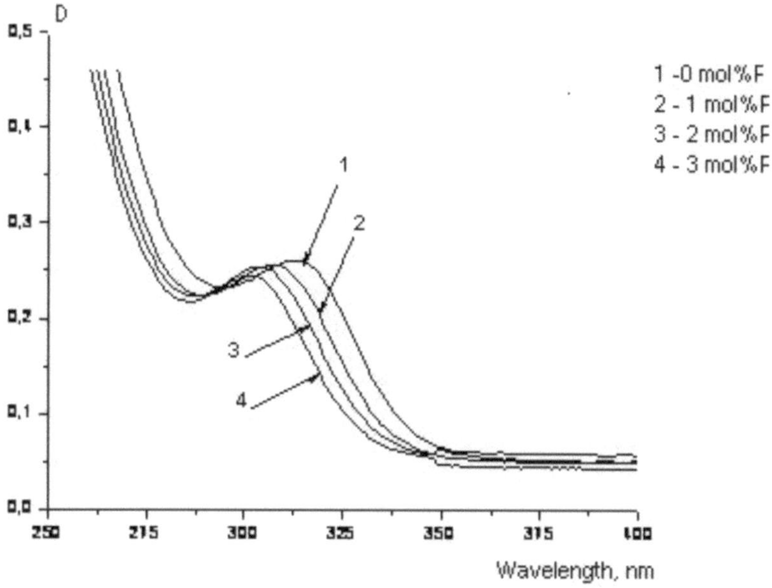

Fig. 1. Absorption spectra of PS glasses for various fluorine concentration.

It is interesting to note that in the case of bromine ions introduction into PS glass composition no halide effect with cerium ions is observed. Within the experiment accuracy variation of bromine ions concentration from 0 to 2.5 molar % does not change the position of the Ce^{3+} ions absorption band.

The freed electrons are trapped. In our case the traps can be comprised by Ag^{+}, Sb^{5+}, Ce^{4+}, by anion vacancies and by other glass defects and admixtures. The following processes are most interesting in our case:

$$Ag^{+} + e^{-} \rightarrow Ag^{0}, \quad (2)$$

$$Sb^{5+} + e^{-} \rightarrow (Sb^{5+})^{-} \quad (3)$$

The electron, trapped by the silver ion comprises the meta-stable center even at the room temperature [10], while the center $(Sb^{5+})^{-}$ is preserved up to the high temperatures [11]. One can see from the absorption spectra, shown in the Fig.2, that UV irradiation of the PS glass results in reducing of the absorption, related to the Ce3+ ions (wavelength 312 nm), while absorption in UV and visible band (350-500 nm) grows up. One can see it in more details in the Fig.3.

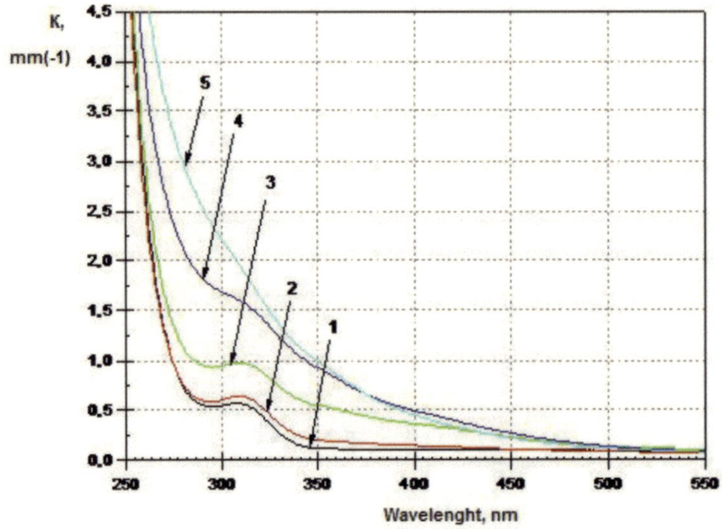

Fig. 2. Absorption spectra of irradiated PS glass (curve 1 - non-irradiated specimen, 2 - irradiation exposure dose - 10 J, 3 - 100 J, 4 - 1000, and 5 - 10000 J).

The reduce of absorption in the range 250 nm during irradiation (Fig.3) makes the basis for assumption that the ions of Ag^+ are responsible for the PS glass absorption band in the range of 250 nm. This assumption exactly coincides with the similar conclusions in the papers [12]. In any case we observe the definite tendency of absorption intensity decrease during irradiation - from ~7.0 mm^{-1} for the starting non-irradiated glass to 4.5 mm^{-1} for 100 J exposure and 3.5 mm^{-1} for 1000 J exposure.

The simplest supposition is that during the irradiation Ag^0 is produced in accordance with reaction (1). However, we have already noted that already at the temperature of 100°C these centers are transformed into the centers of Ag_2^+ and Ag_3^+ [10] due to interaction with the silver ions. The authors of [10] have determined that Ag^0 is oxidized easier than Ag_2^+, while Ag_3^+ and Ag_3^{2+} are yet more stable.

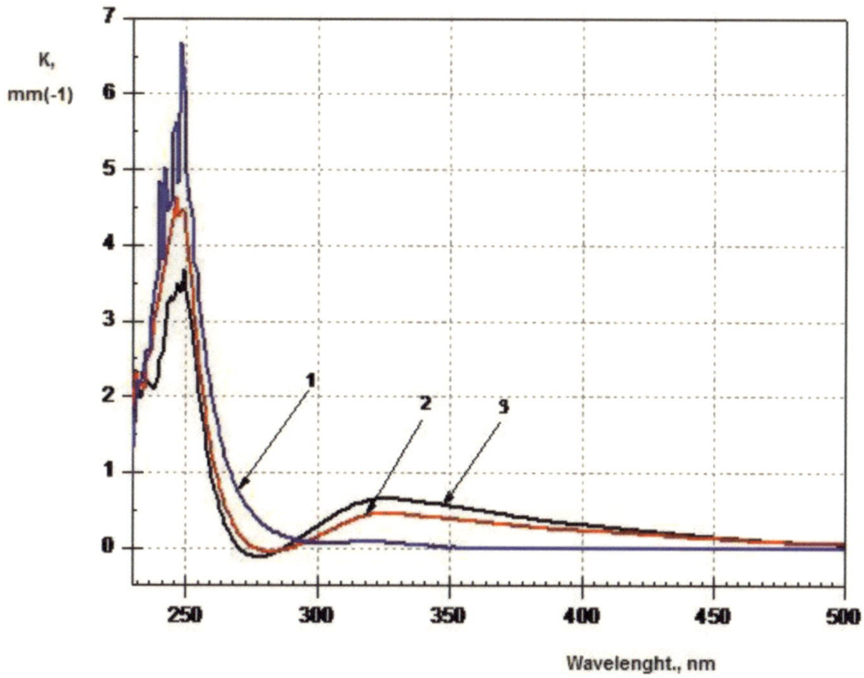

Fig. 3. Absorption spectra of irradiated PS glass (curve 1 - non-irradiated specimen, 2 - difference between spectra of PS after irradiation exposure dose 100 J and starting condition, and 3 - difference between spectra of PS after irradiation exposure dose 1000 J and starting condition).

The conclusion that the redox potential grows up with the increase of size of the molecule nucleus in water solutions provided the authors of [10] the possibility to outline the following stability series:

$$Ag^0 < Ag_2^+ < Ag_3^+/Ag_3^{2+} < ... < Ag_n \quad (4)$$

Hence one can connect the absorption in the range 300-500 nm, increasing with the irradiation dose (Fig.3) with the production of quasi-molecules of $(Ag_2)^+$, $(Ag_3)^+$ etc. According tot he estimations of [13], the molecule Ag_2^+ absorbs radiation with the wavelengths less than 430 nm, Ag_3^+ - 435-525 nm, and Ag_4^+ - 525-705 nm.

Additional evidences for such a statement are based upon the following facts. In the silicate glasses the silver ions are substituting the sodium ones, and the sodium ions, breaking the tie Si-O-Si, are not uniformly distributed across the volume, but are producing the dimers. Hence there exists some possibility that the silver ions can substitute both sodium ions in the dimer, leading thus to production in the glass of the silver dimmer like:

$$\text{Si-O-Ag Ag-O-Si.} \quad (5)$$

Such a structural unit can trap the photoelectron with production of the quasi-molecule of Ag_2^+ without movement of silver ions across the glass. The energy of thermal ionization of the Ag_2^+ quasi-molecule is about ~ 0.8 eV [13]. On the other hand, the wide absorption band from 300 to 500 nm can be related to the significant dispersion of the coupling length of Ag-Ag.

The silver particles are growing up by trapping the electrons on the nuclei, produced on the nucleation stage. These electrons are neutralized by the diffusing Ag^+ ions. Such a mechanism is well known and is used in halide-silver photographic media [14]. At the room temperature the ions of Ag^+ are frozen within the matrix; only for temperatures over, say, 400°C, their diffusion constant is sufficiently high for the measured growth rate.
During the consequent thermal processing the charged centers are losing electrons within the wide range of temperatures. Electron supply can be provided by centers $(Sb^{5+})^-$, which are stable for the temperatures up to 250-300 °C, or silver quasi-molecules. The following reactions can occur under such temperatures:

$$(Sb^{5+})^- \rightarrow e + Sb^{5+} \tag{6}$$

$$Ag_2^+ \rightarrow e + Ag_2^{2+} \tag{7}$$

The relatively low matrix viscosity assists to production of the colloid silver particles Ag_n.

In the Fig.4 are shown the spectra of PS glasses absorption after multiple irradiation by UV light (dose 1000 J) and thermal processing at 450°C during 10 hours. The results, shown in the Fig.4, indicate that for the temperature slightly lower than Tg for these glasses (~500°C), irradiation and thermal processing result in the sufficiently intense absorption in the range 415-426 nm. This peak is typical for the colloid silver particles [15] and is related to the surface plasmon resonance (SPR). The curve (3) in the Fig.4 corresponds to the effect saturation state, i.e. any further repeats of irradiation and thermal processing cycles did not result in modification of either absorption maximum position nor in its intensity.

Peak amplitude of SPR is close to the measurements of the work [16]. Thus one can draw out the conclusion that in the considered specimens, only some 1% of the overall Ag^+ amount is transformed by irradiation and thermal processing into Ag^0, forming the silver nanoparticles. Most probably, the change in maximum position is caused by the size effects. Much larger variation of SPR maximum position is observed in the case of increasing the processing temperature (Fig.5).

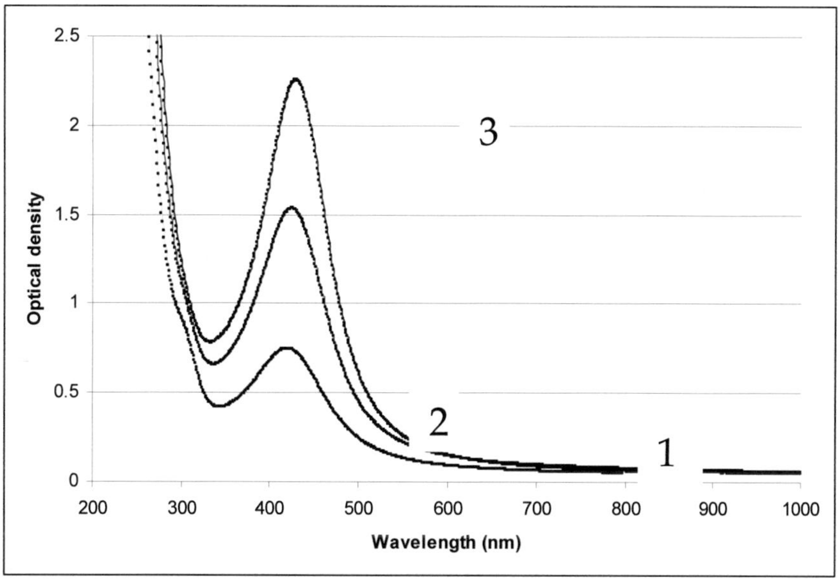

Fig. 4. Absorption spectrum of PS glass specimens after multiple irradiation with exposure 1000 J and processed under the temperature 450°C for 10 hours: 1 - after single time irradiation and thermal processing; 2 - after 3-times processing; and 3 - after 5-times processing.

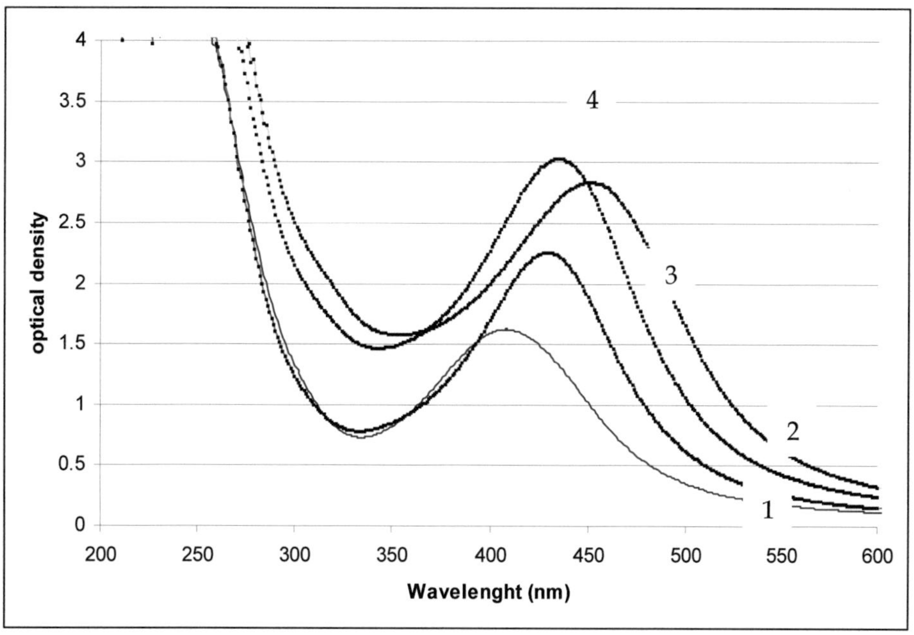

Fig. 5. The absorption spectra of PS glass specimens, multiple irradiated by the dose of 1000 J and processed during 10 hours under the temperature of: 1 - 400°C, 2-450°C, 3-500°C, 4-520°C.

One can see from the Fig.5 that the variation of the processing temperature from 400 to 520°C results not only in shift of the SPR band maximum (from 405 to 460 nm), but also in changing the band halfwidth. Observation of the absorption bands, caused by the silver colloid particles (> 1 nm), under the temperatures, which are much lower than Tg, i.e. under 400-450°C, is the evidence, indicating that, most probably, the silver nanoparticles are produced in the glass cavities, whose size grows up with the temperature increase. For the temperatures higher than Tg the colloid silver deposition can be additionally accompanied by the increase of the internal cavities due to the viscous current.

3. The factors, which influence onto the spectral position of plasmon resonance of silver nanoparticles in glass

Let us consider some possible reasons of the spectral shift of the plasmon resonance band in glass, containing the silver nanoparticles [16]. The following factors can lead to the SPR band shift:

1. modification of the chemical content and crystalline structure of silver nanoparticle, influencing onto its dielectric permeability;
2. modification of the chemical content of the edge nanoparticle - glass, which influences onto the process of scattering the free electrons of the metal on the nanoparticle surface;
3. variation (increase) of the silver nanoparticle size during thermal processing of the glass without modification of its shape;
4. modification of the geometric shape of nanoparticle during irradiation and thermal processing of the glass;
5. modification of the refraction index of matrix, surrounding the silver nanoparticle.

All these factor can influence significantly onto the dielectric permeability of silver nanoparticles even in the case of very small variation of chemical content of nanoparticle.

Let us consider in more details the influence of the last three factors onto the spectral position of absorption band of silver nanoparticles in glass matrix. Let us make the following assumptions: the size of nanoparticles is much less than the radiation wavelength; the concentration of nanoparticles in glass is low; particles do not reveal mutual electromagnetic interaction.

The model, which is based on the Drude-Lorentz theory and is taking into account the scattering of free electrons on the nanoparticle surface, leads to the following relationship for the dielectric permeability [17]:

$$\varepsilon(\omega,r)=\varepsilon_0(\omega)+\frac{\omega_p^2}{\omega^2+i\omega\gamma}-\frac{\omega_p^2}{\omega^2+i\omega(\gamma+Av_F/r)},$$

Here ω is the circular frequency of radiation; r is the nanoparticle radius; ε_0 is the dielectric permeability of the bulk silver; ω_p is the plasmonic frequency of silver; γ and A are the constants, accounting for the processes of scattering of free electrons in metal on electrons, phonons and the nanoparticle surface ($A < 1$); v_F is the Fermi speed of free electrons.

According to the estimations the change of diameter of spherical silver nanoparticle from 3 to 4 nm without shape changes leads to the shift of plasmonic resonance frequency to the longer wavelength side in $\Delta\lambda \approx 10$ nm. This result well correlates with the conclusions of [18], drawn out from the absorption spectra of phosphate glass with silver nanoparticles. One can explain the long wavelength shift after light and thermal processing (Fig.4) and, to smaller extent, after increase of the processing temperature (Fig.5) by the increase of the silver nanoparticles size. However, according to estimations, one cannot explain the observed shift only by changes of silver nanoparticles size.

The geometry shape of silver nanoparticles in glass can differ from the ideal sphere. For example, in the boron-silicate glass the silver nanoparticles can reveal the shape of extended ellipsoid [19, 20]. One can see from the electron microscopy images of silver nanoparticles in phosphate glass, presented in [18, 21], that their shape is also non-spherical. The authors of paper [22], devoted to the polychrome glasses, who have varied the irradiation dose and temperature of processing at the nucleation stage and the stage of nanocrystal growth, have managed to realize the whole gamma of glass colors from yellow to green.

In the Fig.6 is shown the variation of absorption spectra of the PS glass specimens, irradiated by various doses on the nucleation stage. One can note that the increase of the exposure dose on the nucleation stage reduces the deviation of particle shape from spherical, which reveals itself in reduce of the gap between the main bands. Irradiation of PS glass specimens by the dose over 500 J results in observation of only one absorption band, which is an evidence of the spherical shape of colloid nanoparticles.

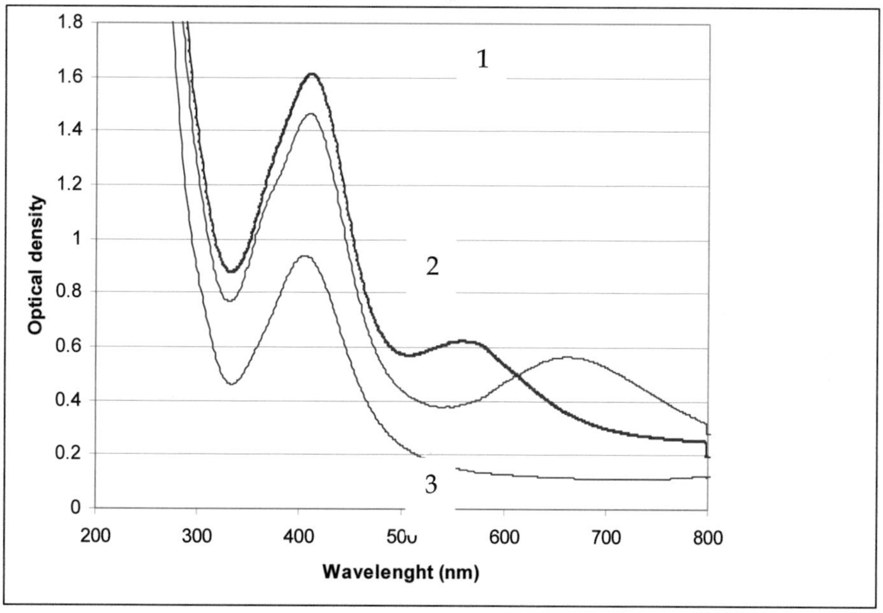

Fig. 6. Absorption spectra of the specimens, whose nucleation took place under UV irradiation by various exposure doses: 1 – 200 J, 2 – 100 J, and 3 – 10 J.

The most interesting results was observed in the case of PS glass irradiation at the stage of nucleation by the dose of 10 J. One can see from the Fig.6 that in this case one can see in the visible range only one band with the maximum around 395 nm. One can see from the Fig.7 that these specimens reveal one more absorption band in the near IR spectral range; the repetition of the irradiation and thermal processing stages results in additional shift of the

second band to the longer wavelength spectral range, i.e. this process is also accompanied by the gap increase.
Since recently photoactivation of PS glasses is very often produced by femtosecond pulsed lasers, emitting in near IR spectral range [23]. In this case it is not necessary to dope the glass by ions of Ce^{3+}, Cu^{+} etc., which work as electron donors during UV irradiation.
Especially interesting is application of short pulse laser radiation for modification of spherical shape of silver nanoparticles in glass [24-28].
Optical microscopy studies [24] have shown that irradiation by circularly polarized femtosecond laser radiation can change the spherical shape of silver nanoparticles in glass. There was outlined the hypothesis that the shape transformation occurs via the stage nanoparticles photoionization by laser irradiation. The rate of dichroism, which can be achieved by laser use, depends upon the initial properties of nanoparticles. For instance, provision of maximum dichroism for the particles with SPR at 413 nm requires irradiation by light with much larger wavelength, and the wavelength has to be increased during irradiation. Such sequence of irradiation at 535 and 670 nm can cause modification of shape from spherical nanoparticle to ellipsoid with the axis ratio >3[28].

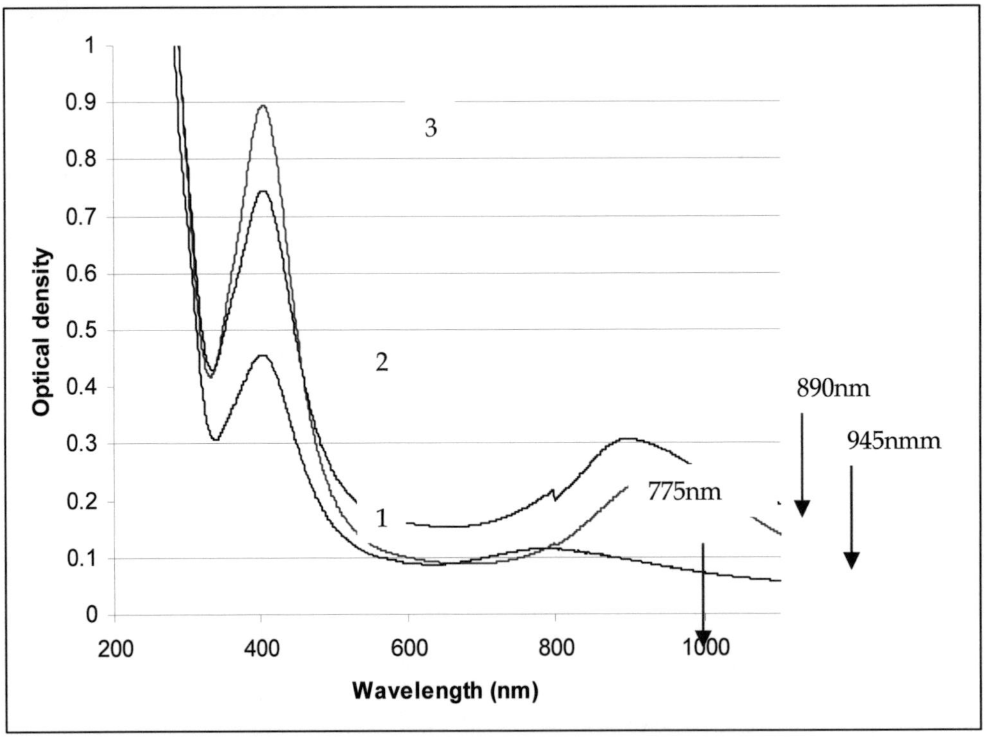

Fig. 7. Absorption spectra of PS specimens, whose nucleation was carried out by UV irradiation at 10 J, while processing occurred in several stages (1, 2 and 3).

We have already noted the SPR maximum position can be strongly influenced by the refraction index of the matrix, surrounding the glass nanoparticle. Naturally, the strongest influence of dielectric environment reveals itself in the spectral position of plasmonic resonance of the silver nanoparticle.

In the dipole approximation the cross-section of absorption by the spherical particle with the shell is described by the following relationship [29]:

$$\sigma_a = \frac{8\pi^2\sqrt{\varepsilon_h}r_s^3}{\lambda}\cdot \mathrm{Im}\left(\delta\right)$$

Here δ is the nanoparticle polarisability:

$$\delta = \frac{\varepsilon_s\varepsilon_a - \varepsilon_h\varepsilon_b}{\varepsilon_s\varepsilon_a + 2\varepsilon_h\varepsilon_b},\ \varepsilon_a = \varepsilon_c\cdot\left(3-2P\right)+2\varepsilon_s P,\ \varepsilon_b = \varepsilon_c + \varepsilon_c\cdot\left(3-P\right)$$

$$P = 1 - \left(\frac{r_c}{r_s}\right)^3$$

ε_h, ε_c, ε_s are the values of the dielectric permeability of medium, core and shell correspondingly, r_c is the core radius, r_s is the shell radius, and λ is the radiation wavelength.

The condition of plasmon resonance is equality to zero of the real part of nanoparticle polarisability denominator [29]:

$$\mathrm{Re}(A) = \mathrm{Re}(\varepsilon_s\varepsilon_a + 2\varepsilon_h\varepsilon_b) = 0\,.$$

One can conclude from these relationships that the thickness of dielectric shell and its dielectric permeability provide the most significant influence onto spectral position of the plasmonic resonance.

The numerical simulation was carried out [30] for the following parameters: refraction index of the medium n_h = 1.5; core material - silver (the optical constants of the silver film were taken from [31] for the spectral range λ=0.4-0.6 μm); the refraction index of the shell n_s = 2.2 (silver bromide was used as the example); core radius was r_c = 2 нм; and the shell thickness $h = r_s - r_c$ varied from 0 to 3 nm.

In the Fig.8 are shown the spectral dependencies of absorption cross-section of spherical nanoparticles with silver core and silver bromide shells of different thickness. One can see from the Figure that the increase of shell thickness from 0 to 3 nm results in shift of plasmonic resonance of nanoparticle to the longer wavelength side in ~100 nm.

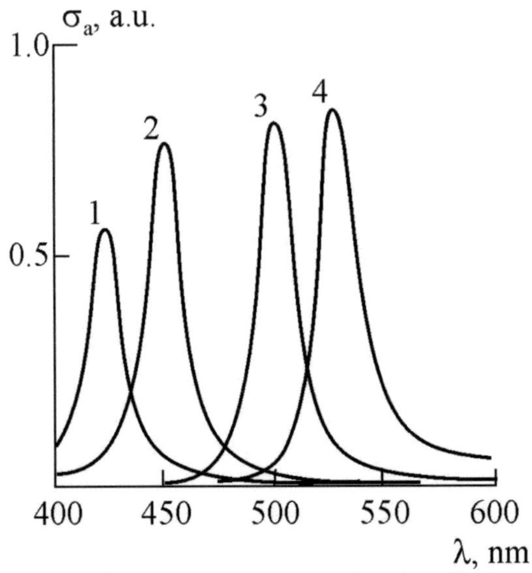

Fig. 8. Spectral dependencies of absorption cross-section for silver nanoparticles with silver bromide shell. 1 - h = 0 nm, 2 - 0.5, 3 - 1.5, 4 - 3 [30].

Let us return to specific media. One can assume that during thermal processing can occur not only variation of thickness of the shell, surrounding the silver nanoparticle, but also variation of this shell refraction index. In the Fig.9 are shown the dependencies of spectral position of nanoparticle plasmonic resonance upon the shell refraction index [30]. One can see from the Figure the increase of the shell refraction index results in shift of plasmonic resonance to the longer wavelength spectral range. The dependence is practically linear. Increase of shell thickness leads to slight increase of slope of dependence of $\lambda_{max}(n_s)$.

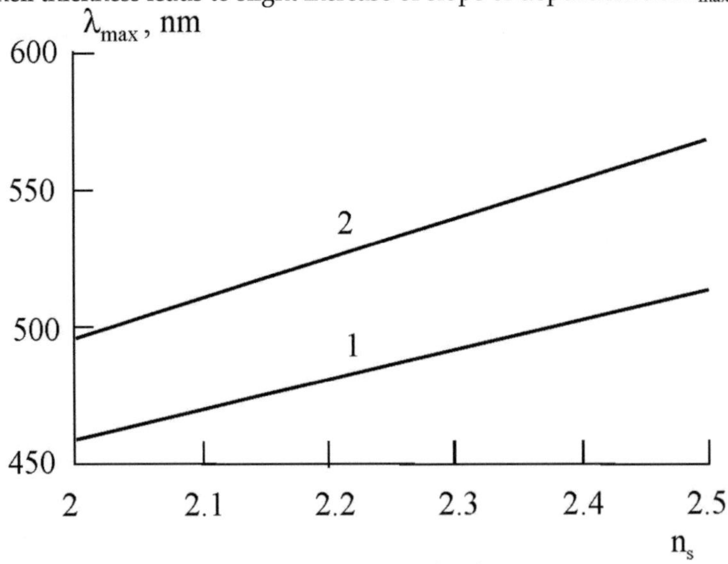

Fig. 9. Dependence of spectral position of nanoparticle plasmonic resonance maximum upon shell refraction index. 1 - h = 1 nm, 2 - 3 [30].

Hence, formation of the dielectric shell on the silver nanoparticle and variation of this shell's thickness during the thermal processing of the glass can provide the rather strong influence onto the spectral position of the plasmon resonance. For instance, in [32] it was noticed that the 3-nm silver particles, produced by γ-irradiation and consequent thermal processing, can be subjected to oxidizing due to interaction with defects on the non-bridged oxygen (NBOs) in the case of temperature growth up to 500°C.
The paper [10] outlines some analogies between silver behavior in glass and its behavior in water solutions from the point of view of tendency of silver particles to be oxidized and thus to reduce their size or disappear with the temperature growth. Let us consider this analogy and note the very low solvability of silver halides in water. Hence one can say that there is a very high probability, especially when the temperature is high, that on the surface of the silver nanoparticle, dissolved in a glass, will be produced the following compounds:

Ag_20, AgF, AgCl, AgBr

The Table presents the values of refraction index (n) of these compounds and their melting temperature (T_m). These compounds may impose significant effect onto spectral position of the plasmon resonance. They are very important for mechanisms of production and behavior of a wide group of light sensitive glasses like photochrome, polychrome, photo-thermo-refractive etc.

Compound	Refraction index (n)	Melting temperature (T_m)
Ag_2O	2.7	decompose 300°C
AgF	~2.03	435°C
AgCl	2.07	455°C
AgBr	2.25	434°C

The above given analysis has shown that the thickness and refraction index of the shell, surrounding the nanoparticle, provides significant influence onto position of the plasmon resonance maximum.

Let us note the relatively low melting temperatures of the compounds, presented in the Table. They are, in any case, lower than that of Tg PS glasses and than the temperatures, usually used for thermal processing with the purpose of Ag nanoparticles formation (450-600°C). Taking also in account the dependence of melting temperature upon the particle size (Thomson equation), one can say, that all these compounds (with the exception of Ag_2O) are liquid during the thermal processing of the PS glasses. This can be the explanation for the following effects:

1. Unusual - for glasses - influence of very small admixtures (like Cu, Cd, Pb etc.) onto the glass properties - the photochrome glasses [33].
2. Dissolution of compounds like NaF and LiF in the liquid phase (photo-thermo-refractive and photosensitive glasses).

True, less than 0.1% Cu_2O is introduced into the halide-silver glasses, providing the drastic changes in glass properties and in properties of the extracted crystalline phase of AgHal

[33]. It would not be so unusual in the case of growing up the monocrystal or synthesis of polycrystalline compound. The glass structure – even with the account for the micro-inhomogeneous content – provides a great number of positions for copper ions, which they can substitute with a high probability. However, in this case, substituting, say, the ions of Na^+, they do not have an opportunity to form their own environment in a solid glass. The situation is, however, different in the case when they (copper ions, for instance) transfer to a liquid phase.

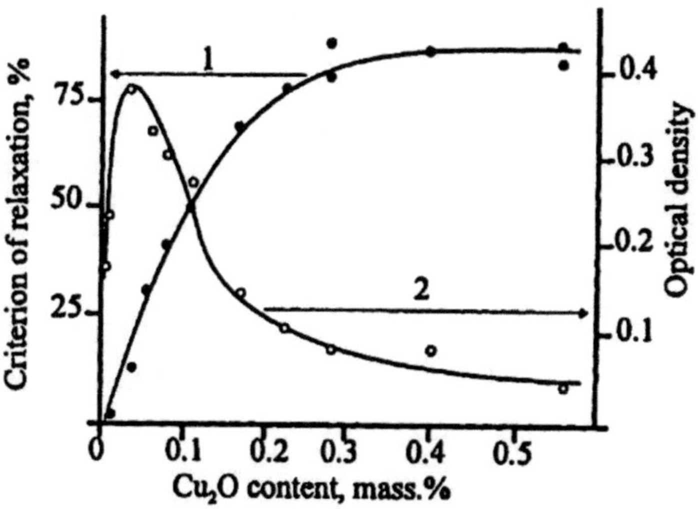

Fig. 10. The influence of copper ions content in the glass onto its photochrome properties (changes in optical density during irradiation and the rate of thermal bleaching (relaxation criterion))

In the Fig.10 is shown the influence of copper ions content in the glass onto the photosensitive properties of the photochrome glasses. One can note the very strong and variable influence. To our opinion, this is an evidence, supporting the hypothesis that during thermal processing the copper ions are entering the liquid silver-halide phase.

Similar ideas can be outlined concerning the mechanism of deposition the crystals of NaF on the photoinduced nanoparticles of Ag in photo-thermo-refractive glasses. It is well known that the sodium fluorine is hardly solvable in silicate glass; it is evaporating from the glass surface even at the comparatively low temperatures. In [34, 35] was shown that in the photo-thermo-refractive glasses no photoinduced deposition of NaF takes place in the absence of KBr in the glass composition. In [35] in such glass was observed the absorption band of colloid silver, covered by the layer of AgBr. Hence, in fact, the sodium fluorine is dissolved in liquid AgBr. Under such temperatures (~500°C) is produced the liquid phase, consisting of AgBr and NaF. The crystalline sodium fluorine is deposited during this phase cooling.

4. Special features of silver nanoparticles formation in PS-glasses under electron beam and thermal action

The classic method of Ag nanoparticles formation in PS glasses, based on UV and thermal action, provides the possibility to obtain the nanoparticles within the overall glass volume. In this case, however, the volume concentration of nanoparticles is low. The reason is that it is impossible to introduce the big number of AG ions into the starting glass composition. In addition, the starting glass has to contain the sensitizing ions of Ce, Sb and Sn. The method of ion implantation [36, 37], which makes it possible to provide high concentration of metal ions in a thin surface layer of glass, is free of the said limitation. In this case under definite conditions the metal nanoparticles are formed already on the implantation stage - without additional thermal processing.

The last years has brought the growing interest to the possibility of modification of the glass surface layers by electron beam. In particular, it was found out that electron irradiation leads to formation of the negatively charged region nearby the glass surface. The latter initiates the field diffusion of the movable positive metal ions - for instance, of K^+ [38]. Redistribution of potassium ions within the glass modifies its chemical composition and provides mechanical stress and refraction index gradient formation. The electrons reveal the emphasized reducing properties, and thus electron irradiation can lead not only to spatial redistribution of glass composition elements, but also to reducing of the positive ions down to the neutral atoms [39, 40] and stimulate growth of nanocrystals in glass [41].

Hence one can post the question - is it possible to form the silver nanoparticles in PS glass under the electron irradiation action, and what are this process peculiarities? We have studied the process of formation of Ag nanoparticles in PS glasses, based on the preliminary irradiation of glass surface by electrons with the energy of 5-30 kV and consequent thermal processing [42, 43]. Our studies have shown that this method makes it possible provide the high concentration of Ag nanoparticles in the surface glass layers even in the case when the starting glass contains the low concentration of silver ions. We have studied in experiment the PS glasses of composition $Na_2O-SiO_2-ZnO-Al_2O_3-Ag_2O$ $-CeO_2-Sb_2O_3-Br-F$; the content of Ag_2O in the starting glass was 0.5% (mass). The irradiation was carried out under the room temperature by electron beam with electron energy 7-30 keV and electron current density 50 $\mu A/cm^2$. Irradiation duration was varied from 100 to 1000 s. The layer of Al with the thickness 100-200 nm was deposited onto the glass surface for charge removal; after irradiation this layer was removed by means of chemical etching. In the Fig.11 are shown the spectra of PS glass optical density before electron irradiation, after it and after several cycles of thermal processing. The irradiation was carried out by electrons with energy 20 keV, and the irradiation dose Q was equal to 100 mC/cm^2. In the Fig.11 is also shown the routine of one thermal processing cycle. One can see from this Figure that irradiation by the electron beam with the said parameters leads to increase of glass absorption in short wavelength spectral range. However, no plasmonic absorption band is observed in this case, indicating thus the lack of metallic silver nanoparticles. Already the first cycle of thermal processing realizes in glass the intense absorption band in the spectral range 390-420 nm, corresponding to the plasmonic resonance of silver nanoparticles. Further thermal processing results in broadening of plasmonic band and modification of its amplitude. Modification of amplitude with the increase of thermal processing cycles' number can be of non-monotonous nature.

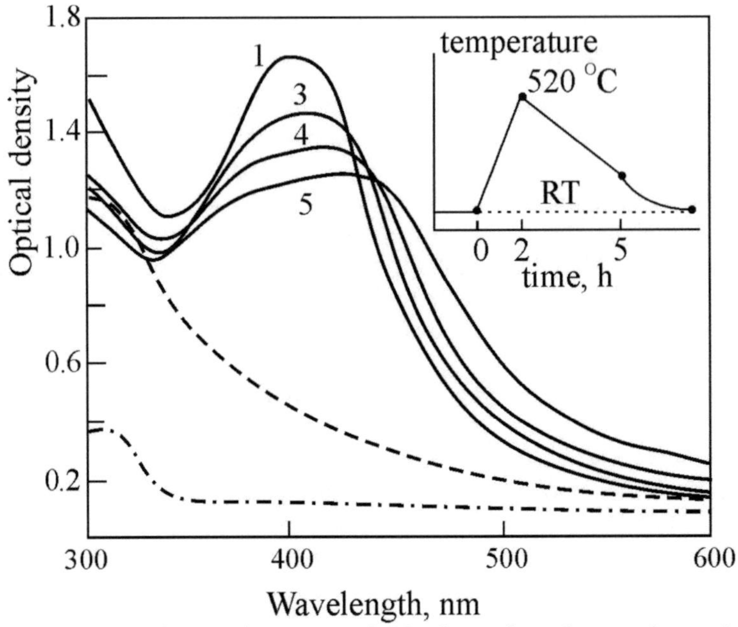

Fig. 11. Transformation of optical spectra of PS glass after electron-beam irradiation an thermal treatment. E_0 = 20 keV, Q = 100 mC/cm^2, irradiation time – 2000 s, (– · – · –) – initial glass; (– – –) - electron-beam irradiated glass; 1…5 – number of thermal treatment. Insert – the mode of thermal treatment.

In the Fig.12 is shown the TEM-image of the PS glass surface layer after electron irradiation and thermal processing (E_0 = 20 keV, Q = 100 mC/cm^2). One can see from the Figure that the average size of silver nanoparticles is 10-12 nm. Their concentration is rather high, and the distance between nanoparticles is sometimes smaller than their diameter. This is an explanation for the absorption spectrum transformation during irradiated glass thermal production. In the case of high concentration of particles and the distance between them less than the radiation wavelength the particles become the electromagnetically coupled. This results in significant increase of plasmonic absorption band [44].

Unlike the classical method of silver nanoparticles formation (UV irradiation and thermal processing), in the case of electron beam irradiation the silver nanoparticles are not uniformly distributed across the volume, but are concentrated in the thin layers nearby the glass surface and parallel to it. The thickness of such layers does not exceed 2 μm, while in some cases it is just approximately 20 nm. In the Fig.13 is sown the image of silver nanoparticles' layer, whose thickness does not exceed two diameters of nanoparticle, while the distance between the particles is 2-5 nm.

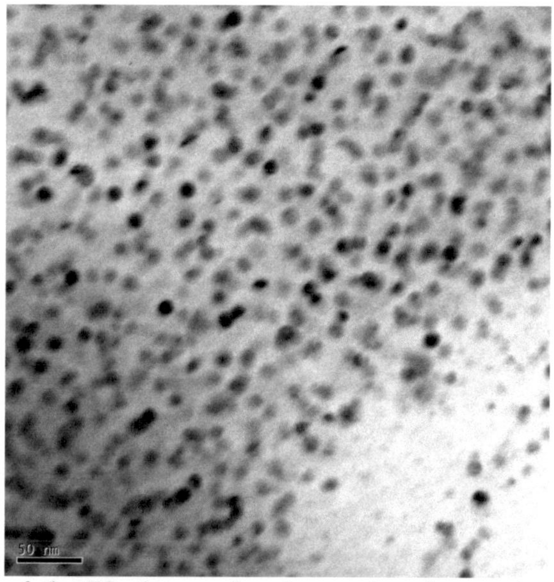

Fig. 12. TEM-image of the PS glass surface layer after electron irradiation and thermal processing. Size of silver nanoparticles 10-12 nm. Scale - 50 nm.

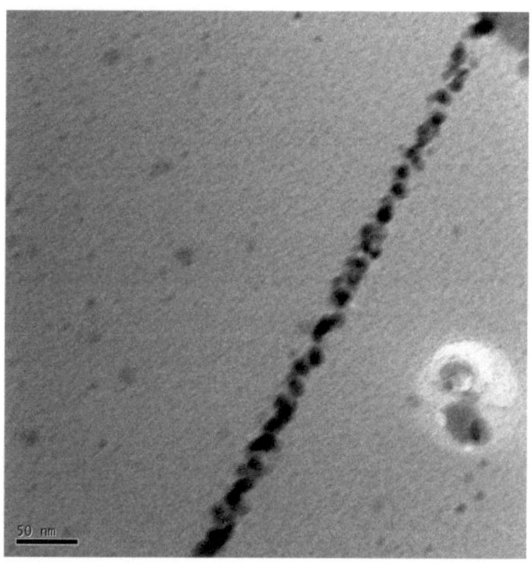

Fig. 13. TEM-image of silver nanoparticles' layer in PS glass. Scale - 50 nm.

One has to note that in the case of the electron energy less than 10 keV and irradiation zone 50-100 mC/cm^2 one can see - just by the naked eye - on the glass surface the silver layer. However, in this case the absorption spectra lack the plasmonic band. It means that in this case the silver layer is continuous or is comprised by the silver nanoparticles, contacting each other. During the thermal processing this layer is gradually disappearing due to silver dissolution in the glass, while the plasmonic band arises in the absorption band.

Let us consider special features of silver nanoparticles formation under electron beam irradiation conditions and the differences of this method from the classical one. In the Fig.14 is shown the geometry of the model, which was used during analysis.

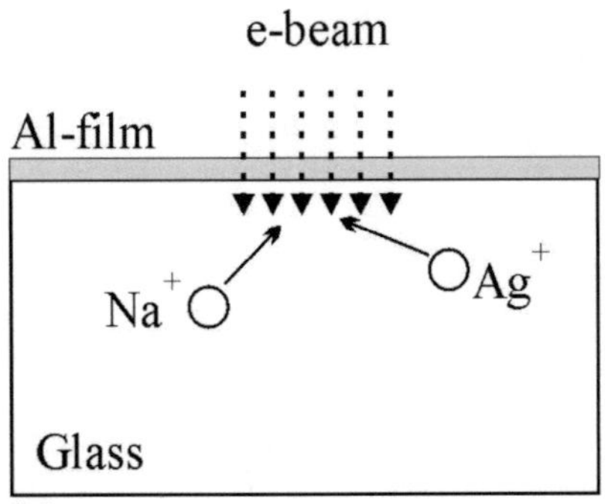

Fig. 14. Geometry of the model.

The depth of electrons' penetration. One can evaluate the depth of electrons' penetration into the glass from spatial distribution of energy losses of electrons. The losses of electron energy are determined by the atomic weight and atomic number of main glass components - of O, Si and Na, and also by glass density. In the Fig.15 are shown the calculated dependencies of electron energy losses for incident beam energies of 5, 20 and 30 keV. One can see from the Figure that for the energy of 5 keV the thickness of the layer, where electron braking takes place, is 250-300 nm, and the losses maximum fills to the depth of 100 nm from the surface. For electrons energy of 20 keV these values are 3 μm and 1 μm correspondingly, while for 30 keV - 6 μm and 25 μm. These layers determine the depth of spatial position in the glass of the volume charge region and of the region of electrons' interaction with the positive metal ions. At the same time the electrons, whose energy still exceeds that of atoms' ionization, can once again ionize the metal atoms. Hence, in these layers two competing processes can take place - reducing and ionization.

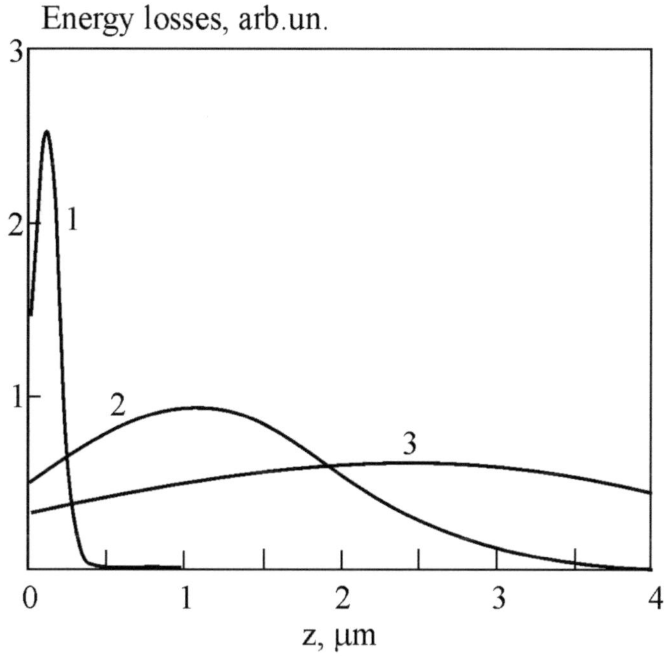

Fig. 15. Distribution of electron energy losses in PS glass. 1- E_0 = 5 keV; 2 - 20; 3 - 30.

Glass heating by electron beam. During glass surface irradiation by the high-energy electrons the major part of their energy is transformed, finally, into heat. The local heating can influence onto the processes of silver ions and atoms diffusion in the irradiated glass zone. Let us consider the influence of glass heating by the electron beam onto the processes of thermal diffusion. The calculation was carried out by means of solving the 3D-problem of thermal conductivity in the cylindrical coordinates for electron beam with the energy 20 keV, beam current density 50 µA/cm^2 and irradiation dose 100 mC/cm^2. The calculation has shown that under such conditions the maximal temperature in the surface layer of the glass with the thickness 1 µm does not exceed 150 ^{o}C. At the distance 20 µm from the surface the temperature is equal to 100 ^{o}C. The coefficient of thermal diffusion of atoms and ions reveals the exponential dependence upon the temperature, and hence the diffusion processes start playing the significant role for the temperatures higher than 300 ^{o}C. So one can draw out the conclusion that in the case of electron irradiation by the beam with the above mentioned parameters the impact of thermal diffusion of silver ions and atoms into the general picture of the process is small.

Tension of the field, produced by volume charge. Electron irradiation of glass produces formation of three electron flows in it - the flow of high-energy electrons of the primary beam, opposite-directed flow of the secondary electrons and the opposite directed flow of the thermalized electrons, which are moving from the specimen to metallic layer on the specimen surface. The solution of the problem of electric conductivity for given conditions made it possible to determine the volume charge, formed inside glass during the electron irradiation, and to evaluate the tension of the electric field, produced by this volume charge. For the electron energy 20 keV and beam current density 50 µA/cm^2 the field tension is 30-

50 kV/cm. At such tension the field diffusion of positive glass ions has to influence significantly onto the process kinetics.

Field diffusion of positive ions. Among all positive ions, presented in the glass combinations, those of sodium and silver are the most movable ones [45]. Spatial distribution of these ions under electron irradiation was determined by solving the field diffusion equation with the account for the jump ion conductivity. The calculation was carried out for the specimen temperature, equal to 150 °C. The results of calculation are shown in the Fig.16. One has to note that the presented curves have the qualitative character, because the values of the coefficients of thermal and field diffusion of sodium and silver ions, presented in literature, are published as the approximate ones, and strongly depend upon the specific glass composition. In addition, the calculation did not took into account the arise in the glass of the opposite directed electric field, related to the negative volume charge, produced by the non-moving negative ions of oxygen and halides.

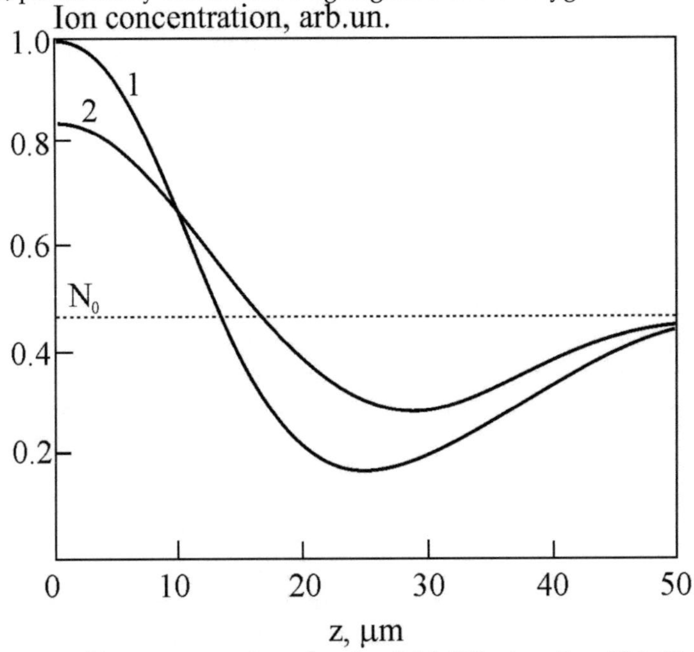

Fig. 16. Distribution of ion concentration after the field diffusion. E_0 = 20 keV. t = 1000 s. 1 - Na^+; 2 - Ag^+. N_0 – initial concentration of ions

One can see from the Figure that the presence of the negative volume charge nearby the glass surface leads to spatial redistribution of sodium and silver ions - nearby the glass surface is formed the region with the high ions' concentration and in the depth - the range with the reduced concentration. Field diffusion increases significantly the concentration of silver ions and atoms in the surface layer of the glass. This is an explanation of the high concentration of silver nanoparticles nearby the surface which is observed in the experiment. Redistribution of silver and sodium ions leads to modification of glass chemical combination in the surface layer and produces the significant mechanical stresses. These two factors can be the reason of silver nanoparticles distribution in thin layers. Other possible reason of layered structures production can be the electric field arising in the glass between the non-moving negative ions and positive ions, which have moved towards the glass surface. When the tension of this field in the glass layer becomes equal to tension of the field, produced by the volume charge of electrons, the diffusion of positive ions in this layer is stopped, but it continues in the gap between this layer and glass surface. At some moment of time the compensating field is produced already at this gap, stopping thus the diffusion etc.

So one can outline the following special features of silver nanoparticles formation in glass under the electron irradiation and the factors, influencing onto this process:

1. The processes of silver ions reduction and silver atoms ionization by electron beam;
2. Formation of volume charge and of electric field with high tension;
3. The processes of field diffusion, increasing the concentration of silver and sodium ions nearby the glass surface;
4. Modification of chemical composition of the glass in its surface layer;
5. Producing the mechanical tensions.

Summarizing one can outline the conclusion that formation of silver nanoparticles in PS glasses by electron irradiation is a prospective method of formation the nano-size metallic structures on the glass surface and nearby it for nanoplasmonics and integrated optics purposes. Easy combining of this method with the electron lithography technology makes it possible to create on the glass the metallic elements of complicated configuration, which can be used for methamaterials production. At the same time the processes, which take place during electron irradiation and consequent thermal processing are still not investigated in details and require further studies.

5. References

1 Weyl W.A. «Coloured Glasses», Sheffield. 1951, 400 p

2. Polavarapu, L.; Qing-Hua Xu; Mohan S.; Dhoni M.S. & Ji, W. (2008) Optical limiting properties of silver nanoprisms Appl. Phys. Letter. 92, 263110

3. Nashchekin, A.V.; Sidorov, A.I.; Usov O.A.; Turoverov K.K.; Tsekhomsky V.A., Kurbatova N.V.; Podsvirov O.A.; Nikonorov N.V.; Nevedomskii V.N. & Sokolov R.V. (2009) SPR biosensors based on Ag nanoparticles in photothermochromic glasses. Topical Problems of Biophotonics (TPB)-2009 2nd Int.Symp., N.Novgorod, Proceedings, pp. 147-148.

4. Borrelli, N.F. & Hall, D.W., in: Ulhmann, D.R. & Kreidl N.J. (Eds.) (1991) Optical Properties of Glass, American Ceramic Society, Westerville, OH p. 87

5. Wang, Y.H.; Peng, S.J.; Lu, J.D.; Wang R.W.; Cheng,Y.G. &. Maob Y.L. (2009) Nonlinear optical properties of Ag nanocluster composite fabricated by 200 keV negative ion implantation. Vacuum 83 pp 412–415.

6. . Beregenoy, A.I. (1966) Sitalls and photositals, M., 348..

7. Pierson, J.E. & Stookey, S.D. (1977) Photosensitive colored glasses. Patent USA N 4017318.

8. Glebov, L.B., Nikonorov, N.V., Panisheva, E.I., Petrovsky, G.T., Savvina, V.V., Tunimanova, I.V. and Tsekhomsky, V.A., (1990) Multichrom glasses - the new material for volume phase hologram recording. Report of AS USSR., 314 4, pp 849-853

9. Kefeli, A.A., (1947) Influence the haloid salts addition on spectral absorption of the glass colored with cobalt. Report of AS USSR, 58 6, pp 1051-1054.
10. Espiau de Lamaestre, R.; Bea, H.; Bernas, H.; Belloni J, &. Marignier, J. L. (2007) Irradiation-induced Ag nanoclaster nucleation in silicate glasses: Analogy with photography. Phys. Rev. 76, pp 205431-1-18
11. Panisheva, E.I.; Solov'eva, N.D. & Tunimanova, I.V. (1993) The influence UV- and γ-irradiation on polychromatic glasse. Fiz. Khim. Stecla 19 1, pp 109-116.
12. Paje, S.E.; Garcia, M.A.; Villegas, M.A. & Llopis, J. (2000) Optical properties of silver ion-exchanged antimony doped glass J.Non-Cryst. Solid, 278 pp 28-136.
13. Latishev, A.N. & Molotsky, I.K., (1969) The theory of thin structure of absorption spectra photo painted silver haloid. J .Sc. A. Ph and M. 14 4, pp264-270.
14. Mott N.F. & Gurney R.W. (1940, Electronic Process in Ionic Crystals, Oxford.
15. Klimov, V.V. (2009) Nanoplasmonic, M. Phismatlit, p 489
16. Nacharov, A.P.; Niranorov N.V.; Sidorov A.I. & Tsekhomsky V.A. (2008) Influence of UV irradiation and heat treatment on morphology of the silver nanoparticles in the photothermorefractiv glasses. Glass Phys. Chem.. 34, 6, pp. 693–699.
17. Kreibig, U. (1976) Small silver particles in photosensitive glass: Their nucleation and growth. Appl. Phys. 10, pp 255-264
18. Hövel, H.; Fritz, S.; Hilger, A. & Volmer, M. (1993) Width of Cluster Plasmon Resonances: Bulk Dielectric Functions and Chemical Interface Dumping. Phys. Rev. B: Condens. Matter 48 24, pp 18178–18188.
19. Uchida, K.; Kaneko, S.;Omi, S; Hata, C.; Tanji, H.; Asahara, Y.; Ikushima, A.J.; Tokizaki, T. & Nakamura, A., (1994) Optical Nonlinearities of a High Concentration of Small Metal Particles Dispersed in Glass: Copper and Silver Particles, J. Opt. Soc. Am. B 11 7, pp 1236–1240
20. Kyoung, M. & Lee, M. (1999) Nonlinear Absorption and Refractive Index Measurements of Silver Nanorods by Z-Scan Technique. Opt. Commun., 171 11, pp 145–148
21. Polizzi, S.; Armigliato, A.; Riello, P.; Borrelli, N.F. & Fagherrazzi, G. (1997) Redrawn Phase-Separated Borosilicate Glasses: A TEM Investigation, Microsc., Microanal.,Microstruct., 8, pp. 157–161.
22. Magruder, R.H.,Witting, J.E. & Zuhr, R.A., (1993) Wavelength Tunability of the Surface Plasmon Resonance of Nanosize Metal Colloids in Glass, J. Non-Cryst. Solids 163 2, pp. 162–168.
23. Stookey S. D.; Beall G. H. & Pierson J. S. (1978) Polychromatic glass. J. Appl. Phys. 49 10, p 5114
24. Qiu, J.; Miura, K. & Hirao, K. 2008. Femtosecond laser-induced micro features in glasses and their applications. J. Non-Cryst. Solid 354 12-13, pp1100–1111
25. Stalmashonak, A.; Graener, H. & Seifert, G. (2009) Transformation of silver nanospheres embedded in glass to nanodisks using circularly polarized femtosecond pulses. Appl. Phys. Letter 94, pp 193111
26. Podlipensky, A; Abdolvand A.; Seifert G. & Graener H. (2005) Femtosecond laser assisted production of dichroitic 3D structures in composite glass containing Ag nanoparticles. Applied Physics A: Materials Science & Processing 80 8, pp 1647-1652

27. Unal, A.; Stalmashonak, Seifert, G. & Graener, H. (2009) Ultrafast dynamics of silver nanoparticle shape transformation studied by femtosecond pulse-pair irradiation. Phys. Rev. B 79, pp115411.
28.Stalmashonak A.; Seifert, G. & Graener H. (2009) .Spectral range extension of laser-induced dichroism in composite glass with silver nanoparticles. J. Opt. A, Pure Appl. Opt. 11, 065001.
29. Boren, C.F. & Hafman, D.R. (1983) Absorption and scattering of light by small particles, John Willey & Sons, New York, 664.
30. Nikonorov N.V.;.Sidorov A.I. ; Tsekhomsky, V.A. & Lasareva K.E.(2009) Effect of dielectric shell of a silver nanoparticle on the spectral position of plasmon resonance of the nanoparticle in photochromic glass. Optics and spectroscopy 107 5, pp 705-707
31. Neeves, A.E., & Birnboim, M.H. (1989) Composite structures for the enhancement of nonlinear-optical susceptibility. JOSA B 6 4, pp787-796.
32. Chen, S.; Akai T.; Kadono, K., & Yazawaa T. (2001) Reversible control of silver nanoparticle generation and dissolution in soda-lime silicate glass through x-ray irradiation and heat treatment. Appl. Phys. Letter. 79 22, pp 3687-3690
33. Dotsenko A.V., Glebov L.B., & Tsekhomsky V.A. (1998). "Physics and Chemistry of Photochromic Glasses". CRS Press New York, p.190.
34. Nikonorov, N.V.; Panisheva, E.I.; Tunimanova, I.V. & Chucharev, A.V. (2001) Effect of the glass composition on the refraction index variation at the photo induced crystallization. Phys. Chem. Glass 27 3, pp365-376.
35.Glebova L.; Lumeau J.; Klimov, M; Zanotto, E.D. & Glebov L.B. (2008) Role of bromine on the thermal and optical properties of photo-thermo-refractive glass. J. Non-Cryst. Solids 354 2-9, pp 456-461
36. Chakraborty P. (1998) Metal nanoclasters in glasses as non-linear photonic materials. J. Mater. Sci. 33, pp 2235-2249.
37. Stepanov A.L., Ganeev R.A., Ryasnyansky A.I., & Usmanov T. (2003) Non-linear optical properties of metal nanoparticles implanted in silicate glass Nucl. Instr. and Meth. In Phys. Res. B 206, pp 624-628.
38. Gedeon O., Zemek J., & Jurek K. (2008) Changes in alkali-silicate glasses induced with electron irradiation J. Non-Cryst. Sol. 354, pp 1169-1173.
39. Jiang N., Qiu J., Gaeta A.L., & Silcox J. (2002) Nanoscale modification of optical properties in Ge-doped SiO2 glass by electron-beam irradiation Appl. Phys. Lett. 80 11, pp 2005-2007.
40. Jiang N., Qiu J., & Spence J.C.H. (2005) Precipitation of Ge nanoparticles from GeO2 glasses in transmission electron microscope Appl. Phys. Lett. 86 143112.
41. Jiang N., Qiu J., & Spence J.C.H. (2007) Precipitation of nanocrystals in glasses by electron irradiation: An alternative path to form glass ceramics? Appl. Phys. Lett. 90 1161909.
42. Nashchekin A.V., Usov O.A., Sidorov A.I., Podsvirov O.A., Kurbatova N.V., Tsekhomsky V.A., & Vostokov A.V. (2009) SPR of Ag nanoparticles in a photothermochromic glasses. Proc. SPIE. 7394, p 73942J.

43. Vostokov A.V., Ignatiev A.I., Nikonorov N.V., Podsvirov O.A., Sidorov A.I., Nashchekin A.V., Sokolov R.V., Usov O.A., & Tsekhomskii V.A. (2009) Effect of electron irradiation on the formation of silver nanoclusters in photothermorefractive glasses Techn. Phys. Lett. 35 9, pp 812-814.
44. Quinten M., Kreibig U. (1993) Absorption and elastic scattering of light by particle aggregates Appl. Opt.. 32 30, pp 6173-6182.
45. Tervonen A., Honkanen S., & Leppihalme M. (1987) Control of ion-exchanged waveguide profiles with Ag thin-film sources J. Appl. Phys. 62, pp 759-763.

11

Silver nanoparticles: sensing and imaging applications

Carlos Caro[1], Paula M. Castillo[1], Rebecca Klippstein[2], David Pozo[2] and Ana P. Zaderenko[1]
[1]*Department of Physical, Chemical and Natural Systems, Pablo de Olavide University*
[2]*CABIMER-Andalusian Center for Molecular Biology and Regenerative Medicine CSIC-University of Seville-UPO-Junta de Andalucia*
Spain

1. Introduction

Recent advances in nanotechnology have allowed the development of robust, and highly sensitive and selective detection methods that are expected to address some deficiencies of conventional detection technologies. Within this context gold and silver nanoparticles have emerged as a powerful tool in sensing and imaging applications due to their surprising optical properties.

Although silver exhibits many advantages over gold, such as higher extinction coefficients, sharper extinction bands, higher ratio of scattering to extinction, and extremely high field enhancements, it has been employed far less in the development of sensors, with the exception of sensors based on surface enhanced spectroscopies. The reason for this is the lower chemical stability of silver nanoparticles when compared to gold. Nevertheless, recent developments include means of protecting efficiently silver nanoparticles that offer far improved chemical stabilities. As a consequence, silver nanoparticles are rapidly gaining in popularity and several research groups have begun to explore alternative strategies for the development of optical sensors and imaging labels based on the extraordinary optical properties of these metal nanoparticles.

In the present chapter, we will focus on recent developments regarding silver nanoparticles and their emerging sensing applications.

2. Surface plasmon

Noble metal nanoparticles display unique optical properties that differentiate them form their bulk counterparts. Probably the most fascinating finding, regarding this peculiar optical performance, is that they often exhibit strong extinction bands in the visible spectrum, and therefore bright and gaudy colours, that are not present in the spectrum of the bulk metal. Although these colours are reminiscent of molecular dyes, it is important to emphasize that their fundamentals are different. While the spectra of molecular dyes can be understood only in terms of quantum mechanics, the extinction spectra of metal nanoparticles can be treated in terms of classical electromagnetism. Moreover, metal nanoparticles also scatter light with high efficiency and, unlike molecular dyes, their extinction spectra are really a combination of both absorption and scattering (Kelly et al., 2002). The interaction of the oscillating electromagnetic field of the light with metal nanoparticles, results in the collective coherent oscillation of the metal conduction electrons with respect to the nanoparticle positive lattice. At a particular frequency of the light this process is resonant, receiving the name of *Localized Surface Plasmon Resonance*, LSPR (Figure 1), and is the responsible of the strong extinction band exhibited by the nanoparticle. Additionally to the extremely high molar extinction coefficients and resonant Rayleigh scattering, LSPR also results in enhanced local electromagnetic fields near the surface of the nanoparticle (Novotny & Hecht, 2006).

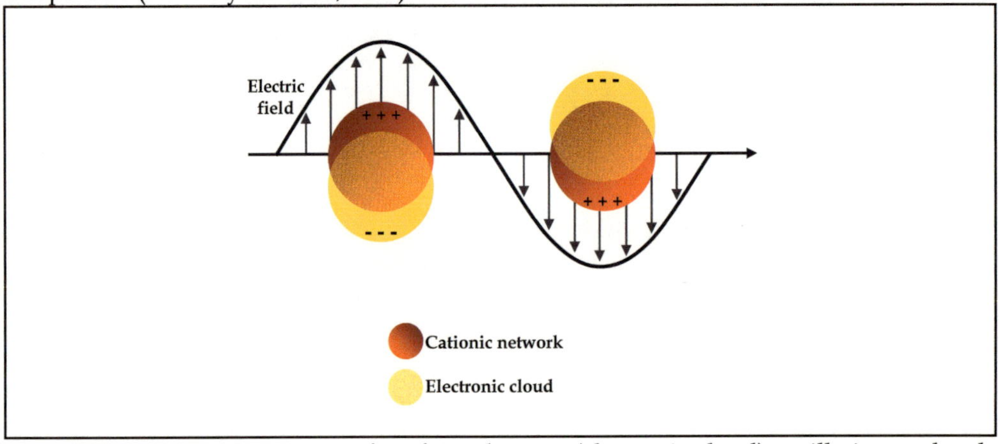

Fig. 1. Schematic representation of surface plasmon (electronic cloud) oscillation under the effect of an electromagnetic field

The first theoretical approach for modelling the optical properties of nanoparticles proposed by Mie, within classical electromagnetic formalisms, is still in common use today. According to the Mie theory, the resonance condition is achieved when the real part of the dielectric function of the metal equals the dielectric function of the surrounding medium (Mie, 1908). Therefore, the LSPR frequency depends both on the nanoparticle itself and on the medium where it is dispersed. Two important consequences arise from this dependency, on the one hand, the LSPR is tunable, i.e., its frequency can be modified through changes in the nanoparticle composition, size and shape (Kelly et al., 2003). On the other hand, metal nanoparticles are sensitive to their local environment, i.e., changes in the dielectric properties of their surroundings results in LSPR shifts that can be measured. Both tunability and sensitivity of LSPR convert metal nanoparticles in materials of choice for optical sensing and imaging applications. The most suitable metals are silver and gold, since the localized plasmon resonance condition mentioned above is satisfied at visible light frequencies. Additional advantages of these metal nanoparticles include simple preparation methods for a wide range of sizes and shapes and easy surface conjugation to a variety of ligands.

Over the last decade, several types of sensors have been developed on the basis of the plasmonic properties of noble metal nanoparticles i.e., extremely high molar extinction coefficients and resonant Rayleigh scattering on one side, and enhanced local electromagnetic fields near the surface of the nanoparticle on the other. These properties are inherent to a given detection mechanism and a given detection technique, enabling the classification of sensors in two main groups depending on the type of interaction involved, and therefore measured, between the metal nanoparticle and the analyte molecule. The first group encompasses sensors involving LSPR frequency shift, due to the interaction between nanoparticle and target molecule. Within this group, two different sensors may be distinguished, depending on the origin of LSPR changes: aggregation sensors and refractive index sensors. In aggregation sensors the LSPR shift is due to the plasmon coupling of nanoparticles in close proximity, in refractive index sensors the LSPR shift is due to changes in the local refractive index of the medium. The second main group of sensors is based on the enormous electromagnetic field enhancement in the vicinity of noble metal nanoparticles, which results in the so called surface enhanced spectroscopies, such as Surface Enhanced Raman Spectroscopy (SERS) and Metal Enhanced Fluorescence (MEF). This simple classification scheme is resumed in Table 1 that collects also the associated measurement techniques.

Sensor	Mechanism		Measurement/technique
Aggregation	LSPR-shift origin:	near-field electromagnetic coupling	Extinction/UV-Vis spectroscopy
Refractive Index		local refractive index changes	Extinction/UV-Vis spectroscopy Elastic Scattering/Dark-field microscopy
SERS	local electric field enhancement		Inelastic Scattering/Raman spectroscopy
MEF			Fluorescence

Table 1. Optical plasmonic sensors

In the following sections of this chapter, the principles of these sensors will be introduced and recent application examples for silver nanoparticles will be given.

3. Chemical sensing based on silver nanoparticles LSPR shift

3.1. Chemical sensing based on interparticle plasmon coupling: Aggregation sensors

The enhancement of electric fields on the nanoparticle surface, owing to plasmon resonance, decays over a distance on the order of the nanoparticle size. This means that when nanoparticles aggregate, i.e., approach each other within this distance, their fields interact leading to interparticle plasmon coupling and, consequently, to coupling-induced LSPR shift (Kreibig & Vollmer, 1995).

Generally, naked silver nanoparticles are stabilized against aggregation by means of anions (e.g., Cl^- or citrate^{3-}) or polymers (e.g., polyvinilpyrrolidone) absorbed on their surfaces. However, changes in dispersant media such as pH and ionic strength can cause nanoparticle destabilization and aggregation. When the nanoparticle aggregates, its LSPR is red-shifted and broadened (Figure 2). This effect has been exploited in the design of so-called aggregation sensors that form the basis of simple, highly sensitive and low cost colorimetric assays, which have been applied to the detection of small molecules, DNA, proteins, toxic metal ions and pollutants.

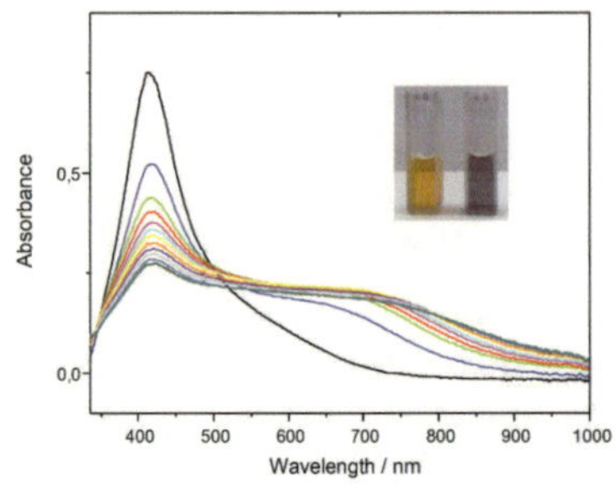

Fig. 2. Absorbance spectra of silver nanoparticles in aqueous solution before (black line) and after aggregation due to increasing salt concentrations (coloured lines). The suspension colour changes from yellow to blue.

In some cases, the proper analyte of interest is able to promote nanoparticle aggregation. An interesting example of this type of analyte-induced detection system has been recently described by Huang and co-workers. The authors have established a colorimetric analytical method for the detection of an anti-inflammatory drug, berberine hydrochloride, by means of citrate stabilized silver nanoparticles. Citrate stabilized silver nanoparticles exhibit a negatively charged surface rendering yellow stable dispersions in water due to the electrostatic repulsion between nanoparticles. In the presence of positively charged berberine hydrochloride, nanoparticles aggregate and suspension colour changes from yellow to green or blue, depending on the aggregation degree. By adjusting pH and ionic strength values of the medium, a good correlation between the absorbance change at the plasmon wavelength of the nanoparticle, and the berberine concentration was achieved, with detections limits of 1.3×10^{-8} M. Assay reliability was validated in berberine contained in commercially available drugs (Ling et al, 2008).

However, this method is restricted to cases where the analyte of interest is able to induce nanoparticle aggregation and no other substance present in the sample interferes by aggregating the nanoparticle. To overcome these limitations, it is necessary to endow aggregation sensors with selectivity. For this purpose, molecular recognition events have been employed to achieve selective colorimetric assays. The general strategy consists in functionalizing the silver nanoparticle surface with a monolayer of molecular recognition ligands. Aggregation is induced by specific recognition and binding to the target (Figure 3). The plasmon band is broadened and red-shifted as a function of the aggregation degree; therefore the change in absorbance is related to the target concentration.

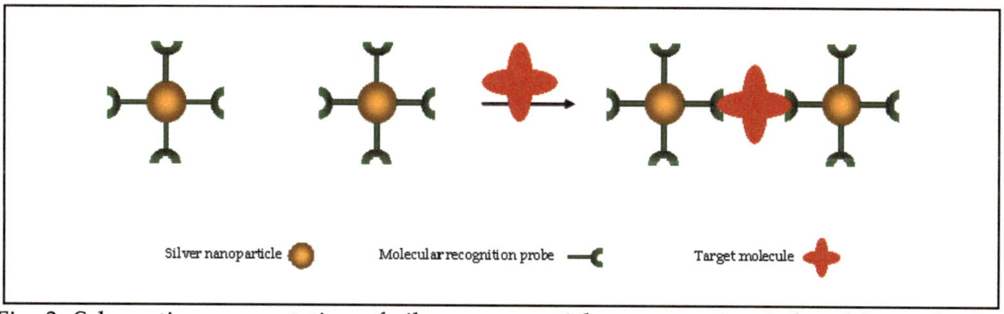

Fig. 3. Schematic representation of silver nanoparticles aggregation induced by molecular recognition events.

The hybridization properties of DNA make this biomolecule an especially suitable candidate in aggregation sensing applications. Silver nanoparticles conjugated to oligonucleotides have recently emerged as powerful tools for the detection of target DNA sequences, and have been used in the design of colorimetric assays based on aggregation induced by sequence-specific hybridization. Different conjugates nanoparticle-oligonucleotide have been proposed to this end (Chen et al., 2004; Liu et al., 2005; Lee et al., 2007).

Additionally, when gold (Jin et al., 2003) and silver (Thompson et al., 2008) nanoparticles are conjugated to oligonucleotides, sharper melting transitions are observed as compared to free oligonucleotides, allowing the differentiation between complementary strands and strands with one or more mismatches by both colorimetric and melting curve analysis. Taking advantage of this property, Mirkin and co-workers developed a "dual" detection method for single-nucleotide-polymorphisms. The authors employed two different types of probe systems, Au and Ag/Au core shell nanoparticles, which offer two different colour signatures for target DNA-directed colorimetric detection. The metal nanoparticles were conjugated to oligonucleotide probe strands, wild-type and single nucleotide mutant, and hybridized to the corresponding DNA targets. After hybridization, nanoparticles aggregated, and these aggregates could be dispersed again by heating them above the melting temperature of the duplex DNA interconnects. Mismatched hybridization, e.g. wild type strand functionalized nanoparticle-mutant DNA target, displayed lower melting points than their matching counterparts, and could be easily differentiated. As in this approach two different colour pairs are available, the method provides a cross-check, enhancing the reliability of the determination (Cao et al., 2005). Further improvement using silver nanoparticles offered higher sensitivity, due to the greater extinction coefficient of silver nanoparticles when compared to their gold analogues (Thompson et al., 2008). Anyway, with the achieved sensitivity, a PCR amplification step is still compulsory.

A completely different strategy, where the molecular recognition is performed away from the nanoparticle enabling the use of naked nanoparticles without loss of selectivity, has been very recently reported (Kanjanawarut & Su, 2009). In this strategy, charge neutral peptide nucleic acids (PNA) are used both as hybridization probe and to induce aggregation of citrate stabilized metal nanoparticles. In the absence of a complementary target DNA, PNA molecules, which remain free in solution, are able to coat nanoparticles inducing aggregation and, consequently, the colour changes in the colloidal solution. When a complementary DNA is present, PNA–DNA complexes are formed and nanoparticles do not aggregate, due to the stabilizing effect of the negative charges of the DNA strands in the

PNA−DNA complexes adsorbed on the nanoparticle surface. Nanoparticle behaviour could be easily followed by UV-Vis spectroscopy. This strategy enabled even single-base mismatch discrimination by means of adding NaCl to accelerate the aggregation process of the nanoparticles. Although both silver and gold citrate capped nanoparticles were employed to carry out the assay, silver rendered higher sensitivity.

Silver aggregation sensors are not restricted to DNA detection, and protein recognition has also been demonstrated. For example, carbohydrate binding proteins have been detected with carbohydrate stabilized nanoparticles. The tetrameric lectin concanavalin A (Con A) was detected by means of silver nanoparticles functionalized with mannose. Each Con A subunit contains a carbohydrate binding site that binds to a mannose ligand on the nanoparticle surface inducing nanoparticle aggregation (Schofield et al., 2006). Conversely, Con A can be utilized to detect a carbohydrate. Silver nanoparticles functionalized with dextran undergo aggregation in the presence of Con A, owing to the interaction of dextran with Con A. Competitive binding of glucose to Con A, is able to restore nanoparticles to solution, permitting the colorimetric detection of glucose. Although the detection limit obtained was below the physiological level of glucose, the specificity of this approach is not nearly sufficient for a diagnostic assay (Zhang et al., 2004a).

Aggregation based chemical sensing has also been used to monitor enzymatic reactions. Unmodified Ag nanoparticles have recently been used to develop a colorimetric assay to detect enzymatic reactions where adenosine triphosphate (ATP) is consumed, such as ATP dephosphorylation by calf intestine alkaline phosphatise and peptide phosphorylation by protein kinase. The assay relies on the capacity of ATP to protect the Ag nanoparticles against salt-induced aggregation. As ATP is consumed in the enzymatic process, nanoparticles begin to aggregate (Wei et al., 2008).

Other applications of protein functionalized silver nanoparticles that have been proposed include a colorimetric pH sensor based on the principle of pH dependent denaturation of proteins. Conformational changes in cytochrome c in response to pH changes could be measured over a wide range of pH values, as a result of plasmon shift in a system formed by silver nanoplates functionalized with the protein (Park et al., 2009).

A different approach to the molecular recognition events consists in functionalizing silver nanoparticles with complexing agents. Aggregation is achieved by formation of inclusion complexes between the functionalized nanoparticle and the guest molecule. Highly sensitive and selective colorimetric assays have thus been accomplished for the detection of histidine (Xiong et al., 2008; Li et al., 2009), pesticides (Xiong & Li, 2008) and Yb^{3+} ions (Han et al., 2009).

A very promising application of silver nanoparticles has been recently described. By means of the interdependency between the extent of nanoparticle coupling-induced LSPR shift and the distance between the interacting nanoparticles, Alivisatos and co-workers have developed a new class of molecular rulers based on plasmon coupling of single gold and silver nanoparticles (Sonnichsen et al., 2005). These "plasmon rulers" where used to study the dynamics of DNA hybridization on the single molecule level, with notorious advantages when compared to conventional rulers based on Fluorescent Resonance Energy Transfer technique (FRET), e.g., scattering measurements offers better sensitivity and lower background signal than FRET, unlike reporter dyes used in FRET, the plasmon ruler neither blinks nor bleaches and affords upper distance limits of 70 nm as compared to approximately 10 nm for FRET.

3.2. Chemical sensing based on plasmon shifts with local refractive index

The silver nanoparticle LSPR is extremely sensitive to changes in local refractive index induced by analyte binding at or near the nanoparticle surface. The nanoparticle response to these changes can be measured as a LSPR shift in the UV-Vis extinction spectrum. This overall LSPR shift depends on surface coverage, i.e., on analyte concentration, allowing the development of refractive index sensors for quantitative determinations. Although other noble metals, such as gold, also exhibit high sensitivity to the local refractive index, the narrower plasmon bandwidth found in silver nanoparticles permits more accurate measurements of the LSPR shift (Lee et al., 2006).

As mentioned above, the extinction spectra of silver nanoparticles are really a combination of both absorption and scattering, and therefore the response of the LSPR to changes in local refractive index can also be measured by means of elastically scattered light. The Rayleigh scattering efficiency of a silver nanoparticle is 10^6 times higher than the fluorescence efficiency of a fluorescent dye molecule (Yguerabide et al., 1998), with the additional advantages that, unlike fluorescent probes, silver nanoparticles neither blink nor bleach. This extremely high scattering quantum yield of silver nanoparticles enables their use in highly sensitive imaging applications using dark-field optical microscopy, e.g. for imaging single living cells (Xu et al., 2004) as well as single receptor molecules on single living cells (Huang et al., 2007).

Apart from high sensitivity, selectivity for the analyte is also required to enable the use of silver nanoparticles in a detection system. For this purpose, the silver nanoparticle surface has to be tailored in such a way that only the target of interest binds to it. As in the previously described aggregation sensors, this can be achieved by functionalizing the nanoparticle surface with ligands capable of specifically binding the target molecule therefore eliminating non-specific surface adsorption (Haes & Van Duyne, 2003; Malinsky et al., 2001). Over the last decade, many functionalization protocols have been developed, nevertheless the most popular is the one based on using a thiolated intermediate linker or directly a thiolated derivative of the ligand. Such a popularity of thiolated compounds is mainly due to their capacity to form stable metal-sulphur bonds with the nanoparticle surface atoms under mild conditions. Moreover, silver nanoparticles modified with alkanethiol self-assembled monolayers exhibit extremely high short-range refractive index sensitivities (Malinsky et al., 2001; Sherry et al., 2006). The incorporation of different functionalities into the terminal position of the alkanethiol provides an easy way to tailor the chemical properties of the nanoparticle surface. These tailored silver nanoparticles have emerged as a powerful tool in the study of adsorption and binding events in biological systems (Ostuni et al., 1999), and have been extensively used in the design of refractive index sensors.

Basically, a refractive index sensor is any device that operates by transducing small changes in local refractive index in the silver nanoparticle surroundings into a measurable wavelength shift response. A widely used configuration for refractive index sensors consists in a surface covered with silver nanoparticles that are coated with a recognition layer designed to bind a specific target molecule.

As first demonstrated by Van Duyne and co-worker, surface-confined triangular silver nanoparticles fabricated by nanosphere lithography on a glass substrate, render highly sensitive refractive index sensors. The recognition layer was constituted by a self-assembled mixed monolayer of alkanethiol molecules, comprising a thiolated compound with a terminal carboxylic acid group that could be used to attach biotin by common coupling chemistry. As a proof of concept, the biotinilated affinity sensor was assayed with streptavidin. Biotin-streptavidin binding induced a plasmon band shift, which was dependent on the streptavidin concentration. Exposure to 100 nM streptavidin, for example, rendered a LSPR red shift of 27 nm. After validation, its use was extended to the more realistic biotin-anti-biotin system with lower binding affinity typical of immunoassays. Limits of detection of 1 pM (streptavidin) and 100pM (anti-biotin) were obtained (Haes & Duyne, 2002). By using distinct ligand modified self-assembled monolayers to functionalize the triangular silver nanoparticles, this affinity sensor scheme has been extended to the optical detection of carbohydrate binding proteins (Yonzon et al., 2004), anti-amyloid-derived diffusible ligands focussed on the diagnosis of Alzheimer's disease (Haes et al., 2004) and small protein toxins (Zhu et al., 2009).

Interestingly, it has been reported that when chromophores are used as ligands, amplified spectral response to target binding can be achieved. For this to happen, the LSPR frequency of the silver nanoparticles has to be close to the molecular resonance of the chromophore, so that they couple (Haes et al., 2006; Zhao et al., 2007). This enhanced sensitivity has been recently applied to detect the binding of low molecular weight substrates and inhibitors (Zhao et al., 2008) as well as different drugs (Das et al., 2009) to human cytochrome P450.

Although the previous examples in this section were based on triangular nanoparticles, many other nanoparticle shapes are usable in refractive index sensors, such as rhombic silver nanoparticles (Zhu et al., 2008), and spherical silver nanoparticles (Gish et al., 2007). Groves and co-workers have recently developed a label-free LSPR sensor constituted by silver nanocubes covered with a self-assembled alkanethiol monolayer and interfaced with a glass-supported model membrane. This sensor was used to monitor and quantify static and dynamic protein binding to the membrane (Galush et al., 2009).

Refractive index sensors are also highly interesting for pollutant detection, for example, very recently a sensor based on silver colloidal nanoparticles has been reported for the detection of volatile organic compounds (Chen & Lu, 2009).

Further exploration of the potential of silver nanoparticles in refractive index sensors showed that a single silver nanoparticle monitored by dark-field microscopy can be used to sense local refractive index changes (McFarland & Van Duyne, 2003). Moreover, single silver nanoparticles functionalized with a single ligand molecule have demonstrated their capacity to constitute an optical sensor by themselves. These "single nanoparticle sensors" permit the target detection at the single molecule resolution and opens the door to studying single molecular interactions of chemical functional groups on the surface of nanoparticles. Sensing and imaging of a single human cytokine molecule, tumor necrosis factor-α (TNFα), have been recently achieved by employing small silver nanoparticles (2.6 nm) functionalized with a single antibody molecule. Such as small nanoparticle size provides intrinsic single molecule detection volumes per nanoparticle, affording high sensitivity and a notably wide dynamic range (0-200 ng/mL) suitable for diagnostic purposes in diseases inducing increased TNFα levels (Huang et al., 2008).

Other examples involving silver nanoparticles in sensing and imaging applications associated to dark field microscopy include the development of a microarray-based DNA hybridization assay to screen for a given polymorphic site in the breast cancer gene BRCA1 (Oldenburg et al., 2002), the study of the action of antibiotics upon the cell membrane of living bacterial cells (Kyriacou et al., 2004; Xu et al., 2004) and imaging nanoparticle binding to fungi (Weinkauf & Brehm-Stecher, 2009).

4. Chemical sensing based on inelastic light scattering: SERS

Raman scattering is an extremely inefficient process. When a molecule interacts with visible light, most of the light is absorbed and only a small fraction is inelastically scattered to render the Raman spectrum of the molecule. As a consequence, Raman cross sections are extremely small, less than $10^{-29}cm^2sr^{-1}$, i.e., about 10 orders of magnitude smaller than that of infrared absorption, and 14 orders of magnitude smaller than that of fluorescent dye molecules. This means that the detection sensitivity of Raman spectroscopy is intrinsically low and, in order to achieve the high sensitivity required in biological or pollutant samples, the scattering intensity should be greatly increased.

Nevertheless, when the Raman spectrum is registered for a molecule located in the vicinity of a metal nanoparticle, or a metallic surface with nanometric roughness, the Raman cross section can be amplified dramatically. This enhancement effect is known as *Surface Enhanced Raman Scattering* or Surface Enhanced Raman Spectroscopy (SERS). SERS is probably one of the most powerful techniques currently available for sensing applications since, additionally to an extremely high sensitivity, it provides valuable structural information on the analyzed molecule.

Although SERS effect has not been completely elucidated, two enhancement mechanisms are generally recognized to contribute to SERS effect, the electromagnetic mechanism and the chemical mechanism. In the electromagnetic mechanism, local electric fields in the surroundings of the metal nanoparticle are enhanced due to the surface plasmon excitation, leading to more intense electronic transitions in molecules located near the nanoparticle, and enhanced Raman scattering (Schatz et al, 2006). The chemical mechanism consists in changes in the polarizability of the molecule, owing to charge transfer interaction between electronic states of the molecule and the metal nanoparticle surface, which results in increased Raman signals (Otto & Futamata, 2006). Even though these two mechanisms contribute to the enhancement, the electromagnetic one plays a major role when compared to the chemical one which is much smaller. Moreover, whereas the chemical mechanism depends on the properties of the analyte molecule, the electromagnetic mechanism only depends on tunable

optical properties of the metal nanoparticles that can be optimized in order to obtain higher SERS enhancements. In this sense, electromagnetic mechanism plays a key role in the development of improved SERS substrates, and current efforts directed to optimize nanostructures will be further discussed within this section.
The most active metal in SERS is silver, followed by gold and copper, in that order. SERS enhancement factors as high as 10^{14}-10^{15} have been described for silver nanoparticles (Kneipp et al., 1997; Xu et al., 1999; Futamata el al., 2002), enabling detection limits as low as a single molecule (Nie & Emory, 1997; Kneipp et al., 1997; Le Ru et al., 2006). The enormous sensitivity achieved in SERS permits the ultrasensitive and ultraselective detection of biomolecules (Ji et al., 2005; Fabris et al., 2008) as well as pollutants (Guerrini et al., 2009a; Abalde-Cela et al., 2009).

The IR signals of molecules nearby silver nanoparticles are also enhanced, giving rise to *Surface Enhanced Infrared Absorption Spectroscopy* (SEIRAS). However, SERS presents the considerable advantage over SEIRAS that aqueous samples may be examined, due to the very weak Raman scattering of water as compared to strong IR absorption. This is extremely significant for biological samples.

The largest Raman scattering enhancements have been described for molecules residing in gaps of a few nanometers between aggregated colloidal silver nanoparticles, the so called "Hot Spots" (Jiang et al., 2003; Michaels et al., 2000). This is attributed to plasmon-coupling between nanoparticles in close proximity (Su et al., 2003; Atay et al., 2004; Fromm et al., 2004; Jana & Pal, 2007). Other factors concerning the nanostructure have also been claimed to influence the enhancement effect, such as the presence of sharp edges and the LSPR coupling to Raman excitation source. On the basis of these considerations, a great deal of the current research effort in SERS focuses on the development of improved SERS active substrates for analytical purposes, by means of controlling composition, size, shape, and interparticle spacing of nanoparticles and their assemblies (Lin et al., 2009).
From the numerous methods actually available to fabricate SERS substrates, aggregation of colloidal nanoparticles is still preferred in applications requiring extremely high sensitivity. Indeed, silver nanoparticle aggregates have been found to be responsible for single-molecule detection in SERS, due to their high content in hot spots (Nie & Emory, 1997; Kneipp et al., 1997).
Aggregation can be achieved by increasing the ionic strength of the medium, (Jana & Pal, 2007). In this salt-induced aggregation strategy, attention has to be paid to salt election, as far as some ions have been claimed to enhance SERS effect, by means of surface activation, while others can even quench it (Doering & Nie, 2002). In fact, the SERS spectra of DNA and RNA mononucleotides may be obtained with high sensitivity when the silver colloid is aggregated with $MgSO_4$ instead of the more commonly used halide ions. (Bell & Sirimuthu, 2006)
Despite being widely utilized, owing to its simplicity and sensitivity, the aggregation method has several drawbacks such as low reproducibility and the lack of stability of the substrates. Very recently Mejías and co-workers have developed a new method for the obtention of highly reproducible films of silver nanoparticle aggregates that are strongly attached to the substrate and exhibit a high hot spots content (Figure 4). Due to the high stability of this SERS substrate, it can even be reused. Moreover, this substrate has shown its potential in quantitative SERS applications (Caro et al., 2008).

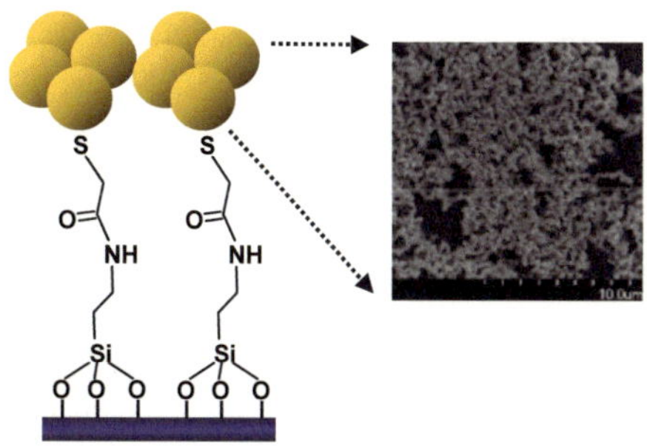

Fig. 4. Schematic representation of the sensor (left). Scanning electron microscopy image of silver nanoparticle aggregates on the sensor surface (right).

Although SERS is a selective technique, given that it provides a vibrational fingerprint of the analyte, distinguishing between closely-related species in complex samples, for example peptides that differ only slightly in sequence, is beyond the Raman discrimination capacity. Furthermore, a necessary condition for obtaining SERS activity is that the analyte molecule possesses affinity for the metal surface to ensure that it will approach the SERS substrate surface. If, to the contrary, the analyte of interest has no affinity for the surface at all, it does not matter how active the substrate is, it will not display SERS effect. These drawbacks have been overcome by coating the nanoparticle surface with a substance that promotes the adsorption or binds specifically to the target molecule. Such surface coated SERS sensors have made possible, for example, the detection of molecules with no affinity for uncoated surfaces, such as polycyclic aromatic hydrocarbons (Guerrini et al., 2009b), and the quantitative *in vivo* glucose measurement (Stuart et al., 2006).

Another interesting example of these modified SERS substrates, consist in a novel and fast method for the qualitative detection of protein kinase activity. To constitute the sensor, silver nanoparticles were deposited on a glass substrate and covalently linked to a target kinase-recognition peptide through a thiol group in the amine terminus of the peptide. After kinase reaction, the covalent attachment of the phosphate group to the peptide tyrosine residue resulted in a prominent change in the Raman spectra, exhibiting a collapse of the 848/828 cm^{-1} doublet to a single peak around 830 cm^{-1}. The reported method proved to be selective even in crude cell lysates (Yue et al., 2009).

In addition to the SERS active substrates, another attractive sensor arrangement has been described, the fiber-optic SERS sensor. In this type of sensor, the substrate is simultaneously part of the excitation and detection mechanism. For example, glass fiber tips coated by silver nanoparticles enabled the recording of spectra of biological samples, such as plant tissues or microbiological cells, with a high spatial resolution and avoiding sample photoinduced destruction due to the considerably reduced laser power required with this sensor (Gessner et al., 2002). In another example, crystal violet molecules, a contaminant in aquaculture, were detected in water far below the ppb detection limit (Lucotti & Zerbi, 2007).

Over the last years a new SERS sensor architecture has emerged as a powerful tool in biological assays, the SERS tags. Briefly, a SERS tag is a chemically encoded labeling agent. The basic structure of the SERS tag is formed by a SERS reporter molecule as labeling agent, i.e., a molecule with a high Raman cross-section, adsorbed on the surface of a silver nanoparticle that can be optionally further encapsulated in a protective shell. Once the basic structure is assembled, it may be conjugated to a recognition element, such as an antibody. In a detection process, the Raman signature of the SERS reporter molecule will reveal whether the labeling agent is present or not. SERS tags represent an excellent alternative to fluorescence-based encoding methods due to the advantages that they present when compared with traditional fluorescence labels, such as narrower emission peaks, increased number of optical signatures, high-level multiplexing, single laser excitation for detection of multiple labels and higher stability.

As mentioned above, a SERS tag can be obtained by simply adsorbing a SERS reporter molecule on the nanoparticle surface, this simple yet effective approach has been employed, for example, to develop a sandwich type immunosensor to detect hepatitis B surface antigen (Ji et al., 2005).

A wide variety of SERS tags have been reported, the most popular being the encapsulated ones, given that encapsulation prevents desorption of the SERS reporter molecule and increases the chemical stability of metal nanoparticles. As encapsulating agent, polymers (Stokes et al., 2006) and silica (Jun et al., 2007) are generally employed. Indeed, silica encapsulated SERS tags are being commercialized.

Exploiting the fact that the SERS effect is stronger in nanoparticle aggregates, Su and co-workers have developed an induced-aggregation method to obtain SERS tags. In this approximation, silver seed nanoparticles are grown, in the presence of appropriate SERS reporters, which are able to induce aggregation. Once the silver clusters reach the desired size, they are stabilized by the addition of bovine serum albumin (BSA). These SERS BSA-coated tags can be easily conjugated to antibodies by direct adsorption, for their use in immunoassays (Su et al., 2005). Covalent binding to antibodies is also possible by employing a cross-linked organic encapsulation with BSA and glutaraldehyde. These SERS tags have been used to map prostate specific antigen in tissue samples (Sun et al., 2007).

A different aggregation strategy has been recently reported by Fabris and co-workers who prepared silver nanoparticle dimers and small aggregates via a dithiol linker, which acts both as a "bridge" between the nanoparticles and as SERS reporter molecule. Further functionalization of these SERS tags with aptamers allowed highly sensitive protein detection (Fabris et al., 2008).

In a complex variation of the SERS tag scheme, Lee and co-workers have conjugated the optical properties of silver nanoparticles with the magnetic properties of iron oxide nanoparticles to obtain SERS encoded multiplex beads that afford both protein separation and identification (Jun et al., 2009). For this purpose sulfonated polystyrene-divinylbenzene beads were successively functionalized with magnetic iron oxide nanoparticles, silver nanoparticles, SERS reporter molecules and, finally, encapsulated with a silica shell. These multiplex beads were tested as separation vehicles on the model system biotin–streptavidin. Multiplex beads retain both magnetic and SERS properties after silica coating and biotin functionalization, offering a wide range of applications in drug screening and combinatorial chemical synthesis. This combination of SERS detection and magnetic separation capacities by means of silver and magnetic nanoparticles, have also been used in a sandwich type immunoassay. In this assay, monoclonal antibody modified silica-coated magnetic nanoparticles, as separation tool, and polyclonal antibody Ag/SiO_2 core-shell nanoparticles embedded with a SERS tag, were employed to bind to target antigen, affording separation and detection. Using this strategy, concentration of human tumor marker α-fetoprotein up to 0.12 μg/ml was detected with a detection limit of 11.5 pg/ml (Gong et al., 2007).

Silver nanoparticles functionalized with SERS reporter molecules exceed the simple molecule detection frontier, and applications such as SERS cellular imaging (Kim et al., 2006; Hu et al., 2007), or intracellular pH sensing (Talley et al., 2004; Wang et al., 2008) are also feasible.

5. Chemical sensing based on metal enhanced fluorescence: MEF

Fluorescence detection still plays a major role in most of the biological assays currently used. Nevertheless, fluorescence detection at the single molecule level has several limitations arising from the use of organic fluorophores, such as low signal to background ratio, poor photostability, and strong photoblinking.

Metal nanoparticles are known to interact with nearby fluorophores affecting their emission intensity. When this interaction results in a fluorescence enhancement, the effect is known as *Metal Enhanced Fluorescence* (MEF). This effect can be understood as arising from two contributions. On the one hand the enhanced emission intensity from the fluorophore can be attributed to the enhanced local fields associated with the metal nanoparticle, as in the case of SERS. On the other hand, the interaction with the nanoparticle LSPR results in increased radiative and nonradiative decay rates and, consequently, increased quantum yield and decreased lifetime (Lakowicz et al., 2002a; Lakowicz, 2005; Cade et al., 2009). Additionally, and as a consequence of the decreased lifetimes, the photostability of the fluorophore increases (Lakowicz et al., 2002b). Considering the potential of MEF, it could therefore improve current fluorescence based techniques.

In contrast to other metals, which typically quench nearby fluorophores (Huang & Murray, 2002), silver nanoparticles enhance the luminescence of fluorophores when they are localized at a distance of 4-10 nm from the silver nanoparticle surface. For example, silver nanoparticles deposited on a glass substrate increase the emission intensity of the common biological label fluorescein by an average of at least three fold (Pugh et al., 2003) and, what is more, fluorescein self-quenching decreases even in highly labeled proteins, permitting the use of ultrabright over-labeled protein as labels (Lakowicz et al., 2003; Lukomska et al., 2004).

Generally, MEF leads to 10-1000 fold intensity enhancements. Some factors affecting the enhancement degree have been described, such as the distance between the silver nanoparticle and the dye (Kühn et al, 2006; Anger et al., 2006), silver nanoparticle aggregation (Zhang et al, 2004b; Zhang et al, 2007a) and the spectral overlap between the LSPR band of the silver nanoparticle and the emission band of the attached fluorophore (Chen et al., 2007).

The detection of DNA hybridization is of interest in numerous biological applications such as microarrays and PCR. To this end, Lakowicz and co-workers have studied in depth the MEF effect applied to DNA. They have found that silver nanoparticles enhance the intrinsic fluorescence of DNA (Lakowicz et al., 2001). Further research demonstrated that when DNA was labeled with a fluorescent probe, silver nanoparticles improved the fluorophore photostability (Malicka et al., 2002). The increased brightness of the fluorophore near silver islands enabled the measurement of the DNA hybridization kinetics as well as the development of the first MEF-based DNA hybridization assay (Malicka et al., 2003). Taking into account that silver nanoparticle LSPR also permits the detection of hybridization events, a further refinement was made and a dual detection method was developed. In this new approach, DNA hybridization was monitored through both nanoparticle absorbance and fluorophore emission spectral changes (Zhang et al., 2006). In these studies, the authors have made use of tiopronin monolayer-protected silver nanoparticles with the additional advantage that these nanoparticles are not cytotoxic (Castillo et al., 2008).

As fluorescent labeling is widely employed in immunoassays, efforts have been made to implement this technique by means of MEF, for example to detect insulin (Lochner et al., 2003) and myoglobin (Barnett et al., 2007).

Biological recognition events, such as antigen-antibody interaction or DNA hybridization, are often kinetically slow, requiring long incubation times. Recently, low-power microwave heating has been employed to accelerate these processes in MEF based assays. The thermal gradient created between the bulk aqueous medium and the metal nanoparticles during microwave heating has been claimed to play a major role in the observed faster biorecognition kinetics in such a microwave-accelerated MEF-based bioassay. This principle has been applied to immunoassays and DNA hybridization assays that could be completed within a few seconds with the benefits of increased sensitivity associated to MEF (Aslan & Geddes, 2008).

Silver nanoparticles have also demonstrated their potential in applications such as the development of solid substrate-based RNA capture assays (Aslan et al., 2006), label-free bioassays by means of enhancement of tryptophan fluorescence in proteins (Szmacinski et al., 2009), the detection of glucose (Aslan et al., 2004) and the development of a MEF ratiometric pH sensor, using a pH-sensitive fluorophore (Aslan et al., 2005).

The development of molecular imaging agents for fluorescent imaging of cells is of great interest in disease diagnosis and elucidation of signalling pathways. Silver nanoparticles, conjugated to target molecules and fluorophores, have emerged as promising tools for cell imaging applications (Zhang et al., 2007b). Advantages of their use as cell imaging agents owing to their increased brightness and decreased lifetimes have been recently reported. Single molecule detection in cells by means of fluorescence requires the utilization of intense laser beams, which cause cell damage. This undesirable effect can be reduced by means of

MEF, which permits the use of weaker laser beams. Furthermore, owing to shorter lifetimes afforded by MEF, which reduces the time that a fluorophore spends in the excited state when it is vulnerable to oxygen attack, photobleaching is also decreased (Borejdo et al., 2008). Moreover, lifetime images generally lack interest given that the lifetimes of most organic fluorophores are close to the value of cell autofluorescence resulting in background interference. However, useful lifetime images can be obtained thanks to the decreased lifetimes obtained in MEF (Zhang et al., 2008).

6. Conclusions

Silver nanoparticles possess many valuable optical properties that have opened the door to new approaches in sensing and imaging applications, offering a wide range of detection modes such as colorimetric, scattering, SERS, and MEF techniques, at extremely low detection limits. Moreover, nanoparticles have made possible the use of scattering imaging techniques and have brought valuable improvements to standard imaging techniques. Silver nanoparticles may be introduced explicitly as labels, or may already be required in the application design to perform other functionalities, such as drug delivery, and simply provide this extremely useful added benefit.

Although LSPR sensing offers many promising features, detection thresholds must be improved for most practical applications. A great deal of current research effort is directed towards the optimization of nanoparticle tunable properties to improve their response in LSPR sensors.

One of the most promising contributions of silver nanoparticles to sensing and imaging applications is SERS. In this highly active research field, hundreds of papers are published each year on both fundamental aspects and applications. SERS has risen to its potential in the ultrasensitive detection of a wide variety of compounds, ranging from chemical pollutants to biomolecules, with detection thresholds as low as a single molecule, and the added advantage to provide structural information on the target molecule. It is to be expected that future improvements in this technique will be directed towards the development of more robust and reproducible sensors with a maximum coverage of hot spots. Additionally to SERS sensors, encapsulated SERS tags have emerged as an alternative to traditional fluorescent labeling offering advantages such as improved stabilities and multiplexing capability. On the other hand, MEF has also shown a great potential to address the deficiencies of traditional fluorescent labeling.

Summarizing, although further research is required, future perspectives of silver nanoparticles include applications in multiplexing assays, massively parallel bioassays for high throughput screening of drugs, diagnostic level sensing for clinical assays, the study of biological interactions at the molecular level and the measurement of intermolecular distances of tens of nm with plasmon rulers.

7. References

Lochner, N.; Lobmaier, C.; Wirth, M.; Leitner, A.; Pittner, F. & F. Gabor. (2003) Silver nanoparticle enhanced immunoassays: one step real time kinetic assay for insulin in serum. *Eur. J. Pharm. Biopharm.*, 56:469-477, ISSN: 0939-6411.

Abalde-Cela, S.; Ho, S.; Rodríguez-González, B.; Correa-Duarte, M. A.; Álvarez-Puebla, R. A.; Liz-Marzán, L. M. & Kotov, N. A. (2009) Loading of Exponentially Grown LBL Films with Silver Nanoparticles and Their Application to Generalized SERS Detection. *Angew. Chem. Int. Ed.*, 48:5326–5329, ISSN: 1521-3773.

Anger, P.; Bharadwaj, P.; Novotny, L. (2006) Enhancement and quenching of single-molecule fluorescence. *Phys. ReV. Lett.* , 96:113002-11300.

Aslan, K.; Zhang, J.; Lakowicz, J. R. & Geddes, C. D. (2004) Saccharide Sensing Using Gold and Silver Nanoparticles-A Review. *J Fluores.*, 14(4):391-400, ISSN 1573-4994.

Aslan, K.; Lakowicz, J.R.; Szmacinski, H. & Geddes, C.D. (2005) Enhanced ratiometric pH sensing using SNAFL-2 on silver island films: metal enhanced fluorescence sensing. *J Fluores.*, 15(1):37–40, ISSN 1573-4994.

Aslan, K.; Huang, J.; Wilson, G. M. & Geddes. C. D. (2006) Metal-Enhanced Fluorescence-Based RNA Sensing. *J. Am. Chem. Soc.*, 128 (13):4206–4207, ISSN: 1520-5126.

Aslan, K. & Geddes, C. D. (2008) A Review of an Ultrafast and Sensitive Bioassay Platform Technology: Microwave-accelerated Metal-enhanced Fluorescence. *Plasmonics*, 3:89–101, ISSN: 1557-1963.

Atay, T.; Song, J. H. & Murmikko, A. V. (2004) Strongly Interacting Plasmon Nanoparticle Pairs: From Dipole−Dipole Interaction to Conductively Coupled Regime. *Nano Lett.*, 4:1627-1631, ISSN: 1530-6992.

Barnett, A.; Matveeva, E. G.; Gryczynski, I.; Gryczynski, Z. & Goldys, E. M. (2007) Coupled plasmon effects for the enhancement of fluorescent immunoassays. *Physica B* 394:297–300, ISSN: 0921-4526.

Bell, S. E. J. & Sirimuthu, N. M. S. (2006) Surface-Enhanced Raman Spectroscopy (SERS) for Sub-Micromolar Detection of DNA/RNA Mononucleotides. *J. Am. Chem. Soc.*, 128:15580-15581, ISSN: 1520-5126.

Borejdo, J., Muthu, P., Talent, J., Gryczynski, Z., Calander, N., Akopova, I., Shtoyko, T., Gryczynski, I. (2008) Reduction of photobleaching and photodamage in single molecule detection: observing single actin monomer in skeletal myofibrils. *J. Biomed. Opt.*, 13:034021-034032, ISSN: 1083-3668.

Cade, N. I.; Ritman-Meer, T.; Kwakwa K. A. & Richards, D. (2009) The plasmonic engineering of metal nanoparticles for enhanced fluorescence and Raman scattering. *Nanotechnology*, 20 285201-285206 , ISSN 1361-6528.

Cao, Y. C.; Jin, R.; Thaxtonm, C. S. & Mirkin, C. A. (2005) A two-color-change, nanoparticle-based method for DNA detection. *Talanta* 67:449–455, ISSN: 0039-9140.

Caro, C.; López-Cartes, C.; Zaderenko A.P. &. Mejías J. A. (2008) Thiol-immobilized silver nanoparticle aggregate films for surface enhanced Raman scattering. *J. Raman Spectrosc.*, 39 (9):1162-1169, ISSN 1097-4555.

Castillo, P. M.; Herrera, J. L.; Fernandez-Montesinos, R.; Caro, C.; Zaderenko, A. P.; Mejias, J. A. & Pozo, D. (2008). Tiopronin monolayer-protected silver nanoparticles modulate IL-6 secretion mediated by Toll-like receptor ligands. *Nanomed*, 3(5):627-635, ISSN 1743-5889.

Chen, Y.; Aveyard, J. & Wilson, R. (2004) Gold and silver nanoparticles functionalized with known numbers of oligonucleotides per particle for DNA detection. *Chem. Commun.*, 24:2804–2805, ISSN 1359-7345.

Chen, Y.; Munechika, K. & Ginger, D. S. (2007) Dependence of Fluorescence Intensity on the Spectral Overlap between Fluorophores and Plasmon Resonant Single Silver Nanoparticles. *Nano Lett.*, 7(3).690-696, ISSN: 1530-6992.

Chen, Y. & Lu, C. (2009) Surface modification on silver nanoparticles for enhancing vapor selectivity of localized surface plasmon resonance sensors. *Sens. Act. B*, 135:492–498, ISSN: 0925-4005.

Das, A.; Zhao, J.; Schatz, G. C.; Sligar, S. G. & Van Duyne, R. P. (2009) Screening of Type I and II Drug Binding to Human Cytochrome P450-3A4 in Nanodiscs by Localized Surface Plasmon Resonance Spectroscopy. *Anal. Chem.*, 81:3754–3759, ISSN: 1520-6882.

Doering, W. E. & Nie, S. (2002) Single-Molecule and Single-Nanoparticle SERS: Examining the Roles of Surface Active Sites and Chemical Enhancement. *J. Phys. Chem. B*, 106:311-317, ISSN 1520-5207.

Fabris, L.; Dante, M.; Nguyen, T.; Tok, J. B.H. & Bazan, G. C. (2008) SERS Aptatags: New Responsive Metallic Nanostructures for Heterogeneous Protein Detection by Surface Enhanced Raman Spectroscopy. *Adv. Funct. Mater.*, 18:2518–2525, ISSN: 1616-3028.

Fromm, D. P.; Sundaramurthy, A.; Schuck, P. J.; Kino, G. & Moerner, W. E. (2004) Gap-Dependent Optical Coupling of Single "Bowtie" Nanoantennas Resonant in the Visible *Nano Lett.*, 4:957-961. ISSN: 1530-6992.

Futamata, M.; Maruyama, Y. & Ishikawa, M. (2002) Microscopic morphology and SERS activity of Ag colloidal particles *Vib. Spectrosc.*, 30:17-23, ISSN: 0924-2031.

Galush, W. J.; Shelby, S. A.; Mulvihill, M. J.; Tao, A.; Yang, P. & Groves, J. T. (2009) A Nanocube Plasmonic Sensor for Molecular Binding on Membrane Surfaces. *Nano Lett.*, 9 (5):2077-2082, ISSN: 1530-6992.

Gessner, R.; Rösch, P.; Kiefer, W. & Popp, J. (2002) Raman Spectroscopy Investigation of Biological Materials by Use of Etched and Silver Coated Glass Fiber Tips. *Biopolymers (Biospectroscopy)*, 67:327–330, ISSN 1520-6343.

Gish, D. A.; Nsiah, F.; McDermott, M. T. & Brett, M. J. (2007) Localized Surface Plasmon Resonance Biosensor Using Silver Nanostructures Fabricated by Glancing Angle Deposition. *Anal. Chem.*, 79:4228-4232, ISSN: 1520-6882.

Gong, J.; Liang, Y.; Huang, Y.; Chen, J.; Jiang, J.; Shen, G. & Yu. R. (2007) Ag/SiO_2 core-shell nanoparticle-based surface-enhanced Raman probes for immunoassay of cancer marker using silica-coated magnetic nanoparticles as separation tools. *Biosensors and Bioelectronics*, 22:1501–1507, ISSN: 0956-5663.

Guerrini, L.; Garcia-Ramos, J. V.; Domingo, C. & Sanchez-Cortes, S. (2009a) Nanosensors Based on Viologen Functionalized Silver Nanoparticles: Few Molecules Surface-Enhanced Raman Spectroscopy Detection of Polycyclic Aromatic Hydrocarbons in Interparticle Hot Spots. *Anal. Chem.*, 81, 1418–1425, ISSN: 1520-6882.

Guerrini, L.; Garcia-Ramos, J. V.; Domingo, C. & Sanchez-Cortes, S. (2009b) Sensing Polycyclic Aromatic Hydrocarbons with Dithiocarbamate-Functionalized Ag Nanoparticles by Surface-Enhanced Raman Scattering. *Anal. Chem.*, 81:953–960, ISSN: 1520-6882.

Haes, A.J. & Van Duyne, R.P.(2002) A Nanoscale Optical Biosensor: Sensitivity and Selectivity of an Approach Based on the Localized Surface Plasmon Resonance Spectroscopy of Triangular Silver Nanoparticles. *J. Am. Chem. Soc.*, 124:10596-10604, ISSN: 1520-5126.

Haes, A.J. & Van Duyne, R.P. (2003) Nanosensors Enable Portable Detectors for Environmental and Medical Applications. *Laser Focus World*, 39:53–156, ISSN: 1043-8092.

Haes, A. J.; Hall, W. P.; Chang, L.; Klein, W. L. & Van Duyne, R. P. (2004)A Localized Surface Plasmon Resonance Biosensor: First Steps toward an Assay for Alzheimer's Disease. *Nano Lett.*, 4 (6):1029–1034, ISSN: 1530-6992.

Haes, A. J.; Zou, S.; Zhao, J.; Schatz, G. C. & Van Duyne., R. P. (2006) Localized surface plasmon resonance spectroscopy near molecular resonances. *J. Am. Chem. Soc.*, 128:10905–10914, ISSN: 1520-5126.

Han, C.; Zhang, L. & Li, H. (2009) Highly selective and sensitive colorimetric probes for Yb^{3+} ions based on supramolecular aggregates assembled from β-cyclodextrin–4,40-dipyridine inclusion complex modified silver nanoparticles. *Chem. Commun.*, 24:3545–3547, ISSN 1359-7345.

Hu, Q. Y.; Tay, L. L.; Noestheden, M. & Pezacki, J. P. (2007) Mammalian Cell Surface Imaging with Nitrile-Functionalized Nanoprobes: Biophysical Characterization of Aggregation and Polarization Anisotropy in SERS Imaging. *J. Am. Chem. Soc.*, 129:14-15, ISSN: 1520-5126.

Huang, T. & Murray, R.W. (2002) Quenching of $[Ru(bpy)_3]^{2+}$ fluorescence by binding to Au nanoparticles, *Langmuir* 18:7077–7081, ISSN: 1520-5827.

Huang, T.; Nallathamby, P. D.; Gillet, D. & Xu, X. N. (2007) Design and Synthesis of Single-Nanoparticle Optical Biosensors for Imaging and Characterization of Single Receptor Molecules on Single Living Cells. *Anal. Chem.*, 79 (20):7708–7718, ISSN: 1520-6882.

Huang, T.; Nallathamby, P. D. & Xu, X. N. (2008) Photostable Single-Molecule Nanoparticle Optical Biosensors for Real-Time Sensing of Single Cytokine Molecules and Their Binding Reactions. *J. Am. Chem. Soc.*, 130:17095–17105, ISSN: 1520-5126.

Jana, N. R. & Pal, T. (2007) Anisotropic Metal Nanoparticles for Use as Surface-Enhanced Raman Substrates. *Adv. Mater.*. 19:1761–1765, ISSN: 1521-4095

Ji, X.; Xu, S.; Wang, L.; Liu, M.; Pan, K.; Yuan, H.; Ma, L.; Xu, W.; Li, J.; Bai, Y. & Li, T. (2005) Immunoassay using the probe-labeled Au/Ag core-shell nanoparticles based on surface-enhanced Raman scattering. *Colloids Surf. A: Physicochem. Eng. Aspects* 257–258, ISSN: 0927-7757.

Jiang, J.; Bosnick, N.; Maillard, M. & Brus, L. (2003) Single molecule Raman spectroscopy at the junctions of large Ag nanocrystals *J. Phys. Chem. B*, 107:9964-9972, ISSN 1520-5207.

Jin, R.; Wu, G.; Li, Z.; Mirkin, C.A. & Schatz, G.C. What controls the melting properties of DNA-linked gold nanoparticle assemblies? (2003) *J. Am. Chem. Soc.*, 125 (6), 1643-1654, ISSN: 1520-5126.

Jun, B.; Kim, J.; Park, H.; Kim, J.; Yu, K.; Lee, S.; Choi, H.; Kwak, S.; Kim, Y.; Jeong, D. H.; Cho, M. & Lee, Y. (2007) Surface-Enhanced Raman Spectroscopic-Encoded Beads for Multiplex Immunoassay. *J. Comb. Chem.*, 9:237-244, ISSN: 1520-4774.

Jun, B.; Noh, M. S.; Kim, G.; Kang, H.; Kim, J.; Chung, W.; Kim, M.; Kim, Y.; Cho, M.; Jeong, D. H. & Lee, Y. (2009) Protein separation and identification using magnetic beads encoded with surface-enhanced Raman spectroscopy. *Analytical Biochem.,* 391:24–30, ISSN 0003-2697.

Kanjanawarut, R. & Su, X. (2009) Colorimetric Detection of DNA Using Unmodified Metallic Nanoparticles and Peptide Nucleic Acid Probes. *Anal. Chem.,* 81 (15):6122–6129, ISSN: 1520-6882.

Kelly, K. L.; Jensen, T. R.; Lazarides A. A. & Schatz G. C. (2002). Modeling metal nanoparticle optical properties, In *Metal Nanoparticles: Synthesis, Characterization, and Applications*, D. L. Feldheim and C. A. Foss, Jr., 89-118, Marcel Dekker, New York, ISBN 0-8247-0604-8.

Kelly, K. L.; Coronado, E.; Zhao, L. L. & Schatz, G. C. (2003) The Optical Properties of Metal Nanoparticles: The Influence of Size, Shape, and Dielectric Environment. *J. Phys. Chem. B*, 107, 668–677, ISSN 1520-5207.

Kim, J.; Kim, J.; Choi, H.; Lee, S.; Jun, B.; Yu, K.; Kuk, E.; Kim, Y.; Jeong, D. H.; Cho, M. & Lee, Y. (2006) Nanoparticle Probes with Surface Enhanced Raman Spectroscopic Tags for Cellular Cancer Targeting. *Anal. Chem.*,78 (19): 6967–6973, ISSN: 1520-6882.

Kneipp, K.; Wang, Y.; Kneipp, H.; Perelman, L. T.; Itzkan, I.; Dasari, R. R. & Feld, M. S.(1997) Single Molecule Detection Using Surface-Enhanced Raman Scattering (SERS). *Phys. Rev. Lett.*, 78:1667-1670, ISSN 1079-7114.

Kneipp, K.; Wang, Y:; Kneipp, H.; Perelman, L.T. & Itzkan, I. (1997) Single molecule detection using surface-enhanced Raman scattering (SERS). *Phys. Rev. Lett.,* 78:1667–1670, ISSN 1079-7114.

Kreibig, U. & Vollmer, M. (1995) *Optical properties of metal clusters* Vol. 25 (Springer-Verlag, Berlin), ISBN: 978-3-540-57836-9.

Kühn, S.; Håkanson, U.; Rogobete, L.& Sandoghdar, V. (2006) Enhancement of single-molecule fluorescence using a gold nanoparticle as an optical *Phys. ReV. Lett.* , 97:017402-017405 , ISSN 1079-7114.

Kyriacou, S. V.; Brownlow, W. J. & Xu, X. N. (2004) Using Nanoparticle Optics Assay for Direct Observation of the Function of Antimicrobial Agents in Single Live Bacterial Cells. *Biochem.*, 43:140-147, ISSN: 1520-4995.

Lakowicz, J.R.; Shen, Y.; Gryczynski, Z.; D'Auria, S. & Gryczynski, I. (2001) Intrinsic fluorescence from DNA can be enhanced by metallic particles. *Biochem. Biophys. Res. Com.*, 286:875-879, ISSN: 0006-291X.

Lakowicz, J.R.; Gryczynski, I.; Malicka, J.; Gryczynski, Z. & Geddes, C.D. (2002a) Enhanced and localized multiphoton excited fluorescence near metallic silver islands: metallic islands can increase probe photostability. *J Fluores.*, 12:299-302, ISSN 1573-4994.

Lakowicz, J.R.; Shen, Y.; D'Auria, S.; Malicka, J. ,Fang, J.; Gryczynski, Z.; Gryczynski, I. (2002b) Radiative decay engineering 2. Effects of silver island films on fluorescence intensity, lifetimes, and resonance energy transfer. *Analytical Biochem.,* 301:261-277, ISSN 0003-2697.

Lakowicz, J.R.; Malicka, J.; D'Auria, S. & Gryczynski, I. (2003) Release of the self-quenching of fluorescence near silver metallic surface. *Analytical Biochem.,* 320:13-20, ISSN 0003-2697.

Lakowicz, J. R. (2005b) Radiative Decay Engineering 5: Metal-Enhanced Fluorescence and Plasmon Emission. *Analytical Biochem.,* 337:171-194, ISSN 0003-2697.

Le Ru, E. C.; Meyer, M. & Etchegoin, P. G. (2006) Proof of Single-Molecule Sensitivity in Surface Enhanced Raman Scattering (SERS) by Means of a Two-Analyte Technique. *J. Phys. Chem. B*, 110:1944-1948, ISSN 1520-5207.

Lee, J.; Lytton-Jean, A. K. R.; Hurst, S. J. & Mirkin, C. A. (2007) Silver Nanoparticle−Oligonucleotide Conjugates Based on DNA with Triple Cyclic Disulfide Moieties. *Nano Lett.*, 7 (7):2112–2115, ISSN: 1530-6992.

Lee, K.S. & El-Sayed, M.A. (2006) Gold and Silver Nanoparticles in Sensing and Imaging: Sensitivity of Plasmon Response to Size, Shape, and Metal Composition. *J. Phys. Chem. B*, 110:19220-19225, ISSN 1520-5207.

Li, H. & Bian, Y. (2009) Selective colorimetric sensing of histidine in aqueous solutions using cysteine modified silver nanoparticles in the presence of Hg^{2+}. *Nanotechnology*, 20:145502-145507 , ISSN 1361-6528.

Lin, X.; Cui, Y.; Xu, Y.; Ren, B. & Tian, Z. (2009) Surface-enhanced Raman spectroscopy: substrate-related issues. *Anal. Bioanal. Chem.*394:1729–1745, ISSN: 1618-2650.

Ling, J.; Sang, Y. & Huang, C. Z. (2008). Visual colorimetric detection of berberine hydrochloride with silver nanoparticles. *J. Pharm. Biomed. Anal.*, 47:860–864, ISSN: 0731-7085.

Liu, S.; Zhang, Z. & Han, M. (2005) Gram-Scale Synthesis and Biofunctionalization of Silica-Coated Silver Nanoparticles for Fast Colorimetric DNA Detection. *Anal. Chem.*, 77 (8):2595–2600, ISSN: 1520-6882.

Lucotti, A. & Zerbi, G. (2007) Fiber-optic SERS sensor with optimized geometry. *Sens. Act. B*, 121:356–364, ISSN: 0925-4005.

Lukomska, J.; Malicka, J.; Gryczynski, I. & Lakowicz, J. R. (2004) Fluorescence Enhancements on Silver Colloid Coated Surfaces. *J Fluores.*, 14:417-423, ISSN 1573-4994.

Malicka, J.; Gryczynski, I.; Fang, J. & Lakowicz, J.R. (2002) Photostability of Cy3 and Cy5-labeled DNA in the presence of metallic silver particles. *J Fluores.*, 12:439–447, ISSN 1573-4994.

Malicka, J.; Gryczynski, I. & Lakowicz, J. R. (2003) DNA hybridization assays using metal-enhanced fluorescence. *Biochem. Biophys. Res. Com.*, 306:213–218, ISSN: 0006-291X.

Malinsky, M.D.; Kelly, K.L.; Schatz, G.C. & Van Duyne, R.P. (2001) Chain Length Dependence and Sensing Capabilities of the Localized Surface Plasmon Resonance of Silver Nanoparticles Chemically Modified with Alkanethiol Self-Assembled Monolayers. *J. Am. Chem. Soc.*, 123:1471–1482, ISSN: 1520-5126.

McFarland, A. D. & Van Duyne, R. P. (2003) Single Silver Nanoparticles as Real-Time Optical Sensors with Zeptomole Sensitivity. *Nano Lett.*, 3 (8):1057-1062, ISSN: 1530-6992.

Michaels, A. M.; Jiang, J. & Brus, L. (2000) Ag Nanocrystal Junctions as the Site for Surface-Enhanced Raman Scattering of Single Rhodamine 6G Molecules *J. Phys. Chem. B*, 104:11965-11971, ISSN 1520-5207.

Mie G. (1908). Contributions to the optics of turbid media, especially colloidal metal solutions. *Ann. Phys.* 25:377–445, ISSN 0003-4916.

Nie, S. & Emory, S.R. (1997) Probing single molecules and single nanoparticles by surface enhanced Raman scattering. *Science*, 275:1102-1106, ISSN: 1095-9203.

Novotny, L. & Hecht, B. (2006). *Principles of nano-optics*, 378-414, Cambridge University Press, United Kingdom, ISBN-13 968-0-521-83224-3.

Oldenburg, S. J.; Genick, C. C.; Clark, K. A. & Schultz, D. A. (2002) Base pair mismatch recognition using plasmon resonant particle labels. *Analytical Biochem.*, 309:109–116, ISSN 0003-2697.

Ostuni, E.; Yan, L. & Whitesides, G. M. The interaction of proteins and cells with self-assembled monolayers of alkanethiolates on gold and silver. *Colloids Surf.B: Biointerfaces* 15:3–30, ISSN: 0927-7765.

Otto, A. & Futamata. (2006). Electronic Mechanisms of SERS, In: Surface-enhanced Raman scattering physics and applications. *Topics in applied Physics* Vol. 103, Kneipp, K.; Moskovits, M. & Kneipp, H., 147-182, Springer-Verlag, Berlin Heidelberg, ISBN-10 3-540-33566-8.

Park, J.; Yoon, D. & Kim, Y. (2009) Applications of silver nanoplates as colorimetric indicators of pH-induced conformational changes in cytochrome c. *Korean J. Chem. Eng.*, 26(1):258-260, ISSN: 1975-7220.

Pugh, V. J.; Szmacinski, H.; Moore, W.E.; Geddes, C.D. & Lakowicz, J.R. (2003) Submicrometer spatial resolution of metal-enhanced fluorescence. *Applied Spectroscopy* 57:1592-1598, ISSN: 0003-7028.

Schatz, G.C., Young, M. A. & Van Duyne R. P. (2006). Electromagnetic theory of SERS, In: Surface-enhanced Raman scattering physics and applications. *Topics in applied Physics* Vol. 103, Kneipp, K.; Moskovits, M. & Kneipp, H., 47-65, Springer-Verlag, Berlin Heidelberg, ISBN-10 3-540-33566-8.

Schofield, C. L.; Haines, A. H.; Field, R. A. & Russell, D. A. (2006) Silver and Gold Glyconanoparticles for Colorimetric Bioassays. *Langmuir*, 22:6707-6711, ISSN: 1520-5827.

Sherry, L. J.; Jin, R.; Mirkin, C. A.; Schatz, G. C. & Van Duyne, R. P. (2006) Localized Surface Plasmon Resonance Spectroscopy of Single Silver Triangular Nanoprisms. *Nano Lett.*, 6 (9):2060-2065, ISSN: 1530-6992.

Sonnichsen, C.; Reinhard, B.M.; Liphardt, J. & Alivisatos, A.P. (2005) A molecular ruler based on plasmon coupling of single gold and silver nanoparticles. *Nat Biotechnol* 23:741-745, ISSN 1087-0156.

Stokes R. J.; Hernandez-Santana, A.; Macaskill, A.; Cormack, P. A. G.; Smith, W. E. & Graham, D. (2006) SERRS-active nanoparticle-polymer beads for ultra-sensitive biodiagnostic applications. *Micro & Nano Lett.*, 1:57-61, ISSN: 1750-0443

Stuart, D. A.; Yuen, J. M.; Shah, N.; Lyandres, O.; Yonzon, C. R.; Glucksberg, M. R.; Walsh J. T. & Van Duyne R. P. (2006) In Vivo Glucose Measurement by Surface-Enhanced Raman Spectroscopy. *Anal. Chem.*,78 (20):7211–7215, ISSN: 1520-6882.

Su, K. H.; Wei, Q. H.; Zhang, X.; Mock, J. J.; Smith, D. R. & Schultz, S. (2003) Interparticle Coupling Effects on Plasmon Resonances of Nanogold Particles *Nano Lett.*, 3:1087-1090, ISSN: 1530-6992.

Su, X.; Zhang, J.; Sun, L.; Koo, T.; Chan, S.; Sundararajan, N.; Yamakawa, M. & Berlin, A. A. (2005) Composite Organic-Inorganic Nanoparticles (COINs) with Chemically Encoded Optical Signatures. *Nano Lett.*, 5(1):49-54, ISSN: 1530-6992.

Sun, L.; Sung, K. B.; Dentinger, C.; Lutz, B.; Nguyen, L.; Zhang, J. W.; Qin, H. Y.; Yamakawa, M.; Cao, M. Q.; Lu, Y.; Chmura, A. J.; Zhu, J.; Su, X.; Berlin, A. A.; S. Chan, & Knudsen, B. (2007) Composite Organic-Inorganic Nanoparticles as Raman Labels for Tissue Analysis. *Nano Lett.*, 7:351-356, ISSN: 1530-6992.

Szmacinski, H.; Ray, K. & Lakowicz, J. R. (2009) Metal-enhanced fluorescence of tryptophan residues in proteins: Application toward label-free bioassays. *Analytical Biochem.*, 385:358–364, ISSN 0003-2697.

Talley, C. E.; Jusinski, L.; Hollars, C. W.; Lane, S. M. & Huser, T. (2004) Intracellular pH Sensors Based on Surface-Enhanced Raman Scattering. *Anal. Chem.*, 76 (23):7064–7068, ISSN: 1520-6882.

Thompson, D. G.; Enright, A.; Faulds, K.; Smith, W. E. & Graham, D.(2008) Ultrasensitive DNA Detection Using Oligonucleotide−Silver Nanoparticle Conjugates. *Anal. Chem.* 80 (8), 2805–2810, ISSN: 1520-6882.

Wang, Z.; Bonoiu, A.; Samoc, M.; Cui, Y. & Prasad, P. N. (2008) Biological pH sensing based on surface enhanced Raman scattering through a 2-aminothiophenol-silver probe. *Biosensors and Bioelectronics,* 23:886–891, ISSN: 0956-5663.

Wei, H.; Chen, C.; Han, B. & Wang, E. (2008) Enzyme Colorimetric Assay Using Unmodified Silver Nanoparticles. *Anal. Chem.*. 80:7051–7055, ISSN: 1520-6882.

Weinkauf, H. & Brehm-Stecher, B. F. (2009) Enhanced dark field microscopy for rapid artifact-free detection of nanoparticle binding to Candida albicans cells and hyphae. *Biotechnol. J.* , 4:871–879, ISSN: 0168-1656.

Xiong, D. & Li, H. (2008) Colorimetric detection of pesticides based on calixarene modified silver nanoparticles in water. *Nanotechnology*, 19:465502-465507 , ISSN 1361-6528.

Xiong, D.; Chen, M. & Li, H. (2008) Synthesis of para-sulfonatocalix[4]arene-modified silver nanoparticles as colorimetric histidine probes. *Chem. Commun.*. 7:880–882, ISSN 1359-7345.

Xu, H.; Bjerneld, E. J.; Käll, M. & Börjesson, L.(1999) Spectroscopy of Single Hemoglobin Molecules by Surface Enhanced Raman Scattering *Phys. Rev. Lett.*, 83:4357-4360, ISSN 1079-7114.

Xu, X. N.; Brownlow, W. J.; Kyriacou, S. V.; Wan, Q. & Viola, J. J. (2004) Real-Time Probing of Membrane Transport in Living Microbial Cells Using Single Nanoparticle Optics and Living Cell Imaging. *Biochem.*, 43 (32):10400–10413, ISSN: 1520-4995.

Yguerabide, J. & E. E. Yguerabide, E. E. (1998) Light-Scattering Submicroscopic Particles as Highly Fluorescent Analogs and Their Use as Tracer Labels in Clinical and Biological Applications. *Anal. Biochem.*262:157–176, ISSN 0003-2697.

Yonzon, C. R.; Jeoung, E.; Zou, S. & Schatz, G. C. (2004) Milan Mrksich and Richard P. Van Duyne. A Comparative Analysis of Localized and Propagating Surface Plasmon Resonance Sensors: The Binding of Concanavalin A to a Monosaccharide Functionalized Self-Assembled Monolayer. *J. Am. Chem. Soc.*, 126:12669-12676 ISSN: 1520-5126.

Yue, Z.; Zhuang, F.; Kumar, R.; Wang, I.; Cronin, S. B. & Liu, Y. (2009) Cell kinase activity assay based on surface enhanced Raman spectroscopy. *Spec. Acta Part A,* 73:226–230, ISSN: 1386-1425.

Zhang, J.; Roll, D.; Geddes, C. D. & Lakowicz, J. R. (2004a) Aggregation of Silver Nanoparticle-Dextran Adducts with Concanavalin A and Competitive Complexation with Glucose. *J. Phys. Chem. B*, 108:12210-12214, ISSN 1520-5207.

Zhang, J.; Malicka, J.; Gryczynski, I. & Lakowicz, J. R. (2004b) Oligonucleotide displaced organic monolayer-protected silver nanoparticles and enhanced luminescence of their salted aggregates. *Analytical Biochem.*, 330:81-86, ISSN 0003-2697.

Zhang, J. & Lakowicz, J. R. (2006) A Model for DNA Detection by Metal-Enhanced Fluorescence from Immobilized Silver Nanoparticles on Solid Substrate. *J. Phys. Chem. B*, 110:2387-2392, ISSN 1520-5207.

Zhang, J.; Fu, Y.; Chowdhury, M. H. & Lakowicz, J. R. (2007a) Metal-Enhanced Single-Molecule Fluorescence on Silver Particle Monomer and Dimer: Coupling Effect between Metal Particles. *Nano Lett.*, 7(7):2101-2107, ISSN: 1530-6992.

Zhang, J.; Fu, Y. & Lakowicz; J. R. (2007b) Single Cell Fluorescence Imaging Using Metal Plasmon-Coupled Probe. *Bioconjugate Chem.*, 18:800-805, ISSN: 1520-4812.

Zhang, J.; Fu, Y.; Liang, D.; Nowaczyk, K.; Zhao, R. Y. & Lakowicz, J. R. (2008) Single-Cell Fluorescence Imaging Using Metal Plasmon-Coupled Probe 2: Single-Molecule Counting on Lifetime Image. *Nano Lett.*, 8(4):1179-1186, ISSN: 1530-6992.

Zhao, J.; Das, A.; Schatz, G. C.; Sligar, S. G. & Van Duyne, R. P. (2008) Resonance Localized Surface Plasmon Spectroscopy: Sensing Substrate and Inhibitor Binding to Cytochrome P450. *J. Phys. Chem.C*, 112:13084–13088, ISSN: 1520-5126.

Zhao, J.; Jensen, L.; Sung, J.; Zou, S.; Schatz, G. C. & Van Duyne, R. P. (2007) Interaction of plasmon and molecular resonances for rhodamine 6G adsorbed on silver nanoparticles. *J. Am. Chem. Soc.*, 129:7647–7656, ISSN: 1520-5126.

Zhu, S.; Du, C. & Fu, Y. (2009) Localized surface plasmon resonance-based hybrid Au–Ag nanoparticles for detection of Staphylococcus aureus enterotoxin B. *Opt. Mat.*, 31:1608–1613, ISSN: 0925-3467.

Zhu, S.; Li, F.; Du, C. & Fu, Y. (2008) A localized surface plasmon resonance nanosensor based on rhombic Ag nanoparticle array. *Sens. Act. B*, 134:193–198, ISSN: 0925-4005.

12

Silver nanoparticles as optical sensors

Chien Wang, Marta Luconi, Adriana Masi and Liliana Fernández
Universidad Nacional de San Luis – INQUISAL-CONICET
Argentina

1. Introduction

The use of silver nanoparticles (AgNPs) as analytical and bioanalytical sensors is receiving significant attention. This relevance arises from its unusual optical, electronic, and chemical properties (Schultz et al., 2000; Taton et al., 2000; Yguerabide & Yguerabide, 1998). The optical excitation of the surface plasmon resonance (SPR) caused by a collective excitation of the conduction band electrons of the nanoparticle, produces an absorption with large molar extinction coefficients and relevant scattering, usually when the particle size is larger than a few tens of nanometers (Haes et al., 2004). Colloid suspensions display brilliant colors as a result of intense light absorption and scattering, a fact first recognized by Faraday (Faraday & Philos, 1857), more than a century ago.

SPR is associated with the coupled oscillation of free electrons on the conduction band accompanying enhanced local electromagnetic field, which is intensely sensitive to surrounding medium conditions. When a nanoparticle is exposed to an electromagnetic wave, the electrons in the particle oscillate at the same frequency as the incident wave. Spectral characteristics of silver nanoparticles are strongly dependent on their size, shape, interparticle spacing and environment (b-Rao et al., 2002). Therefore, the geometry of noble metal nanoparticles would provide important control over linear and nonlinear optical properties (Bruzzone et al., 2005; b-Jiang et al., 2005; Roll et al., 2003).

Optical scattering has shown to be usefull in imaging methods to detect biosystems and has been applied to the diagnostics of cancer cells (Hirsch et al., 2003; El-Sayed et al., 2005).

Other potentialities are related to single-nanoparticle use as chemical and biological sensors (Cognet et al., 2003; McFarland & Van Duyne, 2003); changes in the plasmonic resonance wavelength of maximum absorption or scattering are monitored as a function of changing the chemical and physical environment of the surface of the nanoparticles. The spectral response, SPR or scattering band, must be very sensitives to the changes in the refractive index of the surroundings. Many theoretical and experimental studies (Kelly et al., 2003; Gole & Murphy, 2004) have been made on metal nanoparticles with different geometries with the purpose to find the best nanoparticle configuration and enhance the sensitivity of the plasmon resonance response.

2. Synthesis processes

During the last decades, many synthesis methods have been reported for the preparation of AgNPs.
The processing of nanosized silver particles can be briefly classified as:

- Chemical reduction of silver ions (Fig.1) (Toshima et al., 1993; Liz-Marzan & Philipse, 1995; Rivas et al., 2001; Pyatenko et al., 2005).
- Thermal decomposition in organic solvents (Esumi et al., 1990).
- Reversed micelle processes (Mafuné et al., 2000).
- Photoreduction (b-Zhou et al., 2001).
- Ultrasonic irradiation (Xiong et al., 2002), ^{60}Co-γ-irradiation (Henglein & Giersig, 1999) and microwave irradiation (Pastoriza-Santos & Liz-Marzan. 2002).

No doubt, the main synthetic way is based on the dispersion and condensation processes. Among these methods, AgNPs with spherical, nanowire or nanoprism shapes with tunable sizes have been manufactured (Balan et al., 2007). In most cases, only a low concentration of silver colloids (several millimoles per liter or less) in the presence of suitable stabilizers has been obtained.

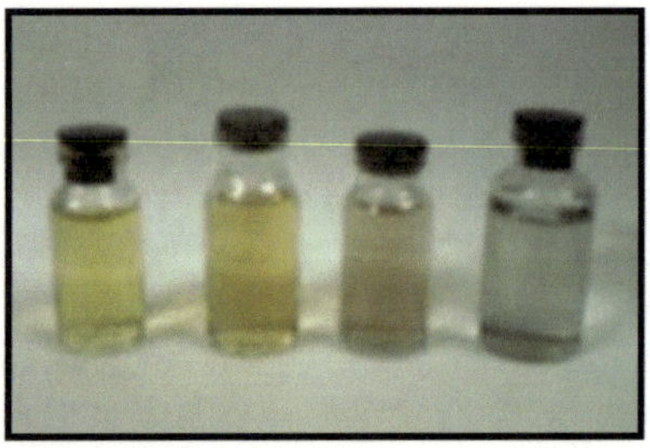

Fig. 1. AgNPs obtained by chemical reduction.

Experimental conditions as reagent concentrations, time and temperature of heating, cleanliness of glassware must be carefully controlled to achieve stable and reproducible colloids. The obtained products tend to form agglomerates and/or change the size or shape during the storage. As AgNPs are fairly unstable in solution (Luo & Sun, 2007), the derivatization step is necessary to obtain monodispersed nanoparticles.
The most common strategy for the formation of stable nanoparticles in liquid is the use of a protective agent, which not only prevents their aggregation, but also results in functionalized particles (a-Rao et al., 2000). The choice of the capping agent is very critical because it determines the stability, solubility, reactivity, and even the size and shape of the nanoparticles during the synthesis. However, the stabilizers such as surfactants and ionic polymers cannot be easily removed from the surfaces of the silver colloids, which unavoidably affect the physicochemical properties of the resulting nanoparticles. Therefore, large-scale synthesis of nanosized silver particles using removable reductants and stabilizers is worthy of investigation.

3. Characterization of Silver Nanoparticles

According to Mie theory (Petit et al, 1993), AgNPs spectra increase in intensity with the increase of size of NPs and decrese of broad of spectral band. Mock (Mock et al., 2002) studied absorption spectra of AgNPs with different morphologies, related with size, shape and wavelength of absorption using TEM; the results made evident that the more assimetric and greater size was the particle, the more bathochromic shift was shown.
Taking into account experimental dates, characterization of the nanoparticles to examine size, shape, and quantity is very important. A number of different measurement techniques can be used for this purpose; UV-visible spectroscopy is a valuable tool for study and structura, associated to scanning and transmission electronic microscopy (TEM), Atomic Force Microscopy (AFM), Scanning Electron Microscopy (SEM) and Dynamic Light Scattering (DLS). These techniques provide the image of a small piece of the sample, and this means that they give information about local properties by characterizing a few nanoparticles at a time (Sosa & Noguez, 2003).
The applications of absorption spectroscopy are the most extensive. However, for examination of nanoparticles, the optical properties require an individually developed theory. Measured absorbance spectrum does not necessarily show the actual absorbance but the extinction of the light. The extinction is both the absorbed and the scattered light from the particles.
SEM is a valuable instrument, when high-resolution images of the surface is desired. The microscope measures the electrons scattered from the sample. Electrons are accelerated by an electric potential varying wavelength at shorter than the one of photons. SEM is capable of magnifying images up to 200.000 times achieving high resolution pictures of the surface and determining the size distribution of nanoparticles.
AFM is an instrument capable of measuring the topography of a given sample. A nanosized tip attached on a cantilever is traced over the sample and a 3D image of the sample topography is generated on a computer.

4. Analytical and bioanalytical sensors

4.1 Colorimetric assays

Colorimetric assays based on the unique surface plasmon resonance properties of metallic nanoparticles have showed to be very useful in the analysis of biomolecular and metal ions; they have received considerable attention due to their simplicity, high sensitivity, low cost, rapidity and they do not need the use of complecated apparatus.
As colorimetric assays, AgNPs have some advantages over gold NPs to a certain degree since they possess higher extinction coefficients relative to gold NPs of the same size. However, little attention has been paid to AgNPs based colorimetric assays, though sensitive and selective colorimetric detection has been reported using AgNPs functionalized with appropriate ligands. These limitations can be attributed to the following facts:

- Functionalization of AgNPs usually cause chemical degradation.
- AgNPs'surface could be easily oxidized.

Colorimetric analytical method using citrate-capped AgNPs has been developed and applied to berberine detection (b-Ling et al., 2008), an important anti-inflammatory drug for heart and intestinal disorders. Citrate-capped AgNPs with negatively charged surface can be symmetrically dispersed in water by the electrostatic repulse interaction of each particle.

However, the presence of positive berberine would induce the aggregation of AgNPs, causing the color change of AgNPs suspension from yellow to green, and then to blue depending on the aggregation degree of AgNPs. The mechanism of color change and the effect of experimental conditions were studied using absorption and light scattering spectrometry. Under the optimum conditions, berberine hydrochloride can be detected from 0.05 μM to 0.4 μM visually based on the color alteration of the solution (LOD = $1.3\ 10^{-8}$ mol L^{-1}). This colorimetric analytical method without use of expensive machines is so convenient, economy and speedy that it has flourishing prospects in analytical chemistry.
A new method for the synthesis of AgNPs based on the incomplete reduction of silver ions by UV light in presence of poly(methacrylic acid) (PMA) has been presented (a-Dubas & Pimpan, 2008). The low power of the UV lamp provides a very slow kinetic of reaction which allows the preparation of purple solution of AgNPs.

UV-vis spectroscopy was used to monitor the kinetic of nanoparticles synthesis; their size and morphology were confirmed by transmission electron microscopy (TEM). The nanoparticles solution displays an absorbance maximum at 515 nm instead of the usually expected at 400 nm for spherical AgNPs. The unreacted silver is thought to be responsible for the color shift by forming an Ag^+/COO^- complex with the PMA stabilizing polyelectrolyte. The resulting solution displays a purple color which can be changed to yellow upon addition of ammonia; there are also medical interests in measuring this analyte in the body as indicator of disorder or disease. Based on this mechanism, an ammonia sensor has been proposed, showing a linear response in the range of 5–100 ppm.

A selective and sensitive method is proposed for detecting cysteine based on the color change of triangular silver nanoprisms at pH 5.5 incubated for 3 min in boiling water bath (Wu et al., 2009). Only cysteine got a color change among all the 20 protein, including alanine, arginine, asparagines, aspartic acid, glutamine, glycine, histidine, isoleucine, leucine, lysine, methionine, phenylalanine, proline, serine, threonine, tryptophan, tyrosine and valine, forming amino acids. With increasing cysteine concentration, blue shift of the dipole plasmon resonance peak was obtained. With the increase of cysteine, the color sequentially changed from blue to bright blue, amaranth, salmon pink and orange. It was found that the blue shift of wavelength is in proportion to the logarithm of concentration of cysteine (log $C_{cysteine}$), fitting the equation of $\lambda_{max} = 471.3 - 93.1 \log C_{cysteine}$ with correlation coefficient 0.9964. Considering that the color of the AgNPs is related to the size, shape, electron density on the colloidal surfaces and the refractive index of the surrounding medium, SEM images of the reaction between triangular silver nanoprisms and cysteine was measured; shape will transform to disk, so the method may serve as a new technique to control the size and shape of nanoparticles.

A sensitive, selective, simple and label-free colorimetric assay using unmodified AgNPs probes has been developed to detect enzymatic reactions (Wei et al., 2008). Enzymatic reactions concerning dephosphorylation and phosphorylation, specifically adenosine triphosphate (ATP) dephosphorylation by calf intestine alkaline phosphatase (CIAP) and peptide phosphorylation by protein kinase A (PKA), were chosen as model systems due to their significant importance to cellular regulation, cellular signaling, and biomedical applications. In the presented method, dephosphorylation and phosphorylation could be readily detected by the color change of AgNPs, with a detection limit of 1 unit mL^{-1} for CIAP and a detection limit of 0.022 unit mL^{-1} for PKA. This work is an important step toward a colorimetric assay using AgNPs and is a promise for enzyme assay in complex systems and for screening of different enzymes inhibitors in future.

Thompson (Thompson et al., 2008) has proposed the synthesis of oligonucleotide-silver nanoparticle (OSN) conjugates and demonstrated their use in a sandwich assay format. The OSN conjugates have practically identical properties to their gold analogues and due to their vastly greater extinction coefficient both visual and absorption analyses can occur at much lower concentrations.

A new one-pot protocol has been proposed for highly stable AgNPs modified with p-sulfonatocalix[n]arene (n = 4, 8) (Xiong & Lil, 2008). The obtained NPs were characterized by transmission electron microscopy, FT-IR and UV–vis spectroscopy. p-sulfonatocalix[4]arene modified AgNPs can be utilized as a novel colorimetric probe for optunal, allowing a rapid quantitative assay of optunal down to a concentration of 10^{-7} mol L^{-1}, showing a great potential for application to real-time in situ detection of optunal and the possible mechanism is discussed.

AgNPs have been synthesized by reduction of silver nitrate in the presence of humic acids (HA) which acted as capping agents (b-Dubas & Pimpan, 2008). In presence of sulfurazon-ethyl herbicide (0, 100, 200, 300, 400, 500 ppm), a variation in color of the nanoparticles solution from yellow to purple is produced. The effect of the humic acid concentration used in the nanoparticles synthesis was studied by varying the [Ag^+:HA] ratio content from [1:1] to [1:100]. UV–vis spectroscopy was used to monitor the extinction spectra of AgNPs after the synthesis and in the herbicide sensing experiments. An average AgNPs size of 5 nm was confirmed by transmission electron microscope (TEM).

A novel beta-cyclodextrin-4,4'-dipyridine supramolecular inclusion complex-modified AgNPs has been synthesized for the colorimetric determination of Yb^{3+} ions in aqueous solution with high sensitivity (LOD 2 10^{-7} mol L^{-1}) (a-Han et al., 2009).

Cysteine modified AgNPs were prepared in aqueous solution, via one-pot protocol, and characterized by transmission electron microscopy (TEM), Fourier transform infrared (FT-IR) and UV–vis spectroscopy (a-Li & Bian, 2009). The nanoparticles provided a simple and rapid strategy to detect visually histidine (His) using Hg^{2+} solution (Fig. 2). The colorimetric sensor allows a rapidly quantitative assay of histidine down to the concentration of 3 10^{-5} mol L^{-1}.

A new spectrophotometric method for the determination of arsenic at a mg g^{-1} level has been developed, based on the formation of AgNPs in anionic micellar medium and consecutive reduction of Ag(I) by arsine (AsH_3) generated from arsenic present in water (Pal & Maji, 2005). The magnitude of absorbance of yellow-coloured silver sol measured at λ_{max} = 395 nm was proportional to arsenic concentration in the sample with linear dynamic range of 0 to 0.5 mg g^{-1} (R^2 = 0.998). Molar absorptivity equalled 4.98 10^{-3} L mol^{-1} cm^{-1}, and Sandell´s sensitivity was 1.50 10^{-2} μg cm^{-2}. The proposed method is simple and provides reproducible results with standard deviations within ± 5%, free of interferences for Fe^{2+}/ Fe^{3+}, Ca^{2+}, Mg^{2+}, Sb(III), Se(IV), PO_4^{3-}, $SiO_3^{=}$, NO_3^{-}, Cl^{-}, $SO_4^{=}$, humic acid, pesticides/herbicides (such as 2,4-D, endosulfan, atrazine), etc.

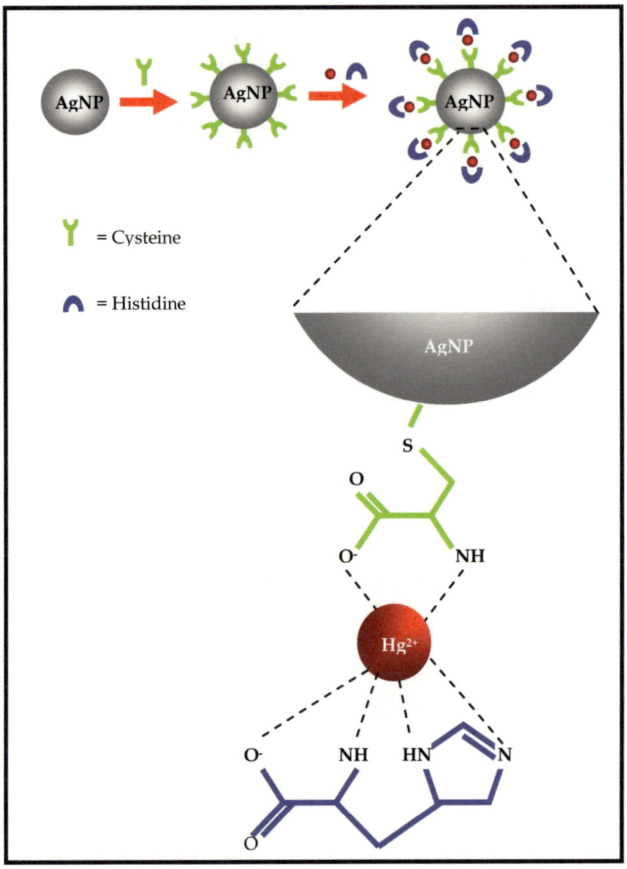

Fig. 2. Schematic representation of colorimetric sensor for histidine determination.

4.2 Spectrofluorimetric Analysis

Fluorescence detection is emerging as a powerful tool with wide applications in life sciences. This methodology allows in situ studies including binding of ligands such as a drug or hormone to a receptor, measurement of concentration of metabolites, study of structure, orientation, fluidity and distances between molecules. Yet in practical situations, where complex samples of biological origin must be used, such as in proteomics or clinical medicine, the presence of interfering background makes fluorescence detection more difficult and the need arises for the fluorescence signals to be amplified. Fortunately a physical nanoscale phenomenon exists which enables such amplification (a-Liu et al., 2006).

A good strategy to achieve improved sensitivity is the application of metal nanostructures in experimental protocols; these materials generally ensure high and/or spatially homogenous fluorescence enhancement. With the progress of nanotechnologies and the spectral theories, the luminescence properties of metal nanoparticles have been studied. But literatures about the nanoparticles-sensitized fluorescence are scarce (Zhao et al., 2008; Goldys & Xie, 2008).

Among the noble metal nanoparticles, AgNPs have attracted more attention because of their advantage on various studies such as photosensitive components, catalysts, and surface-enhanced Raman spectroscopy.

Liu et al. (b-Liu et al, 2007) prepared a kind of AgNPs by reaction of silver nitrate solution and ammonia gas. The strong fluorescence signal of the AgNPs solution can be quenched after the adding of the calf thymus DNA (ct-DNA) solution. Two excitation peaks appeared at the wavelengths of 239 and 314 nm respectively by emitting at 384 nm. The synthetic method proposed, was based on the following reactions:

$$Ag^{+} + OH^{-} \longrightarrow AgOH\downarrow$$

$$2AgOH \longrightarrow Ag_2O + H_2O$$

$$Ag_2O \xrightarrow{hv} Ag$$

The quenched NPs fluorescence intensity was linear with the concentration of ct-DNA in the range of 0.5 to 5.0 104 ng mL^{-1} with a detection limit of 0.3 ng mL^{-1}.
The interferences of bovine serum albumin, amino acid and glucose were very weak. And among tested ions, Mg^{2+}, Zn^{2+}, Fe^{3+}, Ca^{2+}, $SO_4^{=}$, Cl^{-} and I^{-} can be allowed in relatively high concentrations, but Cu^{2+} and PO_4^{3-} ions can only be allowed in relatively low concentrations.

Zhao et al. (Zhao et al., 2008) studied the fluorescence of terbium(III) when it is excited in the presence of ciprofloxacin (CPLX) in aqueous solution. CPLX can form complex with Tb(III) ion and the intramolecular energy transfer from CPLX to Tb(III) takes place when excited. The maximum fluorescence emission band of the terbium ion is located at 545 nm with λ_{ex} = 272 nm, but this intensity obviously increases when the AgNPs are added to the Tb(III)-CPLX system. This result indicates that AgNPs can promote the energy transfer between Tb(III) and CPLX. The size of the AgNPs affects the total intensity and is most enhanced by them at pH 6.0. The relative intensity is proportional to the concentration of CPLX. Based on this phenomenon, a new method for the determination of CPLX was developed by using a common spectrofluorometer to measure the intensity of fluorescence. The calibration graph for CPLX is linear in the range of 3.0 10^{-9} to 1.0 10^{-5} mol L^{-1}. The detection limit is 8.5 10^{-10} mol L^{-1}. The method was applied satisfactorily to the determination of CPLX in tablets and capsules. Their results showed that AgNPs with certain size and concentration can enhance the fluorescence.

Actually, luminescence methods for determining polycyclic aromatic hydrocarbons (PAHs), including adsorption–luminescence ones, are among the most sensitive methods. Nanosized organized media, including surfactant-modified noble metal nanoparticles, deserve special attention. Surfactants are most frequently used in the synthesis of noble metal nanoparticles for their stabilization and for the characteristic of modifying not only the properties of the test solution, but also those of the adsorbent surface A procedure for the synthesis of hydrophobic AgNPs in two-phase water organic emulsions was developed by Olenin et al.

(a-Olenin et al., 2008; b-Olenin et al., 2009), in which silver nitrate was reduced with sodium borohydrate in an aqueous solution followed by their transfer into the organic phase using CTAB as a phase transfer catalyst. When small amounts (few μL) of the chemically modified AgNPs sol were added to a dilute pyrene solution, the violet fluorescence intensity of pyrene decreased proportionally to the concentration of the nanoparticles added, and fluorescence with λ_{max}= 440 and 470 nm typical for pyrene excimers was observed in the long-wavelength region. The chemically modified AgNPs can be used as an adsorbent for preconcentrating pyrene from dilute n-hexane solutions followed by the direct luminescence determination, at room temperature, in the adsorbent matrix at concentrations below 0.01g mL^{-1} in test solutions.

Shang et al. (Shang et al., 2009) developed a sensitive fluorescent method for detecting cyanide based on the inner filter effect (IFE) of AgNPs (Fig. 3). The method proposes to use the powerful absorption of AgNPs to produce a tune in the emission and excitation of an isolated fluorescence indicator, which occurs in the IFE-based fluorescent assays. In the presence of cyanide, the absorber AgNPs dissolves gradually and leads then to recovery the IFE-decreased emission of the fluorophore. The method can detect cyanide with good selectivity over other common anions and ranging from 5.0 10^{-7} to 6.0 10^{-4} mol L^{-1} with a detection limit of 2.5 10^{-7} mol L^{-1}. It compares favorably with other reported fluorescent methods and has further application in cyanide-spiked water samples with a recovery between 98.2 and 101.4%.

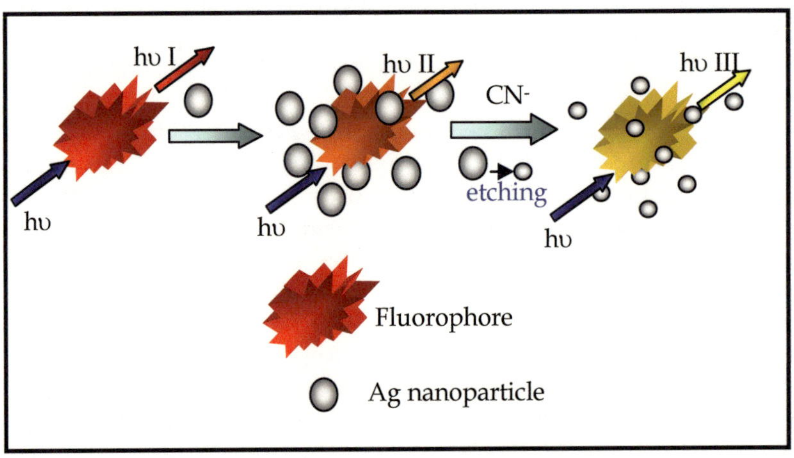

Fig. 3. Schematic representation of AgNPs-cyanide determination by IFE.

A sensitive and selective fluorimetric sensor for the assay of ascorbic acid using AgNPs as emission reagent was proposed by Park et al. (Park et al., 2009). AgNPs were prepared by reaction of silver nitrate solution and ammonia gas in aqueous-gaseous phase. AgNPs were used as a fluorescence probe for the assay of ascorbic acid using its quenching effect on the emission of AgNPs. They proposed that this effect was due to the complexation between ascorbic acid and AgNPs and established the quenching mechanism by Stern-Volmer law. The quenched fluorescence intensity was linear with the concentration of ascorbic acid in the range of 4.1 10^{-6} to 1.0 10^{-4} mol L^{-1} (r = 0.9985) with a detection limit of 1.0 10^{-7} mol L^{-1}.

The RSD for repeatability of the sensor for the assay of ascorbic acid concentration of 3.0 10^{-5} and 4.0 10^{-6} mol L^{-1} was found to be 1.5 and 1.3%, respectively. The method was applied to the determination of ascorbic acid in vegetables and vitamin C tablets.

4.3 Methodologies based on scattering phenomena

Investigations on light-scattering properties of nanomaterials and their application in analysis have been going on for more than 10 years (a-Ling et al., 2009). Yguerabide and Yguerabide (Yguerabide & Yguerabide, 1998) anticipated the potential applications of these particles in biochemical assay and cell biology. Under appropriate light, AgNPs of 40 nm in diameter scatter blue light. These particles could potentially be used in biochemical assay and cell imaging as fluorescent analogs. The light-scattering properties of new nanomaterials (van Dijk et al., 2006; a-Pan et al., 2007) in different composition, size and shape have greatly attracted the attention of analysts, so the applications of nanomaterials for analytical purposes have grown dramatically. This topic constitutes an atractive approach which is in incipient development because of the unique advantages of NPs of colorful light-scattering properties, comparable to optical probes with various fluorescent dyes.

AgNPs open exciting new ways to create efficient optical probes based on the strongly enhanced spectroscopic signals that can occur in their local optical fields. One of the most impressive effects associated with local optical fields is surface-enhanced Raman scattering (SERS) (Otto, 1984; Moskovits, 1985; Campion & Kambhampati, 1998; b-Kneipp et al., 1999). This technique is a form of Raman spectroscopy based on the interaction of the incident light with a molecule adsorbed onto the roughened metal surface. Derived from Raman spectroscopy, there are other interesting techniques employing AgNPs as substrate, such as surface-enhanced resonance Raman scattering (SERRS), surface-enhanced Raman optical activity (SEROA) and tip-enhanced Raman spectroscopy (TERS) (Pettinger et al., 2002; a-Kneipp et al., 2006).

Similar to IR, Raman spectroscopy (RS) yields detailed information about molecular vibrations. These kind of molecular motions are very sensitive to strength and types of chemical bonds. Therefore vibrational spectroscopy techniques are useful not only in identifying molecules but also reflect changes in the surrounding of the molecules and are thus, helpful in studying intra- and inter molecular interactions.

While infrared spectroscopy is based on absorption, reflection and emission of light, RS is based on inelastic scattering of radiation usually in the visible of near-infrared region by a solid, liquid or gaseous sample. In this context, scattering occurs due to collisions between photons and molecules (Fig. 4).

Irradiation of light with the frequency υ_o upon a certain molecule transfers a number of photons with the energy $E = h\upsilon_o$ (Fig. 5). For instance, laser light having a wavelength of 500 nm and an optical output of 1W emits approximately 2.5 10^{18} photons per second. These photons include photons colliding with molecules as well as those that pass without interacting with molecules. When carbon tetrachloride, which is a transparent liquid, is irradiated, it was found that about 10^{13} through 10^{15} photons collide with a molecule and change their directions among a total of 2.5 10^{18} photons. Most photons colliding with molecules do not change their energy after the collision (elastic collision) and the ensuing radiation is called Rayleigh scattering. Therefore, Rayleigh scattering consists of scattered photons which have the same frequency as the incident light, discussed in further section.

Only a very small number of the photons that impact on the molecule, exchanges energy with them during the course of the collision (inelastic collision). Scattering phenomenon in which incident photons exchange energy with a molecule is known as RS. Depending on frequency having the scattered light after colliding with molecule it could be Stokes RS or anti-Stokes RS. If an incident photon delivers a quantum ($h\upsilon$) of energy to the molecule, and the energy of the scattered photon is reduced to $h(\upsilon_o - \upsilon)$, the scattering phenomenon is known as Stokes RS. Otherwise, when an incident photon receives the $h\upsilon$ energy from the molecule, the energy of the scattering photon rises to $h(\upsilon_o+\upsilon)$, it is known as anti-Stokes RS.

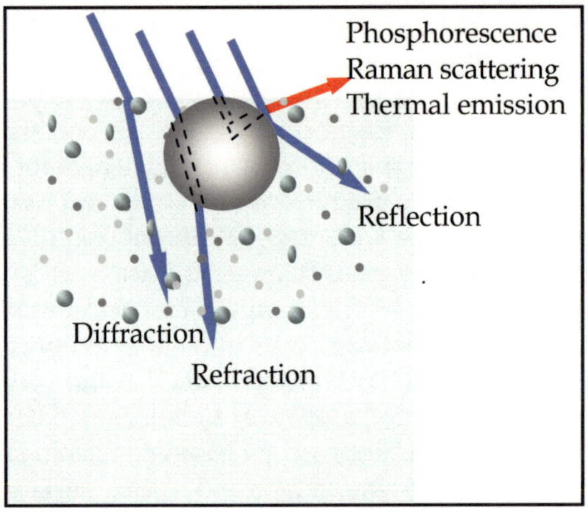

Fig. 4 Optical phenomena in the interphase air/aqueous AgNPs solution.

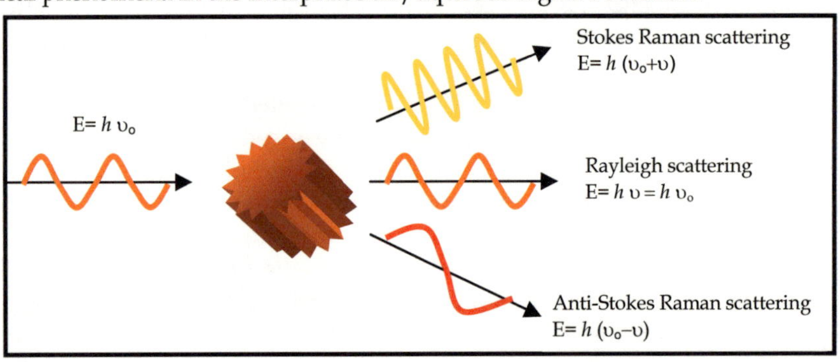

Fig. 5. Scattering phenomena.

The change in wavelength that is observed when a photon undergoes RS is attributed to the excitation (or relaxation) of vibrational modes of a molecule. Because of the diverse functional groups have different characteristic vibrational energies; every molecule has a unique Raman spectrum. Therefore, RS is a vibrational spectroscopic technique complementary to IR spectroscopy, but excitation and emission involve higer-energy photon similar to those used in electronic absorption and fluorescence spectroscopy.
It should be stressed that the basic feature of RS is scattering, which can occur at any wavelength. There is a change in the direction of the light, but no photon annihilation takes place as in an electronic absorption or fluorescence process. The selection rules are very

different from those in IR: Raman signals are associated with vibrations that cause a change in polarizability.Even so, RS cross sections are typically 14 orders of magnitude smaller than those of fluorescence; therefore, the Raman signal is several orders of magnitude weaker than the fluorescence emission in most cases. In fact, many materials are not readily suited to conventional RS. These include dyes that fluoresce in the same spectral range as their Raman emission, so that the useful Raman spectrum is masked overwhelmingly.
Due to the inherently small intensity of the Raman signals, the limited sensitivity of available detectors, the intensity of the excitation sources, and the potential interferences from the substances which fluoresce, the applicability of RS has been restricted for many years. However, its utility as an analytical technique improved with the advent of the laser and the evolution of photon detection technology. Another approaching of the technique is the selection of excitation wavelength close to an electronic absorption band of the chromophore. When the overlapping of molecular absorption band with the laser wavelength occurs, the intensity of the normal Raman scattering can be enhanced by several orders of magnitude. This is known as Resonance Raman Scattering (RRS); while smaller is the frequency difference between laser and electronic transition, stronger is the RRS intensity. Since the unique expected resonance effects, are thouse produced when the laser line coincides with an electronic absorption, RRS combines sensitivity and selectivity. In principle, it enables one to selectively observe a chromophoric solute in a dilute solution or study a particular Raman scatter in a complex matrix.

4.3.1 Surface-enhanced Raman Scattering

In 1977, Jeanmaire and Van Duyne demonstrated that the magnitude of the RS signal can be greatly enhanced when the scatterer is placed on or near a roughened noble-metal substrate (Jeanmaire & Van Duyne, 1977). This enhanced scattering process is known as surface-enhanced Raman (SER) scattering, a term that emphasizes the key role of the noble metal substrate in this phenomenon. A number of different metals such as silver, gold, copper and aluminium can be used to provide the optically active surface (Campion & Kambhampati, 1998; b-Kneipp et al., 1999); however, silver tends to give the largest enhancement factors using visible excitation wavelengths. Even though the exact mechanism of the enhancement effect of SERS is still a matter of debate in the literature, there are two primary theories which explain the phenomenon.
The first one uses plasma resonance model related to the optical properties of free-electron-like metals. Known as electromagnetic theory, it relies upon the excitation of localized surface plasmons (LSP) on nanoscale roughness features of metal substrate when it is irradiated by light. The excited LSP provides electromagnetic fields in order that when the Raman scatterer is subjected to these fields, the magnitude of the induced dipole increases. Taking account that the selection rules of Raman effect for a molecule are determined by its polarizability, accordingly, the intensity of the inelastic scattering increases (Raman scattering) when its polarizability increases.
The field enhancement becomes greatest when the plasmons frequency (ω_p) is in resonance with the radiation. In this instance, the LSPR (localized surface plasmons resonance) provides strong electromagnetic fields which leads to great enhanced Raman scattering from those molecules close to the metal surface. Although it is difficult to determine the extent of the enhanced fields from the nanostructured surface, theoretical calculations for well-characterized nanoparticles suggest that these fields degrade exponentially with a

characteristic decay length of ~2 nm (Schatz & Van Duyne, 2002). The three major strategies for confining a SERS analyte within the electromagnetic fields are physisorption, chemisorption, and partitioning via a self-assembled monolayer (Fig. 6).

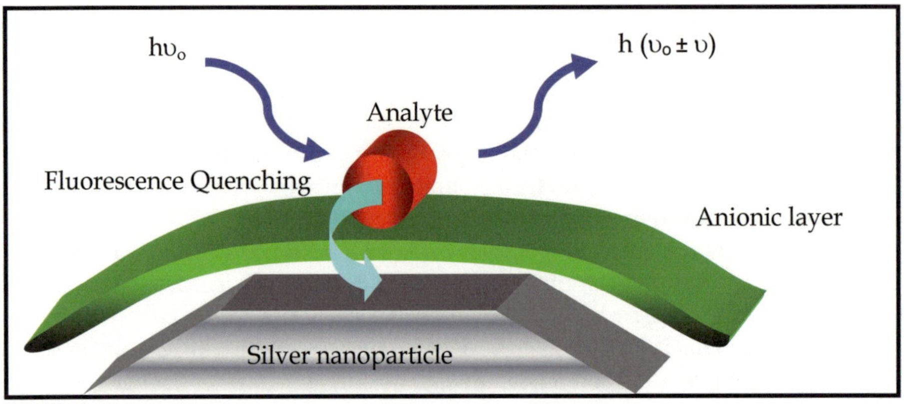

Fig. 6. SERS phenomena.

While the electromagnetic theory of enhancement can be applied regardless of the molecule studied, it does not fully explain the magnitude of the enhancement observed in many systems. Therefore, a second approach is based on the concept of the "active sites" at the metal surface. This model is known as chemical theory, due to the fact that for many molecules, often those with alone pair of electrons, with which the molecules can bond to the surface, a distinctly different mechanism of enhancement has been described, which does not involve surface plasmons. Through the formation of charge-transfer complexes between the metal and the molecules occupying these active sites induces enhanced Raman scattering effect.

Explanation using molecular orbital theory, the HOMO (Highest occupied molecular orbital) to LUMO (Lowest unoccupied molecular orbital) transition for many molecules requires much more energy than the infrared or visible light which is typically involved in Raman experiments. When the HOMO and LUMO of the adsorbate fall symmetrically about the Fermi level (chemical potential) of the metal surface, light of half energy can be employed to make the transition. This phenomenon takes place because the metal acts as a charge-transfer intermediate. Thus, a spectroscopic transition that might normally take place in the UV can be excited by visible light. However, chemical theory only applies for species which have formed a bond with the surface, so it clearly cannot explain the observed signal enhancement in all cases, while the electromagnetic theory can be applied even for molecules physisorpbed at a distance of more than 10 Å from the surface (Campion & Kambhampati, 1998).

The SERS technique has many advantages: because of it is a vibrational spectroscopy technique, a SER spectrum provides much more information about molecular structure and the local environment than an electronic spectroscopy technique, such as fluorescence. As the SERS analyte must be on or near a noble metal, which provides nonradiative pathways for the decay of excited states, fluorescence interference is strongly quenched. This becomes a solution for analysis of many fluorescent chromofores which were not possible to be analyzed by traditional Raman spectroscopy. Also, minor changes in the orientation of an

adsorbate can be discerned, because slight variations yield measurable shifts in the locations of SERS spectral peaks.
Furthermore, the abrupt decay of the electromagnetic fields ensures that only adsorbed molecules on or near the noble-metal substrate (within 4 nm) are probed. This quality makes SERS an ideal tool for surface studies, trace analyses, or biomolecular interactions.
Anothers distinctive advantages are the fact that water exhibits extremely weak SERS signals and AgNPs substrates are compatible to aqueous-phase. This makes the SERS technique employing AgNPs well suited for analysis performed on molecules in aqueous media. Therefore, for *in vivo* and *in vitro* biological studies it is found an increasingly interest as bioanalytical and bioimaging tool.
One of the disadvantages on SERS is the fact that not all analytes are good Raman scatterer. Therefore, this technique encounters an important limitation due to the great number of compound which gives a poor Raman signal. However, this hitch is resolved easily by employing SERRS as an alternative surface-enhanced Raman scattering technique.
To achieve SERRS, a laser excitation frequency is chosen to coincide with the absorption frequency of the chromophore (electronic absorption band). The molecular resonance is given as result of the wavelength overlapping between molecular electronic absorption band and the laser frequency, giving RRS phenomenon. This combination of RRS with resonance surface-enhanced fenomenon provides to SERRS the extreme sensitivity. A number of chromophores, which possess high Raman cross sections, can be attached to analytes with poor Raman signal. The formed complex is then adsorbed onto a suitably roughened metal surface and a laser source is chosen to create the respective resonances, giving a spectral enhancement with minimum background. Accordingly, this technique can be applied for analytes which in otherwise, cannot be determined by RS or SERS. Compared to SERS, SERRS appears to operate more effectively at low concentrations. However, SERS is more molecularly specific and nearly as effective at enhancing the scattering process at higher concentrations.
Total SERS/SERRS enhancement factors can reach 14 orders of magnitude, which produce nonresonant/resonant surface-enhanced Raman signals to a level comparably to or even better than fluorescence.
Unlike fluorescence, which produces relatively broad bands, Raman scattering as a vibrational, i.e., structure-specific method yields a unique spectrum composed of several narrow spectral lines, resulting in well-distinguishable spectra even for similar molecules. This lead to the development of SEROA, a novel method based on Raman optical activity (ROA) combined to surface-enhanced phenomenon for the analysis of enantiomers. ROA effect provides important information on molecular structure on the basis of differences in Raman spectra generated by left and right circularly polarized light. ROA is particularly sensitive to chirality, a molecular property that can be used to characterize polymeric assemblies as well as small molecules. Hence, the secondary structures of proteins and polyribonucleotides and also their building blocks can be discriminated. This includes individual amino acids, nucleosides, and many other biosynthetic precursors and small molecules of biological relevance which are optically active. Therefore, SEROA provide high spectral specificity, multiplex capabilities, and photostability for enantiomers determinations (Cao et al., 2002; Docherty et al., 2004).
Recently, SERS employing AgNPs as substrate to provide active optical surface has been used extensively as a signal transduction mechanism in biological and chemical sensing. Examples are cited in Table 1 and 2.

Methodology description	Analyte(s)	Analytical performance	References
Label-free protein determination	Lysozyme Catalase	LOD 5 μg mL^{-1} 0.05 μg mL^{-1}	c-Han et. al., 2009
Colloidal silver staining for Western Blot label-free protein determination	Bovine serum albumin	LOD 2 ng/band	d-Han et al., 2008
Intracellular pH sensor	H^+ (range 6-8)	---	Talley et al., 2004
Multiplex imunoassay for cellular proteins uing fuorescent-SERS dots by fluorescence bright field imagine	CD34, Sca-1, and SP-C	Identification	Woo et al., 2009
Specific virus DNA sequence determination	Human immunodeficiency virus type 1.	Identification	Wabuyele et al., 2005
In situ measurements for chemical analysis of biofilms	Multispecies biofilms; multiplex components	---	Ivleva et al., 2008
Kinetics monitoring endospore germination by Mirror sandwich SERS substrates	L-alanine; different temperatures	---	Jacquitta et al., 2006
Ratio-method using isotopically labeled internal standard	Creatinine	LOD 11.4 μg mL^{-1}	Stosch et al., 2005
bis-acridinium dication lucigenine as molecular assembler	Organochlorine pesticide endosulfan	LOD 20 μg L^{-1}	Guerrini et al., 2008
Ion-sorption on surfaces	Uranyl ion	LOD 20 ng mL^{-1}	Bhandari D. et al., 2009
Heat-induced SERS sensing	Glutathione	LOD 50 nM	Huang et al., 2009
SERS microflow cell for study of pigment within cyanobacteria from samples of rock	Scytonemin pigment	LOD 2 nM	Wilson et al., 2007

Table 1. Analytical applications of SERS phenomenon.

For detecting some biomolecules, a protocol for adding an aggregation agent to induce strong SERS has been used (Fig. 7). At pH lower than the protein isoelectric point, all the target proteins carry net positive charges, so these proteins can be adsorb on silver surfaces by electrostatic interactions and other interactions already mentioned. Accordingly, multiple interaction sites of an individual protein may bridge two or more silver nanoparticles and induce the subsequent SERS.

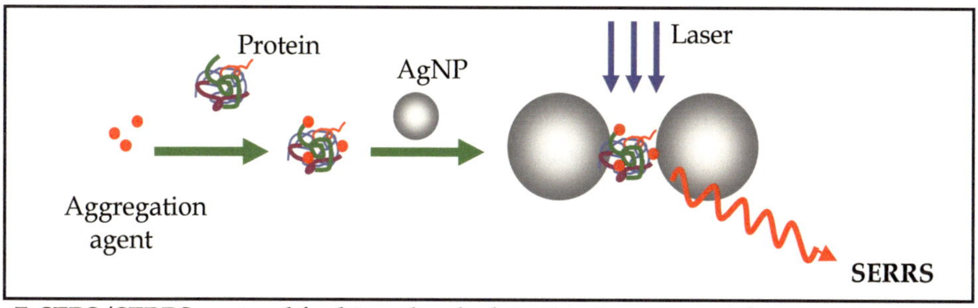

Fig. 7. SERS/SERRS protocol for biomolecule detection in presence of aggregation agent.

Moreover, immobilized silver colloidal nanoparticles have been used as a SERS substrate to perform microspectroscopic imaging. This technique was successfully applied for cancer cells imaging and other bio-molecules. As biosensor, is a highly sensitive tool for *in-vivo* study of many living system. The most critical aspect of performing a SERS experiment is the fabrication of the noble-metal substrates.

Methodology description	Analyte(s)	Analytical performance	References
Fluorescein isothiocyanate linked immunoabsorbent assay	Human IgG	LOD 0.2 ng mL^{-1}	b-Han et al., 2008
Multiplex detection of DNA sequences	Methicillin-resistant *Staphylococcus aureus*	LOD 10^{-12} mol mL^{-1}	MacAskill et al., 2009
Colloidal silver Staining for Western Blot label-free protein determination	Myoglobin	LOD 4 ng/band	d-Han et al., 2008
Rhodamine-labeled oligonucleotide sequence in a bead-based lab-on-a-chip format	Nucleic acid sequences of *Chlamydia trachomatis*	LOD 20.6 nM	Monaghan et al., 2007
Overtones and combinations in single-molecule spectra	Perylenetetracar boxylic diimides	Single molecule level	Goulet et al., 2003

Detection and characterization of structurally similar monosaccharides	D-glucose D-galactose D-mannose D-ribose D-arabinose D-xylose D-lyxose	Identification by "fingerprints" with samples 5 μL (10^{-2} M) each	Mrozek & Weaver, 2002

Table 2. Analytical applications of SERRS phenomenon.

Single-molecule SERS and the future

The early single-molecule detection with SERS was achieved by Nie and Kneipp groups independently at different experimental conditions. The Nie study included a correlated topographical and optical characterization of unaggregated AgNPs dosed with Rhodamine 6G (R6G) molecules. They hypothesized that only 1 of every 100–1000 nanoparticles is "optically hot" and that only 1 of every 10,000 surface sites on a "hot particle" has efficient enhancement (Nie et al., 1997). Accordingly, the single-molecule enhancement is 10^6–10^7 larger than the population-averaged enhancement.

Kneipp research group, on the other hand, probed small (100–150nm) silver colloid aggregates dosed with crystal violet molecules. The large single-molecule enhancement (10^{14}) is hypothetically attributed to large electromagnetic fields generated by fractal-pattern clusters of silver colloid nanoparticles (d-Kneipp et al., 1997).

Since these two pioneering experiments, SERS has been used to detect single molecules of biologically significant compounds, such as adenosine monophosphate (c-Kneipp et al., 1998) and hemoglobin (Xu et al., 1999).

Although the entire SERS community is excited by the recent development of single-molecule SERS, a new controversy surrounds the huge enhancement factors. Current hypotheses regarding the single-molecule enhancement mechanism center on the concept that SERS substrates have a small number of "hot spots", which are thought to occur at the junctions between two nanoparticles. Correlated single molecule SERS and topographical studies have clearly demonstrated that single-molecule SERS occurs only in compact aggregates of nanoparticles where these junctions may exist (Michaels et al., 2000). The large electromagnetic fields at the junction act as an optical trap for the Raman-active molecule and yield large scattering signals as well as enhanced photochemistry. The single-molecule SER spectra fluctuate on the time scale of 1 s; this is a signature of the molecule moving in and out of the hot spot and changing its orientation. Theoretical modeling of the electromagnetic field between two nanoparticles separated by ≤ 1 nm reveals a surface junction excitation and the efficient interaction of the molecular wave function with the wave function of the excited metal surface (a-Jiang et al., 2003).

An encouraging method for generalizing SERS to a wide variety of substrates is the development of tip-enhanced Raman spectroscopy (TERS) (Pettinger et al., 2002). In this technique, the electromagnetic field enhancement is provided by the excitation of the LSPR of a scanning probe. This eliminates the need to use noble-metal substrates to observe SERS. The probe can be a scanning tunneling microscopy probe, a metal-coated atomic force microscopy probe, a tapered optical fiber with a nanoparticle or thin metal film at the tip, or any other nanoscale-sharpened metallic object. Because of the strongest field enhancement is localized to a small volume very close to the apex of the probe, TERS can potentially provide chemical information at a spatial resolution below the diffraction limit. Theoretical modeling of TERS has suggested that enhancement factors of 10^7 and a spatial resolution of < 5 nm can be achieved (Richards et al., 2003).

4.3.2 Rayleigh light scattering (RLS)

Rayleigh (a-Rayleigh, 1871; b-Rayleigh, 1871), developed a theory of light scattering for small spherical particles which scatter but do not absorb light. Later, Mie (Mie, 1908), developed a more general theory of light scattering that applies to spherical particles of any size that can absorb as well as scatter light.
Rayleigh theory can be expressed as follow:

$$I = \frac{16\pi^4 a^6 n_{med}^4 I_0}{r^2 \lambda_0^4} \left| \frac{m^2 - 1}{m^2 + 2} \right|^2 \sin^2(\alpha) \qquad (1)$$

where I_0 is the intensity of incident monochromatic light, a is the angle between the detection direction **r** and the direction of polarization of the incident beam, n_{med} is the refractive index of the medium surrounding the particle, and m is the relative refractive index of the bulk particle material (refractive index of a macroscopic piece of the material of which the particle is composed divided by n_{med}) (Yguerabide & Yguerabide, 1998).
The wavelength of the scattered light is the same as that of the incident light beam. The scattered light intensity I depends on wavelength, and in general both m and n_{med} also depend on wavelength; r is the distance between the particle and the position where the scattered light is detected.
There are strong analogies between fluorescence and light scattering, but also there are differences that must be taken into consideration, namely only ones:

- Light produced by a scattering process has the same wavelength as the incident light; fluorescence occurs at higher wavelengths than excitation process and the fluorescence spectrum profile is independent of incident light wavelength.
- Light scattered by a small spherical particle is completely polarizad; light emitted by a fluorescent solution is only partially polarizad.
- The excited state lifetime for light scattering is usually less than 10^{-11} s whereas fluorescence lifetimes are usually in the range of 1 to 300 ns.
- Scattering intensity is not noticeably quenched; fluorescence intensity is sensitive to a variety of quenchers.

A simple and sensitive method for the determination of nucleic acids has been established (a-Zhou et al., 2009) based on the further enhancement effect of resonance light scattering of AgNPs-fsDNA by Al(III). The results for the determination of plasmid DNA in actual samples are satisfactory. The interaction mechanism investigation indicates that fsDNA and AgNPs combine with Al(III) through electrostatic attraction and adsorption bridging action to form AgNPs-Al(III)-DNA aggregations. This method is very simple, rapid and effective for determination of nucleic acids,which may be suggested for further uses in biology and nanoscience.

The RLS spectra were obtained by simultaneously scanning the excitation and emission monochromators over the range of 250–600 nm (i.e. $\Delta\lambda = 0$ nm). The intensity of resonance light scattering was measured at the maximum wavelength (398 nm) in a 1 cm quartz cell, with the slit width at 10 nm for the excitation and emission. The enhanced RLS intensity of AgNPs–Al(III)-nucleic acids system was represented as $\Delta I = I_R - I_R^0$, where I_R and I_R^0 were the RLS intensities of the systems with and without nucleic acids.
Under optimum conditions, there were linear relationships between the enhancing extent of RLS and the concentration of nucleic acids in the range of 1.0 10^{-9} - 1.0 10^{-7} g mL^{-1}, 1.0 10^{-7} - 2.0 10^{-6} g mL^{-1} for fish spermDNA(fsDNA), 1.0 10^{-9} - 7.0 10^{-8} g mL^{-1} for calf thymusDNA (ctDNA) and 1.0 10^{-9} - 1.0 10^{-7} g mL^{-1} for yeast RNA (yRNA). The detection limits (S/N= 3) of fsDNA, ctDNA and yRNA were 4.1 10^{-10} g mL^{-1}, 4.0 10^{-10} g mL^{-1} and 4.5 10^{-10} g mL^{-1}, respectively.

4.3.3 Second order scattering (SOS)

Second-order scattering (SOS), a phenomenon associated to fluorimetric determination, appears at the double wavelength of the exciting light. As SOS may interfere with fluorimetric measurement, it is always minimised off and eliminated as an unwanted phenomenon.
When a beam of parallel monochromatic light passes though a solution, an intensive radiation can also appear at half wavelength of the incident light, named "anti-double scattering" or "frequency doubling scattering" (FDS). FDS and SOS have been successfully applied to the determination of cations and cationic surfactants (b-Li et al., 2002).
The scattering intensity has been calculated to be:

$$I(\lambda,\lambda_{ex}) = K\, b\, c\, f(d)\, f(\Delta\lambda)\, E_m(\lambda)\, E_x(\lambda_{ex}) \qquad (2)$$

where K is the ratio coefficient, c the concentration of the scattering particle, b the thickness of the cell, d the size of the scattering particle, $\Delta\lambda$ the difference of excitation and emission wavelengths, $E_m(\lambda)$ the intensity distribution of the scattering light, $E_x(\lambda_{ex})$ is the excitation spectrum of the scattering light (Jiang et al., 2001).
When $\lambda_{ex} = (1/2)\,\lambda$, and $\lambda_{ex} = 2\lambda$, the $I\lambda$ relation curves are just the SOS and FDS spectra. That is:

$$I_{SOS} = K\, b\, c\, f(d)\, f(\Delta\lambda)\, E_m(\lambda)\, E_x[(1/2)\,\lambda] \qquad (3)$$

$$I_{FDS} = K\, b\, c\, f(d)\, f(\Delta\lambda)\, E_m(\lambda)\, E_x(2\lambda) \qquad (4)$$

Ding et al. (Ding et al., 2006) reported at first time, the effects of AgNPs on the SOS and fluorescence properties of Tb(III)–quinolones (Qs) complexes in solution. The new determination method of quinolones **was** developed using a common spectrofluorometer. This field is in the beginnings, and many new opportunities for nanoparticles will arise in the coming decades.
The SOS intensity reached the maximum at 545 nm with λ_{ex} = 274 nm. All the SOS and fluorescence intensities (ΔI) were measured against the blank which was prepared by the same way, but without quinolones, and thus $\Delta I = I - I_0$ was obtained.
The size of the AgNPs was found that affect the SOS intensity and the energy transfer between Tb(III) and quinolones. When the size of the AgNPs in the system increases, the

scattering intensity increases. Thus the selection of the appropriate diameter of AgNPs is very important. The results showed that:

- when the diameter of AgNPs was about 25 nm, the SOS intensity was very low;
- when the diameter was approximate 70 nm, the scattering of AgNPs was in the highest light and the SOS intensity was not proportional to the amount of quinolones (Qs);
- when the diameter was 45 nm, the intensity was greatly enhanced and the relative intensity was in the linear range.

Therefore in the next experiments the size of 45 nm was chosen, that is, the concentration of reduced $AgNO_3$ was $1.0\ 10^{-4}$ mol L^{-1}.
The concentration of AgNPs is associated with the aggregation degree. Thus AgNPs concentration directly affects the intensity of SOS and fluorescence. So selection of the concentration of AgNPs is necessary. The effect of AgNPs concentration on the intensity in the concentration range $1.0\ 10^{-8}$ - $1.0\ 10^{-4}$ mol L^{-1} was studied. The results indicated that the largest enhancement appeared when $1.0\ 10^{-6}$ mol L^{-1} AgNPs was used. Species such as vitamin B_1, uric acid, hemoglobin, myoglobin, Fe(III) and Al(III) have a relatively high interference. Dilution could be considered to minimize the interference in the application of urine samples.
The proposed method was applied to the determination of pipemidic acid (PPA) and lomefloxacin (LMFX) in tablets and urine of subjects over pharmacological treatment with these drugs with successfully results.

A new modified EDTA-AgNPs have been synthesized and characterized (Fig 8) (SEM, UV-vis and fluorescence spectroscopy) (Wang et al., 2009). Through SEM images it could be confirmed that these EDTA-AgNPs were monodispersed in aqueous medium, instead of the agglomerated forms present before derivatized reaction. The fluorescent quenching and SOS decrease of these nanoparticles produced by presence of nitrate was studied. Quenching mechanism was proposed for SOS phenomenon. An innovative methodology was developed for ultra-trace nitrates quantification and successfully applied for its determination in commercial parenteral

solutions. The main advantage of the proposed method is the possibility of direct nitrates determination with very good accuracy, sensitivity and tolerance, without the need of previous reduction to nitrite, neither any previous treatment for samples used. The obtained results showed that AgNPs can be applied as optical sensor for nitrate determination.

During the synthesis of the AgNPs by conventional heating it was observed that after adding citrate, the color of solution changed from colorless to yellow, making evident the formation of colloidal silver (Fig. 9). The optimal reaction time was fixed at 6 min; when the time exceed the optimal, the reaction mix change to greenish grey.
SEM images show that, in this instance, the particles which were nearly spherical were not well separated and stuck together to form many small groups. This typical behavior has been reported by other researchers (He et al., 2004; Jiang et al., 2005).

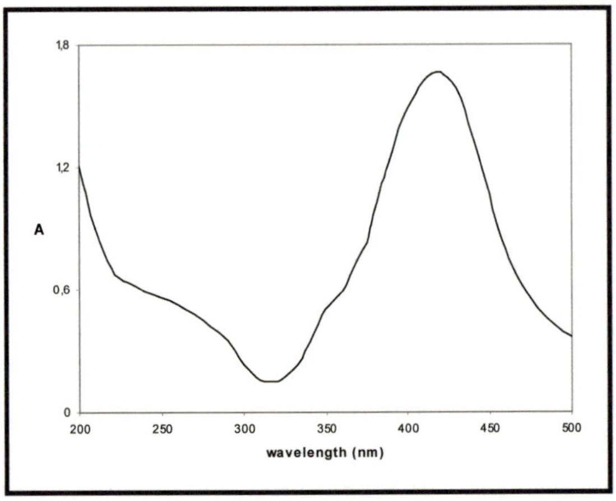

Fig. 8. Absorption spectrum of synthesized AgNPs-EDTA (λ_{max} = 415 nm).

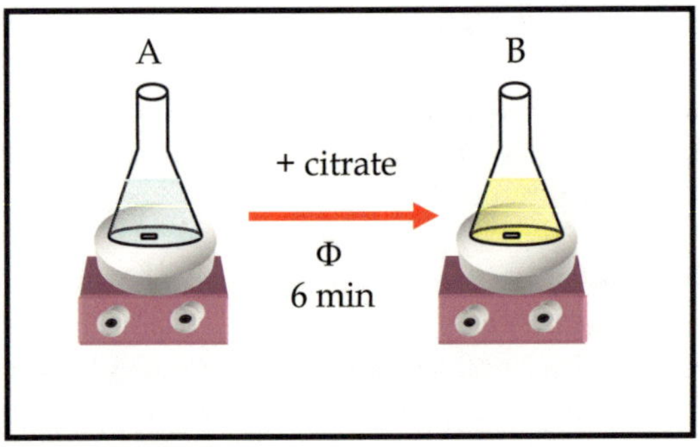

Fig. 9. Synthesis of AgNPs by citrate reduction.
A: $AgNO_3$ solution (colorless); B: AgNPs obtention (yellow).

The chelating reagent EDTA has been used as derivatizing agent for obtaining stable nanoparticles (b-Pan et al., 2007). This coater suffers a chemisorption process onto the surface of AgNPs through carboxylate groups, which are coordinated symmetrically to the Ag atoms. The surface of AgNPs remains negatively charged and, in presence of counter ions, acquires an electrostatic double layer, that it provides a repulsive force enabling to silver colloid to be stable in aqueous solution (Creighton et al., 1979; Heard et al., 1983; Vickova et al., 1996).

Considering that the possibility of using EDTA-AgNPs with analytical purposes depends on obtaining nanomaterial of high pure grade, it was necessary to add a purification step to eliminate all excess of reactant. Concentrated NaOH aqueous solution was added to EDTA-AgNPs solution and, after 15 min of reacting time, the flocculation process took place (He et al., 2004). It was observed that the solution color changed immediately from yellow to violet. The dramatical diminution of Z potential led to form the big flaks of nanoparticles, which facilitated the phase's separation by flocculation (Fig. 10).

This phenomenon is reversible once the hydroxide is removed by subsequent wash steps, obtaining a transparent solution of monodispersed nanoparticles (Fig. 11) with average size of 40 nm (Fig. 12A, B and C).

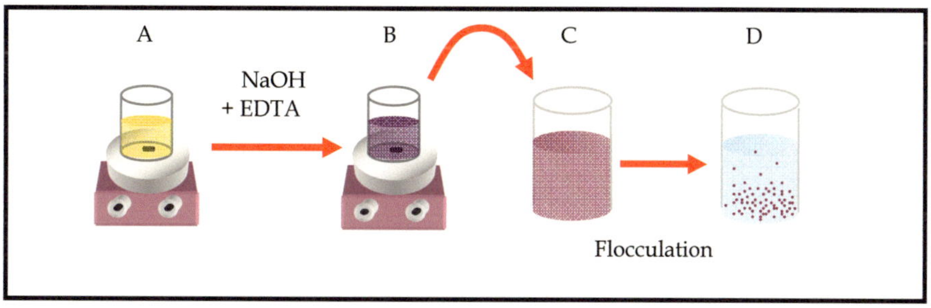

Fig. 10. Derivatization of AgNPs and flocculation process.
A: Recently synthesized AgNPs solution; B: Capped EDTA reaction; C: EDTA-AgNPs product; D: Spontaneous flocculation of EDTA-AgNPs.

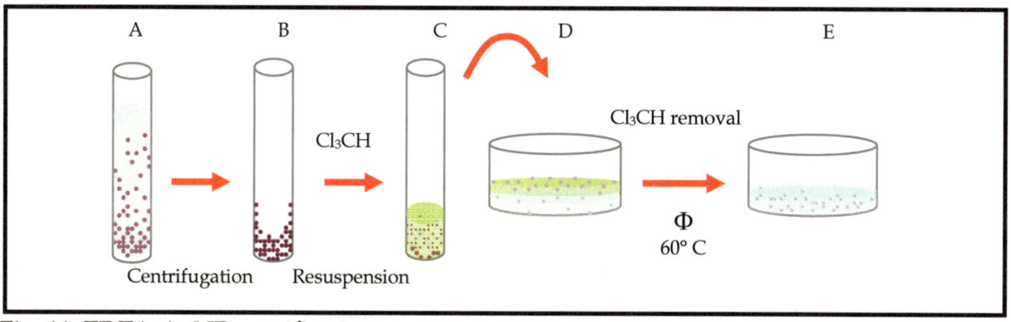

Fig. 11. EDTA-AgNPs purification process.
A: EDTA-AgNPs floccules; B: Solvent remotion; C: EDTA-AgNPs chloroform suspension; D: Chloroform remotion process; E: EDTA-AgNPs monodisperse solution.

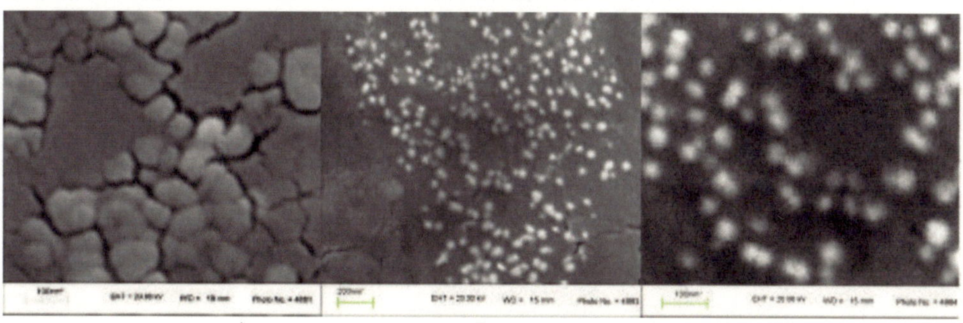

A B C

Fig. 12. SEM images of AgNPs. A: Before derivatization step. B and C: After EDTA derivatization step.

There exist a wide variety of substances which act as quenchers of fluorescence as well as different types of quenching process. Nitrate is considered an electron deficient. This kind of

quencher probably involves donation of electrons from the surface of nanoparticles to the quencher, deactiving the excitate state responsible of fluorescence (Lakowicz, 1999). It is well known that fluorescent emission and SOS are associated phenomena; consequently the fluorescence quenching of EDTA-AgNPs by nitrates affects directly to its SOS signals.
In order to determine the quenching type, a study of the K_{sv} (Stern–Volmer constant) from the modified Stern–Volmer equation (Eqs. (1) and (2)) was carried out at different temperature.

$$F_o / F = 1 + K_{sv} C_q \qquad (5)$$

Stern-Volmer equation for fluorescent quenching, where F_o and F are fluorescent emissions of the fluorophore in absence and presence of the quencher respectively; K_{sv} is the Stern–Volmer constant; and C_q is the concentration of the quencher.
Replacing the term Fo/F by Io/I:

$$I_o / I = 1 + K_{sv} C_q \qquad (6)$$

modified Stern–Volmer equation for SOS decrease; where I_o and I are SOS intensities of EDTA-AgNPs in absence and presence of nitrates, respectively.

Table 3 shows the obtained K_{sv} values. The linearity of the Stern–Volmer plot, as the value of $K_{sv,}$ which enhanced with increasing temperature (Fig. 13), indicated that the quenching mechanism of Ag-EDTA NPs by presence of nitrate is a single dynamic quenching (Lakowicz, 1999).

When the time of AgNPs synthesis process is extended, a larger size of nanoparticles is obtained and in the fluorescence spectrum, the huge dispersion of the light produces an increase of the noise. After derivatization, the noise of fluorescent spectrum was minimized; nevertheless, the decreasing of SOS and fluorescence quenching for

these nanoparticles in presence of nitrate was not efficient. This phenomenon may be due to the superficial activity diminution produced by the increase
of the particles size; simultaneously, it is reduced the capability of nitrate to produce the deactivation process of the excited state.

Temperature (K)	K_{SV} value (μg mL^{-1})
293	0.736
303	0.991
313	1.213
323	1.456

Table 3. K_{SV} values for different experimental temperatures.

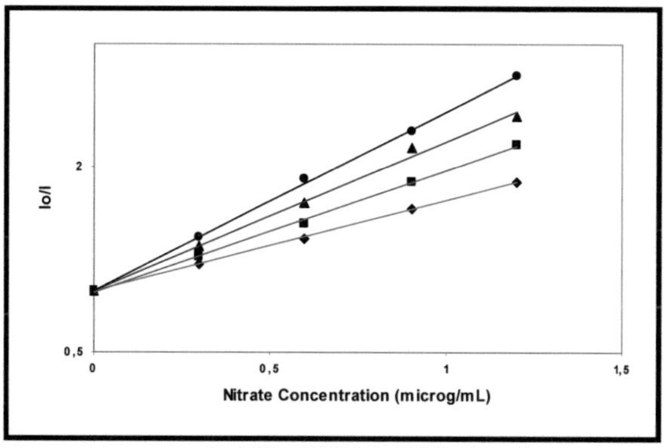

Fig. 13. Influence of temperature on AgNPs-EDTA SOS signals in presence of nitrate. [NO_3^-]: 0.00; 0.30; 0.60; 0.90; 1.20 μg mL^{-1}. Instrument conditions: λ_{ex}= 225 nm; λem= 450 nm.

Under the optimal experimental conditions, calibration curves for the determination of nitrate by SOS and fluorescence were obtained. By SOS decrease, the results showed a good linear relationship over the range 6.4 10^{-4} to 3.0 μg mL^{-1}. The linear regression equation was ΔI_{sos} = 228.50 C (μg mL^{-1}) + 0.84 with regression coefficient r = 0.997. The LOD estimated was 1.8 10^{-4} μg mL^{-1}; while for quenching of fluorescence emission the equation of calibration curve was ΔF = 34.40 C (μg mL^{-1}) + 1, value means lost of sensitivity compared with SOS technique.
The obtained results showed that the EDTA-AgNPs can be applied as sensor for nitrate (Table 4).

Samples	Base value	Added (μg mL^{-1})	Found (μg mL^{-1})	Recovery [b] ± RSD[c] (%)
A	--	0.000	0.006	--
	0.006	0.250	0.252	98.43 ± 1.89
	0.006	0.500	0.503	99.40 ± 1.77
	0.006	0.750	0.768	101.50 ± 2.21
B	--	0.000	0.009	--
	0.009	0.250	0.263	101.54 ± 1.45
	0.009	0.500	0.507	99.60 ± 2.10
	0.009	0.750	0.749	98.68 ± 1.87

Table 4. Nitrate determination in parenteral solutions. Recovery study[a].

[a] AgNPs-EDTA systems prepared and measured as described in general procedure.
[b] Recovery= 100 x [(found-base)/added]. [c](n=6).

The effect of AgNPs on the fluorescence and SOS intensity of Tb(III) - ciprofloxacin (CPLX) complex in aqueous solution has been reported (Zhao et al., 2008) and a new method was developed to determine the concentration of CPLX by the SOS and fluorescence intensity of the system, which expands the field of analytical application of AgNPs.
To a 10 mL calibrated tube were added in the order of 2 mL of HAc–NaAc buffer solution, Tb(III) ion solution, CPLX, and AgNPs. And the mixture was diluted to 10 mL with doubly distilled water, shaken thoroughly, and left to stand for 30 min. In order to choose the optimal excitation and emission wavelengths, the different exciting wavelengths from 220 to 350 nm were used to record the signal peaking at 440–700 nm ($\lambda_{SOS} = 2\lambda_{ex}$), correspondingly. The intensity reached the maximum at 545 nm with λ_{ex} = 272 nm. All data were obtained by using the slit-width of excitation and emission of the spectrofluorometer at 5 nm. The relative intensities of fluorescence and SOS ΔI (the difference between the intensity of CPLX solution and that of the blank solution without CPLX) were proportional to the concentration of CPLX.
Under the optimum condition defined, the calibration graphs for CPLX were obtained. The relative standard deviation was 1.6 % for nine determinations of 1.0 10^{-6} mol L^{-1} CPLX. The linear range of the calibration graph is from 3.0 10^{-9} to 1.0 10^{-5} mol L^{-1}, and the detection limit calculated from the standard deviation of the blank (the reagent blank without CPLX, n = 19) (3σ) is 8.5 10^{-10} mol L^{-1}.
The proposed method was applied to the determination of CPLX in capsules and tablets, and compared with UV–vis method without significant difference between the labeled content and that obtained by this method.

5. Conclusions

In this chapter, some representative researchs based on the application of AgNPs to chemical and biological sensors using absorption, emission and scattering phenomena, have been commented. The inusual optical properties mainly related to their high extinction coefficient and tunable particle shapes, have turned to AgNPs in a very attractive and special usefull analytical tool: their have been applied to trace determination of different nature analytes with sucessfully results.
Due to its high efficiency and low cost, the light scattering techniques based on AgNPs shows particular promise in the development of noninvasive in vivo sensing techniques for the study of biological systems, particular to immunoassays and to monitoring the interaction between proteins and enzymes in living cells. Light-scattering signals could have widespread applications in detection and quality control of new developed nanodrugs.
AgNPs have a promising future in designing of optical sensors. Their utilization will be driven by the need for smaller detection devices with lower limits of detection. Continual advances in nanofabrication technology and optical characterization techniques are expected to overcome the difficulties encountered in the development of TERS. One major challenge is improve the collection and detection efficiency of the traditional SERS instrumentation. Another challenge is the fabrication of an optical probe that incorporates a SERS substrate. The modification of fiber-optic probes with metal films or nanoparticles provides a promising solution, but great care must be taken to ensure long-term stability of the metal substrate and to permit reuse of the probe.

The success of new applications of AgNPs depends on improvement in the understanding of the properties of SPR. Further efforts and systematic study must be assumed in order to offer new developments to extend the analytical applications field of AgNPs.

6. References

Balan, L.; Malval, J.; Schneider, R. & Burget, D. (2007). Silver nanoparticles: New synthesis, characterization and photophysical properties. *Mat. Chem. Phys.*, 104, 2-3, (August 2007), 417–421, ISSN 0254-0584.

Bhandari, D.; Wells, S.; Retterer, S. & Sepaniak, M. (2009). Characterization and detection of uranyl ion sorption on silver surfaces using surface enhanced Raman spectroscopy. *Anal. Chem.*, DOI: 10.1021/ac901266f, ISSN 0003-2700.

Bruzzone, S.; Malvaldi, M.; Arrighini, G. & Guidotti C. (2005). Theoretical study of electromagnetic scattering by metal nanoparticles. *J. Phys. Chem. B*, 109, (February 2005), 3807- 3812, ISSN 0022-3654.

Campion, A. & Kambhampati, P. (1998). Surface-enhanced Raman scattering. *Chem. Soc. Rev.*, 27, (February 1998), 241−250, ISSN 0306-0012.

Cao, Y.; Jin, R. & Mirkin, C. (2002) Nanoparticles with Raman spectroscopic fingerprints for DNA and RNA detection. *Science*, 297, (August 2002), 1536−1540, ISSN 0036-8075.

Cognet, L.; Tardin, C.; Boyer, D.; Choquet, D.; Tamarat, P. & Lounis, B. (2003). Single metallic nanoparticle imaging for protein detection in cells. *Proc. Natl. Acad. Sci. U.S.A.*, 100, 20, (September 2003), 11350-11355, ISSN 0027-8424.

Creighton, J.; Blatchford, C. & Albrecht, M. (1979). Plasma resonance enhancement of Raman scattering by pyridine adsorbed on silver or gold sol particles of size comparable to the excitation wavelength. *J. Chem. Soc. Faraday Trans.* 2, 75, 790-798, ISSN 0300-9238. DOI: 10.1039/F29797500790.

Daniels, J.; Caldwell, T.; Christensen, K. & Chumanov, G. (2006). Monitoring the kinetics of *Bacillus subtilis* endospore germination via surface-enhanced Raman scattering. *Anal. Chem.*, 78, 5, (Febrary 2006), 1724-1729, ISSN 0003-2700.

Ding, F.; Zhao, H.; Jin, L. & Zheng, D. (2006). Study of the influence of silver nanoparticles on the second-order scattering and the fluorescence of the complexes of Tb(III) with quinolones and determination of the quinolones. *Anal. Chim. Acta,* 566, 1 (April 2006) 136–143, ISSN 0003-2670.

Docherty, F.; Clark, M.; McNay, G.; Graham, D. & Smith, W. (2004). Multiple labelled nanoparticles for bio detection. *Faraday Discuss.*, 126, (October 2003), 281−288, ISSN 1364-5498.

a) Dubas, S. & Pimpan, V. (2008). Green synthesis of silver nanoparticles for ammonia sensing. *Talanta*, 76, (June 2008), 29-33. ISSN 0039-9140.

b) Dubas, S. & Pimpan, V. (2008). Humic acid assisted synthesis of silver nanoparticles and its application to herbicide detection. *Mater. Lett.*, 62, (June 2008), 2661-2663, ISSN 0167-577X.

El-Sayed, I.; Huang, X. & El-Sayed, M. (2005). Surface plasmon resonance scattering and absorption of anti-EGFR antibody conjugated gold nanoparticles in cancer diagnostics: applications in oral cancer. *Nano Lett.*, 5, 5, (May 2005), 829-834, ISSN 1530-6984.

Esumi, K.; Tano, T.; Torigae, K. & Meguro K. (1990). Preparation and characterization of bimetallic palladium-copper colloids by thermal decomposition of their acetate compounds in organic solvents. *Chem. Mater.*, 5, (September 1990) 564-567, ISSN 0897-4756.

Faraday, M. (1857). The Bakerian lecture: experimental relations of gold (and other metals) to light. *Phil. Trans. R. Soc. Lond.* , 147 (January 1857) 145-181, ISSN 0261-0523.

Goldys, E. & Xie, F. (2008). Metallic nanomaterials for sensitivity enhancement of fluorescence detection. *Sensors*, 8, (February 2008), 886-896, ISSN 1424-8220.

Gole, A. & Murphy, C. (2004). Seed-mediated synthesis of gold nanorods: role of size and nature of the seed. *Chem. Mater.*, 16, 19, (August 2004), 3633-3640, ISSN 0897-4756.

Goulet, P.; Pieczonka, N. & Aroca, R. (2003). Electrokinetic focusing injection methods on microfluidic devices. *Anal. Chem.*, 75, 8, (March 2003), 1918-1923, ISSN 0003-2700.

Guerrini, L.; Aliaga, A.; Cárcamo, J.; Gómez-Jeria, J.; Sanchez-Cortes, S.; Campos-Vallette, M. & García-Ramos, J. (2008). Functionalization of Ag nanoparticles with the bis-acridinium lucigenin as a chemical assembler in the detection of persistent organic pollutants by surface-enhanced Raman scattering. *Anal. Chim. Acta,* 624, 2, (August 2008), 286-293, ISSN 0003-2670.

Haes, A.; Zou, S.; Schatz, G. & Van Duyne, R. (2004). Nanoscale optical biosensor: short range distance dependence of the localized surface plasmon resonance of noble metal nanoparticles. *J. Phys. Chem. B*, 108, (March 2004), 6961-6968 ISSN 0022-3654.

a) Han, C; Zhang, L. & Li, H. (2009). Highly selective and sensitive colorimetric probes for Yb($^{3+}$) ions based on supramolecular aggregates assembled from beta-cyclodextrin-4,4'-dipyridine inclusion complex modified silver nanoparticles. *Chem. Commun.*, 24, (June 2009), 3545-3547, ISSN 1359-7345.

b) Han, X.; Cai, L..; Guo, J.; Wang, C.; Ruan W.; Han, W.; Xu, W.; Zhao, B. & Ozaki, Y. (2008). Fluorescein Isothiocyanate linked immunoabsorbent assay based on surface-enhanced Raman scattering. *Anal. Chem.,* 80, 8, (March 2008), 3020-3024, ISSN 0003-2700.

c) Han, X.; Huang, G. ; Zhao B.; & Ozaki, Y. (2009). Label-free highly sensitive detection of proteins in aqueous solutions using surface-enhanced Raman scattering. *Anal. Chem.,* 81, 9, (March 2009), 3329-3333, ISSN 0003-2700.

d) Han, X.; Jia, H.; Wang, Y.; Lu, Z.; Wang, C.; Xu, W.; Zhao, B. & Ozaki, Y. (2008). Analytical technique for label-free multi-protein detection based on Western Blot and surface-enhanced Raman scattering. *Anal. Chem.,* 80, 8, (February 2008), 2799-2804, ISSN 0003-2700.

Haynes, C. & Van Duyne, R. (2001) Nanosphere lithography: a versatile nanofabrication tool for studies of size-dependent nanoparticle optics. *J. Phys. Chem.* B, 105, (June 2001), 5599–5611, ISSN 1089-5647.

He, B.; Tan, J.; Liew, K. & Liu, H. (2004). Synthesis of size controlled Ag nanoparticles. *J. Mol. Catal. A Chem.*, 221, 1-2 (November 2004) 121-126, ISSN 1381-1169.

Heard, S.; Grieser, F. & Barraclough, C. (1983). The characterization of Ag sols by electron microscopy, optical absorption, and electrophoresis. *J. Colloid Interface Sci.* 93, 2 (June 1983) 545-555, ISSN 0021-9797.

Henglein, A. & Giersig, M._(1999). Formation of colloidal silver nanoparticles: capping action of citrate. *J. Phys. Chem. B.*, 103, 44, (November 1999), 9533-9539, ISSN 1520-6106.

Hildebrandt, P. & Stockburger, M. (1984). Surface-enhanced resonance Raman spectroscopy of Rhodamine 6G adsorbed on colloidal silver. *J. Phys. Chem.*, 88, (November 1984), 5935–5944, ISSN 0022-3654

Hirsch, L.; Stafford, R.; Bankson, J.; Sershen, S.; Rivera, B.; Price, R.; Hazle, J.; Halas, N. & West, J. (2003). Nanoshell-mediated near-infrared thermal therapy of tumors under magnetic resonance guidance. *Proc. Natl. Acad. Sci. U.S.A.*, 100, 23, (November 2003), 13549-13554, ISSN 0027-8424.

Huang, G.; Han, X.; Hossain, M. & Ozaki, Y. (2009). Development of a heat-induced surface-enhanced Raman scattering sensing method for rapid detection of glutathione in aqueous solutions. *Anal. Chem.*, 81, 14, (June 2009), 5881-5888, ISSN 0003-2700.

Ivleva, N.; Wagner, M.; Horn, H.; Niessner, R. & Haisch, C. (2008). In situ surface-enhanced Raman scattering analysis of biofilm. *Anal. Chem*, 80, 22, (October 2008), 8538-8544, ISSN 0003-2700.

Jeanmaire, D. & Van Duyne, R. (1977). Surface Raman spectroelectrochemistry: Part I. Heterocyclic, aromatic, and aliphatic amines adsorbed on the anodized silver electrode. *J. Electroanal. Chem.*, 84, (November 1977), 1–20, ISSN 0022-0728.

a) Jiang, J.; Bosnick, K.; Maillard, M. & Brus, L. (2003). Single molecule Raman spectroscopy at the junctions of large Ag nanocrystalsJ. *Phys. Chem. B*, 107, (September 2003), 9964–9972, ISSN 1089-5647.

b) Jiang, Z.; Yuan, W. & Pan, H. (2005). Luminescence effect of silver nanoparticle in water phase. *Spectrochim. Acta Part A*, 61, 11-12 (September 2005) 2488-2494, ISSN 1386-1425.

Kelly, K.; Coronado, E.; Zhao, L. & Schatz, G. (2004). The optical properties of metal nanoparticles: the influence of size, shape, and dielectric environment. *J. Phys. Chem. B*, 107, 3, (December 2002), 668-677, ISSN 1089-5647.

a) Kneipp, H.; Kneipp, J. & Kneipp, K. (2006). Surface-enhanced Raman Optical Activity on Adenine in Silver Colloidal Solution. *Anal. Chem.*, 78, 4, (January 2006), 1363-1366, ISSN 0003-2700.

b) Kneipp, K.; Kneipp, H.; Itzkan, I.; Dasari, R. & Feld, M. (1999). Ultrasensitive chemical analysis by Raman spectroscopy. *Chem. Rev.*, 99, (September 1999), 2957−2975, ISSN 0009-2665.

c) Kneipp, K.; Kneipp, H.; Kartha, V.; Manoharan, R.; Deinum, G.; Itzkan, I.; Dasari, R. & Feld, M. (1998). Detection and identification of a single DNA base molecule using surface-enhanced Raman scattering (SERS). *Phys. Rev. E*, 57, 6, (June 1998), R6281–R6284, ISSN 1063-651X.

d) Kneipp, K.; Wang, Y.; Kneipp, H.; Perelman, L.; Itzkan, I.; Dasari, R. & Feld, M. (1997). Single Molecule Detection Using Surface-Enhanced Raman Scattering (SERS). *Phys. Rev. Lett.*, 78, (March 1997), 1667–1670, ISSN 0031-9007.

Lakowicz, J. (1999). *Principles of fluorescence spectroscopy*, 2nd Ed., 239–240, Plenum Press, ISBN-10: 0306460939. ISBN-13: 978-0306460937, New York.

a) Li, H. & Bian, Y. (2009). Selective colorimetric sensing of histidine in aqueous solutions using cysteine modified silver nanoparticles in the presence of Hg^{2+}.* Nanotechnol., *20, (April 2009), 145502 (6pp), ISSN 0957-4484.

b) Li, N.; Liu, S. & Luo, H. (2002). Frequency doubling scattering and second-order scattering technology as novel methods for the determination of the inclusion constant of procaine hydrochloride to β-cyclodextrin. a) *Anal. Chim. Acta*, 472, 1-2 (November 2002) 89–98, ISSN 0003-2670.

a) Ling, J.; Huang, C.; Li, Y.; Zhang, L.; Chen, L. & Zhen, S. (2009). Light-scattering signals from nanoparticles in biochemical assay, pharmaceutical analysis and biological imaging. *Trends in Anal. Chem.*, 28, (April 2009), 447-453, ISSN 0167-2940.

b) Ling, J.; Sang, Y. & Huang, C. (2008). Visual colorimetric detection of berberine hydrochloride with silver nanoparticles. *J. Pharmaceut. Biomed. Anal.*, 47, (August 2008), 860-864, ISSN 0731-7085.

a) Liu, C.; Li, Z.; Du, B.; Duan, X. & Wang, Y. (2006). Silver nanoparticle-based ultrasensitive chemiluminescent detection of DNA hybridization and single-nucleotide polymorphisms. *Anal. Chem.*, 78, (June 2006), 3738-3744 , ISSN 0003-2700.

b) Liu, C.; Yang, X.; Yuan, H.; Zhou, Z. & Xiao, D. (2007) Preparation of silver nanoparticle and its application to the determination of ct-DNA. *Sensors*, 7, (May 2007), 708-718, ISSN 1424-8220.

Liz-Marzan, L. & Philipse, A. (1995). Stable hydrosols of metallic and bimetallic nanoparticles immobilized on imogolite fibers. *J. Phys. Chem.*, 99, (October 1995) 15120–15128, ISSN 0022-3654.

Luo, Y. & Sun, X. (2007). Rapid, single-step preparation of dendrimer-protected silver nanoparticles through a microwave-based thermal process. *Mater. Lett.*, 61, 8-9, (April 2007), 1622-1624, ISSN 0167-577X.

MacAskill, A.; Crawford, D.; Graham, D. & Faulds, K. (2009). DNA sequence detection using surface-enhanced resonance Raman spectroscopy in a homogeneous multiplexed assay. *Anal. Chem.*, DOI: 10.1021/ac901361b, ISSN 0003-2700.

Mafuné, F.; Kohnok, J.; Takeda, Y. & Kondow, T. (2000). Structure and stability of silver nanoparticles in aqueous solution produced by Laser ablation. *J. Phys. Chem. B.*, 104, 35, (August 2000) 8333–8337, ISSN 1520-6106.

Mc Farland, A. & Van Duyne, R. (2003). Single silver nanoparticles as real-time optical sensors with zeptomole sensitivity. *Nano Lett.*, 3, 8, (August 2003), 1057-1062, ISSN 1530-6984.

Michaels, A.; Jiang, J. & Brus, L. (2000). Ag nanocrystal junctions as the site for surface-enhanced Raman scattering of single Rhodamine 6G molecules. *J. Phys. Chem. B*, 104, (December 2000), 11965–11971, ISSN 1089-5647.

Mie, G. (1908). Beiträge zur Optik trüber Medien speziell kolloidaler Goldlösungen (contributions to the optics of diffuse media, especially colloidal metal solutions). *Ann. Phys.*, 25, (March 1907), 377–445, ISSN 0003-3804.

Mock, J.; Barbic, M.; Smith, D.; Schultz, D. & Schultz, S. (2002). Shape effects in plasmon resonance of individual colloidal silver nanoparticles. *J. Chem. Phys.*, 116., 15, (April 2002), 6755-6759, ISSN 0021-9606.

Monaghan, P.; McCarney, K.; Ricketts, A.; Littleford, R.; Docherty, F.; Smith, E.; Graham, D. & Cooper, J. (2007). Bead-based DNA diagnostic assay for Chlamydia using nanoparticle-mediated surface-enhanced resonance Raman scattering detection within a lab-on-a-chip format. *Anal. Chem.*, 79, 7, (February 2007), 2844-2849, ISSN 0003-2700.

Moskovits, M. (1985). Surface-enhanced spectroscopy. *Rev. Mod. Phys.*, 57, 3, (July 1985), 783–826, ISSN 00346861.

Mrozek, M. & Weaver, M. (2002). Detection and identification of aqueous saccharides by using surface-enhanced Raman spectroscopy. *Anal. Chem.*, 74, 16, (June 2002), 4069-4075, ISSN 0003-2700.

Nie, S.; Emory, S. R. (1997). Probing single molecules and single nanoparticles by surface-enhanced Raman scattering. *Science*, 275, (February 1997), 1102–1106, ISSN 0036-8075.

a) Olenin, A.; Romanovskaya, G.; Krutyakov, Y.; Lisichkin, G. & Zuev, V. (2008). Sensitized fluorescence of silver nanoparticles in the presence of Pyrene. *Doklady Chem.* (April 2008), 419, 91-94, ISSN 0012-5008.

b) Olenin, A.; Romanovskaya, G.; Krutyakov, Y.; Vasil'eva, S.; Kudrinskii, A. & Lisichkin, G. (2009). Synthesis and adsorption and luminescence properties of hydrophobic silver nanoparticles in the presence of pyrene. *J. Anal. Chem.*, 64, (January 2009), 26–30, ISSN 1061-9348.

Otto, A. (1984). Electronic scattering, spin effects, SERS and morphic effects, In : *Light scattering in solids IV*, Cardona, M., & Guntherodt, G., (Ed.), 289–418, Springer-Verlag:, ISBN 3540119426, Berlin, Germany.

Pal, A. & Maji, S. (2005). Arsine-induced formation of silver nanoparticles in micellar medium. application to spectrophotometric determination of arsenic. *Chem. Anal. (Warsaw)*, 50, 6, (November - December 2005), 1077- 1086, ISSN 0009-2223.

a) Pan, H.; Liang, F.; Mao, C. & Zhu, J. (2007). Sonochemical synthesis and resonance light scattering effect of Zn(II)bis(1-(2-pyridylazo)-2-naphthol) nanorods. *Nanotechnol.*, 18, (May 2007), 195606, ISSN 0957-4484.

b) Pan, H.; Tao, X.; Mao, C.; Zhu, J. & Liang, F. (2007). Aminopolycarboxyl-modified Ag_2S nanoparticles: Synthesis, characterization and resonance light scattering sensing for bovine serum albumin. *Talanta*, 71, 1 (January 2007) 276-281, ISSN 0039-9140.

Park, H.; Alam, S.; Lee, S.; Karim, M.; Wabaidur, S.; Kang, M. & Choi, J. (2009). Optical ascorbic acid sensor based on the fluorescence quenching of silver nanoparticles. *Luminescence*, DOI: 10.1002/bio.1160.

Pastoriza-Santos, I. & Liz-Marzan, L. (2002). Formation of PVP-protected metal nanoparticles in DMF. *Langmuir*, 18, 7, (February 2002) 2888–2894, ISSN 0743-7463.

Petit, C.; Lixon, P. & Pileni, M. (1993). Synthesis in situ of silver nanocluster in AOT reverse micelles. *J. Phys. Chem.*, 97, 49 (December 1993) 12974-12983, ISSN 0022-3654.

Pettinger, B.; Picardi, G.; Schuster, R. & Ertl, G. (2002). Surface-enhanced and STM-tip-enhanced Raman spectroscopy at metal surfaces. *Single Mol.*, 3, 5-6 (November 2002), 285–294, ISSN 1438-5163.

Pyatenko, A.; Yamaguchi, M. & Suzuki, M. (2005). Laser photolysis of silver colloid prepared by citric acid reduction method. *J. Phys. Chem. B*, 109, 46, (November 2005) 21608-21611, ISSN 1520-6106.

a) Rao, C.; Kulkarni, G.; Thomas, P. & Edwards, P. (2000). Metal nanoparticles and their assemblies. *Chem. Soc. Rev.*, 29, (January 2000) 27- 35, ISSN 0306-0012.

b) Rao, C.; Kulkarni, G.; Thomas, P. & Edwards, P. (2002). Size-Dependent Chemistry: Properties of Nanocrystals, *Chem. Eur. J.*, 8, 1, (December 2001), 28-35, ISSN 0947-6539.

a) Rayleigh, J. (1871). On the light from the sky, its polarisation and colour. I. *Phil. Mag.*, 41, (April 1871), 107–120, ISSN 1364-2812.

b) Rayleigh, J. (1871). On the light from the sky, its polarisation and colour. II. *Phil. Mag.*, 41, (April 1871), 274–27, ISSN 1364-2812.

Richards, D.; Milner, R.; Huang, F. & Festy, F. (2003) Tip-enhanced Raman microscopy: practicalities and limitations. *J. Raman Spectrosc.*, 34, (September 2003), 663–667, ISSN

Rivas, L.; Sanchez-Cortes, S.; Garcia-Ramos, J. & Morcillo, G. (2001). Growth of silver colloidal particles obtained by citrate reduction to increase the Raman enhancement factor. *Langmuir*, 17, (January 2001) 574–577, ISSN 0743-7463.

Roll, D.; Malicka, J.; Gryczynski, I.; Gryczynski, Z. & Lakowicz, J. (2003). Metallic colloid wavelength-ratiometric scattering sensors. *Anal. Chem.*, 75, 14, (July 2003), 3440-3445, ISSN 0003-2700.

Schatz, G. & Van Duyne, R. (2002). Vol. 1. In *Handbook of vibrational spectroscopy*, Chalmers, J. & Griffiths, P., (Ed.), 759–774, John Wiley & Sons, ISBN 978-0-471-98847-2, New York.

Schultz, S.; Smith, D.; Mock, J. & Schultz, D. (2000). Single-target molecule detection with nonbleaching multicolor optical immunolabels. *Proc. Natl. Acad. Sci. U.S.A.*, 97, 3, (February 2000), 996-1001, ISSN 0027-8424.

Shang, L.; Qin, C.; Jin, L.; Wang, L. & Dong, S. (2009). Turn-on fluorescent detection of cyanide based on the inner filter effect of silver nanoparticles. *Analyst*, 134, 7, (July 2009), 1477-1482, ISSN 0003-2654.

Sosa, I. & Noguez, C. (2003). Optical properties of metal nanoparticles with arbitrary shape. *J. Phys. Chem.B,* 107, (June 2003), 6269-6275, ISSN 1089-5647.

Stosch, R.; Henrion, A.; Schiel, D. & Güttler, B. (2005). Surface-enhanced Raman scattering based approach for quantitative determination of creatinine in human serum. *Anal. Chem.*, 77, 22, (October 2005), 7386-7392, ISSN 0003-2700.

Talley, C.; Jusinski, L.; Hollars, C.; Lane, S. & Huser, T. (2004). Intracellular pH sensors based on surface-enhanced Raman scattering. *Anal. Chem.*, 76, 23, (October 2004), 7064-7068, ISSN 0003-2700.

Taton, T.; Mirkin, C. & Letsinger, R. (2000). Scanometric DNA array detection with nanoparticle probes. *Science*, 289, 5485, (September 2000), 1757-1760, ISSN 0036-8075.

Thompson, D.; Enright, A.; Faulds, K.; Smith, W. & Graham, D. (2008). Ultrasensitive DNA detection using oligonucleotide-silver nanoparticle conjugates. *Anal. Chem.*, 80, (April 2008), 2805-2810, ISSN 0003-2700.

Toshima, N.; Yonezawa, T. & Kushihashi K. (1993). Polymer-protected palladium–platinum bimetallic clusters: preparation, catalytic properties and structural considerations. *J. Chem. Soc. Faraday Trans.*, 89, 14, (July 1993) 2537-2543, ISSN 0956-5000.

Van Dijk, M.; Tchebotareva, A.; Orrit, M.; Lippitz, M.; Berciaud, S.; Lasne, D.; Cognet, L. & Lounis, B. (2006). Absorption and scattering microscopy of single metal nanoparticles. *Phys. Chem. Chem. Phys.*, 8, (August 2006), 3486-3495, ISSN 1463-9076.

Vickova, B.; Tsai, D.; Gu, X. & Moskovits M. (1996). A microscopic surface-enhanced raman study of a single adsorbate-covered colloidal silver aggregate. *J. Phys. Chem.*, 100, 8 (February1996) 3169–3174, ISSN 0022-3654.

Wabuyele, M. & Vo-Dinh, T. (2005). Detection of human immunodeficiency virus type 1 DNA sequence using plasmonics nanoprobes. *Anal. Chem.*, 77, 23, (October 2005), 7810-7815, ISSN 0003-2700.

Wang, C.; Luconi, M.; Masi, A. & Fernández, L. (2009). Derivatized silver nanoparticles as sensor for ultra-trace nitrate determination based on light scattering phenomenon. *Talanta*, 77, 3 (January 2009) 1238–1243. 0039-9140.

Wei, H.; Chen, C.; Han, B. & Wang, E. (2008). Enzyme colorimetric assay using unmodified silver nanoparticles. *Anal. Chem*, 80, (September 2008), 7051–7055, ISSN 0003-2700.

Wilson, R.; Monaghan, P.; Bowden, S.; Parnell, J. & Copper, J. (2007). Surface-enhanced Raman signatures of pigmentation of cyanobacteria from within geological samples in a spectroscopic-microfluidic flow cell. *Anal. Chem.*, 79, 18, (August 2007), 7036-7041, ISSN 0003-2700.

Woo, M.; Lee, S.; Kim, G.; Baek, J.; Noh, M.; Kim, J.;Park, S.; Tehrani, A.; Park, S.; Seo, Y. Kim, Y.; Lee, Y.; Jeon, D. & Cho, M. (2009). Multiplex immunoassay using fluorescent-surface enhanced Raman spectroscopy dots for the detection of bronchioalveolar Stem cells in murine lung, *Anal. Chem.*, 81, 3, (December 2009), 1008-1013, ISSN 0003-2700.

Wu, T.; Li, Y. & Huang, C. (2009). Selectively colorimetric detection of cysteine with triangular silver nanoprisms. *Chin. Chem. Lett.*, 20, (May 2009), 611-614, ISSN 1001-8417.

Xiong, D. & Li, H. (2008). Colorimetric detection of pesticides based on calixarene modified silver nanoparticles in water. *Nanotechnol.*, 19, (November 2008), 465502 (6pp), ISSN 0957-4484.

Xiong, X.; Xie, Y.; Du, G.; Liu, X. & Tian, X. (2002). Ultrasound-assisted self-regulation route to Ag nanorods. *Chem. Lett.*, 102, 1, (January 2002), 98-99, ISSN 0897-4756.

Xu, H.; Bjerneld, E.; Käll, M. & Börjesson, L. (1999). Spectroscopy of single hemoglobin molecules by surface enhanced Raman scattering. *Phys. Rev. Lett.*, 83, (November 1999), 4357–4360, ISSN 0031-9007.

Yguerabide, J. & Yguerabide, E. (1998). Light-scattering submicroscopic particles as highly fluorescent analogs and their use as tracer labels in clinical and biological applications: II. Experimental characterization. *Anal. Biochem.*, 262, 2, (September 1998), 157-176. ISSN 0003-2697.

Zhao, H.; Ding, F.; Wang, X.; Ju, H. ; Li, A. & Jin, L. (2008). A study on silver nanoparticles-sensitized fluorescence and second-order scattering of the complexes of Tb(III) with ciprofloxacin and its applications. *Spectrochim. Acta* Part A, 70, (July 2008) 332–336, ISSN 1386-1425.

a) Zhou, H.; Wu, X. & Yang J. (2009). Study on the interaction of nucleic acids with silver nanoparticles-Al(III) by resonance light scattering technique and its analytical application. *Talanta,* 78, 3, (May 2009) 809–813. 0039-9140.

b) Zhou, Y.; Hao, L.; Hu, Y.; Zhu, Y. & Chen Z. (2001). Synthesis of nanowires and coral-shaped nanostructures of Ag by an ultraviolet photo-reduction technique at room temperature. *Chem. Lett.*, 30, 11, (November 2001) 1192-1993, ISSN 0366-7022.

13

Biosynthesis and application of silver and gold nanoparticles

Zygmunt Sadowski
Wroclaw University of Technology
Poland

1. Introduction

Nanotechnology has dynamically developed as an important field of modern research with potential effects in electronic and medicine (Glomm 2005, Chan 2006, Boisselier and Astruc 2009). Nanotechnology can be defined as a research for the design, synthesis, and manipulation of structure of particles with dimension smaller than 100nm. A new branch of nanotechnology is nanobiotechnology. Nanobiotechnology combines biological principles with physical and chemical procedures to generate nano-sized particles with specific functions. Nanobiotechnology represents an economic alternative for chemical and physical methods of nanopaticles formation. These methods of synthesis can be divided on intra cellular and extracellular (Ahmad et al. 2005).
This integration of nanoparticles with biological molecules has lead to the development of diagnostic devices, contrast agents, and important tools in cancel therapy. Nanobiotechnology describes an application of biological systems for the production of new functional material such as nanoparticles. Biosynthetic methods can employed either microorganism cells or plant extract for nanoparticles production. Biosynthesis of nanoparticles is an exciting recent addition to the large repertoire of nanoparticles synthesis methods and now, nanoparticle have entered a commercial exploration period. Gold and silver nanoparticles are presently under intensive study for applications in optoelectronic devices, ultrasensitive chemical and biological sensors and as catalysts. This chapter is devoted to biosynthesis and application of gold and silver nanoparticles

2. Biosynthesis of silver and gold nanoparticles by microorganisms

2.1 Synthesis of nanoparticles by bacteria

An important part of work in nanobiotechnology concerns the synthesis of nanoparticles of different chemical compositions, sizes, shapes, and polydispersity. Many microorganism produce inorganic materials ether intra- or extracellularly. Well-known example is magnetotactic bacteria which able to synthesize magnetic nanoparticles (Bazylinski and Frankel 2004). Magnetotactic bacteria are motile, prokaryotes that move along geometric field lines. They produce magnetosomes, unique intracellular structure contains a magnetic

particle, in narrow range of very low oxygen concentration. Magnetotactic bacteria usually mineralize either oxide magnetite Fe_3O_4 or iron sulfide Fe_3S_4 - greigite.
An extracellularly preparation of metal nanoparticles generally involves the reduction of metal ions in solution.
The formation of extracellular and intracellular silver nanoparticles by bacteria (*Pseudomonas stulzeri, Escherichia coli, Vibrio cholerae, Pseudomonas aeruginosa, Salmonells typlus,* and *Staphylococcus currens* has been investigated (Lengke et al. 2007).
Various microbes are known to reduce metal ions to the metals. The formation of extracellular silver nanoparticles by photoautotrophic cyanobacterium *Plectonema boryanum* had been described (Langke et al. 2007). The procedure of synthesis was as follows: 5 mL of silver solution (approximately 560 mg/L) was added to 5 mL of washed cyanobacteria culture (approximately 10 mg dry weight). The synthesis was conducted at 25, 60 and 100^{0}C for up to 28 days in the dark. Only, at 100^{0}C, the soluble silver was completely precipitated from solutions within 28 days. A greyish-black silver particles adhered to bacterial cells were observed macroscopically. The reaction products were analyzed using transmission electron microscopy (TEM) and X-ray photoelectron spectroscopy (XPS). The addition of $AgNO_3$ caused the precipitation both inside and outside the microbial cells. At 60^{0}C, silver nanoparticles were deposited at the cell surface. At 100^{0}C, the cyanobacterial cells were incrusted by silver nanoparticles. The size of nanoparticles inside the cell was ranging from 1 to 40 nm. The size of nanoparticles of silver which were precipitated outside the bacteria cells was in the range of 1 -200nm.

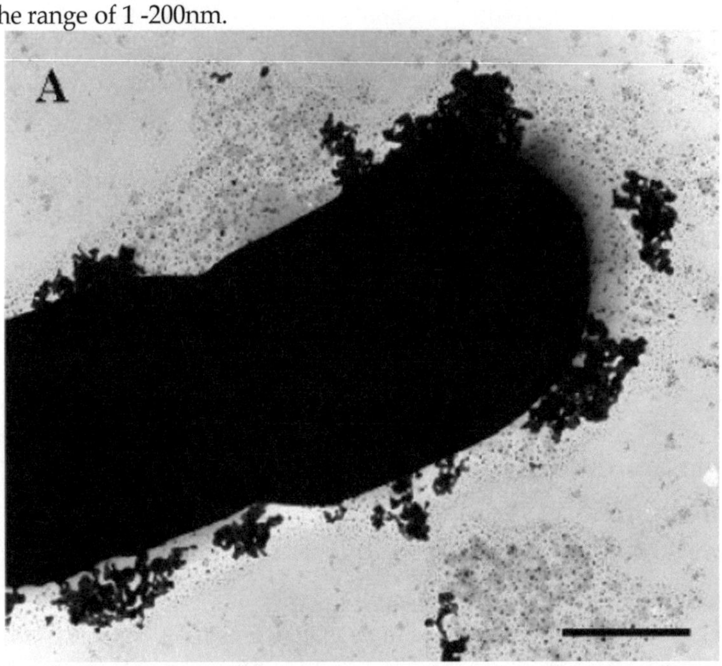

Fig. 1. Precipitated silver nanoparticles on the cyanobacteria cell surface (from Langke et al. 2007)

The bioreduction of the Ag^+ ions could be associated with metabolic processes utilizing nitrate by reducing nitrate to nitrile and ammonium (Langke et al. 2007).

Cyanobacteria commonly use nitrate as the major source of nitrogen. Nitrate was reduced by cyanobacteria metabolic process.

$$NO_3^- + 2H^+ + 2e^- = NO_2^- + H_2O$$

$$NO_2^- + 8H^+ + 6e^- = NH_4^+ + 2H_2O$$

It suggests that Ag^+ ions could be reduced by an intracellular electron donor (Lengke et al., 2007).

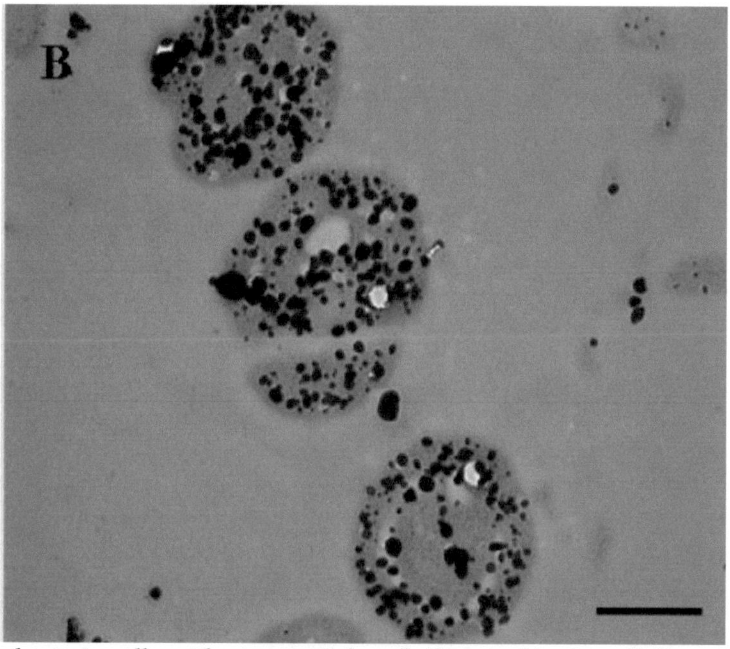

Fig. 2. Cyanobacteria cells with nanoparticles of silver inside the cell (from Langke et al. 2007

The intracellular recovery of gold by microbial reduction of $AuCl_4^-$ ions using the anaerobic bacterium *Shewanella algae* has been investigated (Konishi et al. 2006). The solution turned light yellow after 1h, indicating the initial formation of gold nanoparticles.

Silver nanoparticles in the range of 50 nm were synthesized by supernatant of *Bacillus licheniformis* (Kalishwaralal et al. 2008). *Bacillus licheniformis* is a gram positive, thermophilic bacterium, commonly found in the soil. As was showed previously, during the visual observation culture supernatant incubated with silver nitrate showed a color change from yellow to bran. The apparence of brown color suggested the formation of silver nanoparticles. The XRD pattern shows four characteristic peaks in the whole spectrum. The peaks at 2Θ values of 38.48°, 44.48°, 64.69° and 77.62° corresponding to 111, 200, 220 and 311 planes for silver crystal, respectively (Fig. 3).

Recently, a rapid method for synthesizing silver nanoparticles by treating the aqueous silver nitrate solution with culture supernatants of different strains of *Enterobacteria* such as *Klebsiella pneumonia* has been described (Shahverdi et al. 2007 and Mokhtari et al. 2009). The process of synthesis was quite fast. As it has been presented, the silver nanoparticles were formed within 5 min of the silver ions coming in contact with the culture supernatant . The particle size histogram of silver nanoparticles showed the particle range in size from 28.2 nm to 122nm with the average size value of 52.5 nm.

Enterobacteria is a Gram-negative bacteria, usually associated with intestinal infections. The investigations which have been realized by Mokhtari and coworkers showed that piperitone (3 methyl-6-1 methylethyl)-2 cyclohexan-1-one) can be responsible for the silver ions reduction to metal (Mokhtari 2009). This conclusion supports the hypothesis that nitroreductase enzymes may be involved in silver ions reduction process.

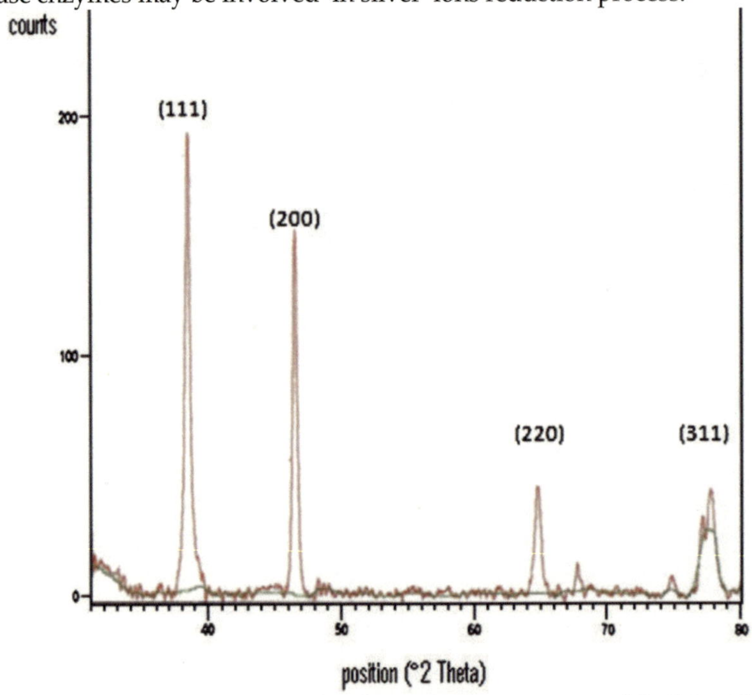

Fig. 3. XRD pattern of silver nanoparticles formed after reaction of *B. licheniformis* culture supernatant with $AgNO_3(1\ 10^{-3}\ M)$ (from Kalishwaralal et al. 2008)

The effect of visible-light irradiation on the synthesis of silver nanoparticles has been recently investigated (Mokhtari et al., 2009). The following procedure was used for the silver nanoparticles formation using a supernatant of *B. pneumonia* in the presence of light. The first step was silver chloride suspension preparation. The sodium chloride solution (50mL, 140 mg/L) was added to 50 mL of silver nitrate solution (340 mg/L) in a dark pale. The sediment fraction of AgCl was separated, cleaned and redispersed in distilled water. Then 1 mL of culture supernatant of *K. pneumonia* was added to the suspension. The silver nanoparticles fabrication was realized in the presence of various visible light intensities, generated by a 75 W halogen lamp. The experimental results confirmed a proposed mechanism involving the conversion of AgCl into Ag nanoparticles. This conversion of AgCl to silver nanoparticles by culture supernatant of *K. pneumonia* in a bright condition is presented at Fig. 4. Generally, AgCl is treated as the main intermediate compound during the bioreduction of the silver ions.

Kalimuthu and coworkers (Kalimuthu et al. 2008) have investigated the process of synthesis silver nanoparticles using bacteria *Bacillus lichemiformis* and sonification of reacting mixture. *Bacillus licheniformis* were isolated from sewage collected from municipal wastes. Ultrasonic destruction of bacteria cell was carried out with ultrasonic processor over three 15 s periods.

The interval between periods was 45 s. Fabricated silver nanoparticles had a size ranging from 2 nm to 100 nm. The average particle size was found to be around 50 nm.

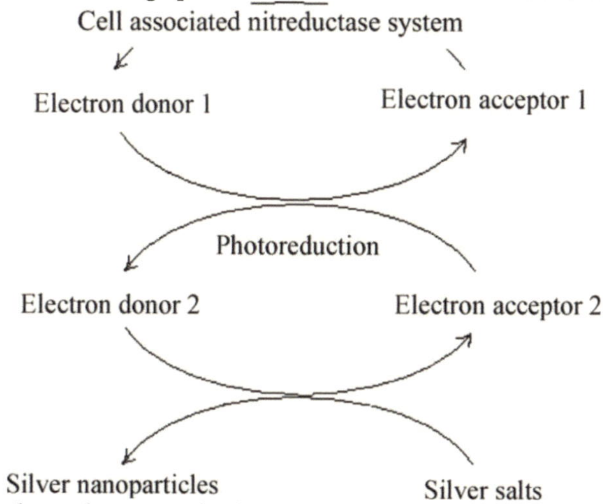

Fig. 4. Hypothetical mechanism of silver nanoparticles synthesis using the culture of *B.lichemiformis* (from Mokhtari et al. 2009)

The enzyme involved in the fabrication of nanoparticles can belong to nitrate reductasa, presented in *B..licheniformis.* This enzyme reduces the silver ions to metallic silver. It is know that NADH is dependent nitrate reductaze enzyme are important factor in the biosynthesic of metal nanoparticles. The possible mechanisms of reduction of silver ions is using nitrate reductasa, as it was presented in Fig.5.

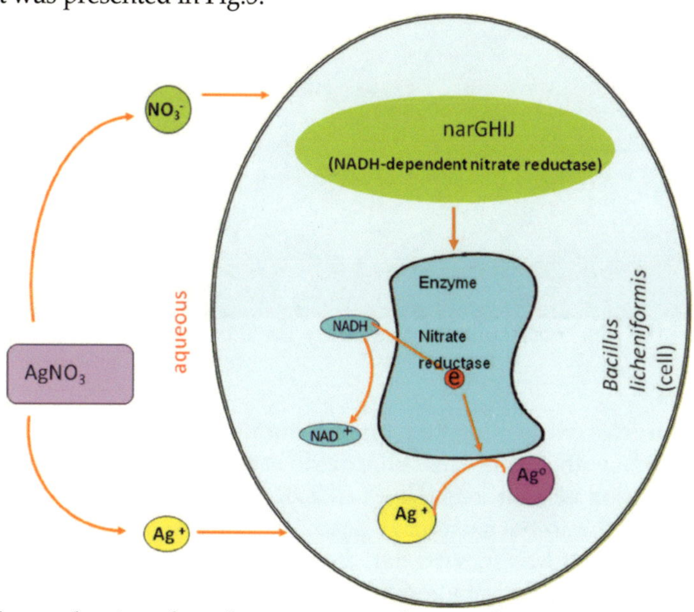

Fig. 5. Possible mechanism for silver nanoparticles synthesis using *Bacillus licheniformis* (from Kalimuthu et al. 2008)

Minaeian and coworkers (Minaeian et al. 2008) used different cultures which were sterilized and inoculated with fresh culture of the strains (*Bacillus subtilis, Lactobacillus acidophilus, Klebsiella pheumoniae, Escherichia coli, Enterobacterdoacae, Staphylococcus aureus, Candida albicans*). The biosynthesized silver nanoparticles have the size range 50-100 nm.

Novel method of biosynthesis of silver nanoaparticles using a combination of culture supernatant of *Bacillus subtilis* and microwave irradiation was proposed by Saifuddin and coworkers (Saifuddin et al. 2009). The formation of nanoparticles by this method was extremely rapid. The samples (supernatant and $AgNO_3$ solution) were subjected to several short burst of microwave irradiation at the frequency of 2.45 GHz, at power output of about 100 W in a following cyclic mode on 15 s off 15 s to prevent of overheating. The synthesized nanoparticles had the size range of 5 -50 nm.

The gold nanoparticles were synthesized using similar procedure. The two isolated strains of *Pseudomonas aerruginosa* were adopted to synthesis of gold nanoparticles (Husseiny et al. 2007). The synthesis of stable gold nanocubes by the reduction of aqueous $AgCl_4^-$ by *Bacillus licheniformis* has been described (Kalishwaralal et al. 2009).

Fig. 6. SEM image of gold nanocubesfabricated by *Bacillus licheniformis* (from Kalishwaralal 2009)

The gold nanoparticles were prepared on the surface of bacteria cells as a result of incubation of bacteria with $AuCl_4^-$ ions. Procariotic bacteria *Rhodopseudomans capsulate* were adapted to bioreduction of gold ions (He et al. 2007). The aqueous chloroaurate ions were reduced during exposure to the bacteria *R. capsulate* biomass. The reaction was completed after 48 h of incubation. It was showed that the shape of gold nanoparticles was controlled by pH of solution. For the explanation of this behavior, the following mechanism was discussed. The aqueous chloroaurate ions were reduced during took a contact with bacterial cell groups. These groups such as amino, sulfhydryl and carboxylic had a positive charge These positive charges depend on the solution pH. The adsorption of $AuCl_4^-$ ions onto the cell surface was supposed. Bioreduction of gold ions seems to be initiated by electron transfer from NADH by NADH-dependent reductase as electron carrier. Then, the gold ions obtained electrons reduced to gold.

2.2 Synthesis of nanoparticles by the fungal systems

The fungi are extremely good candidates in the synthesis of metal nanoparticles. The synthesis of silver particles using two *Aspergillus niger* strains was investigated (Sadowski et al. 2008 A and B). These strains were isolated from a soil. Inoculated fungi were prepared in Petri dishes at the room temperature using 2% malt extract with 0.5% yeast extract. Fungal biomass preparation was grown aerobically in the liquid medium containing (g/L): KH_2PO_4 7.0; K_2HPO_4 2.0; $MgSO_4$ $7H_2O$ 0.1; $(NH_4)_2SO_4$ 1.0; yeast extract 0.6 and glucose 10.0. After the incubation, the biomass was separated and extensively washed with distilled water. Fresh and clean biomass was collected with 100 mL of Milli-Q deionized water and new incubation was carry out. After the incubation, the supernatant was obtained by passing suspension through Whatman filter paper No. 1. For synthesis of silver nanoparticles, $AgNO_3$ 1mM solution of the final concentration was mixed with 50mL of cell filtrate in an Erlenmeyer flask and agitated at 25^0 C in the dark. Sample 1 mL was with drown at different time intervals and absorbance was measured using UV-visible spectrophotometer. The spectra recorded at different times of biosynthesis is presented in Figure 7.

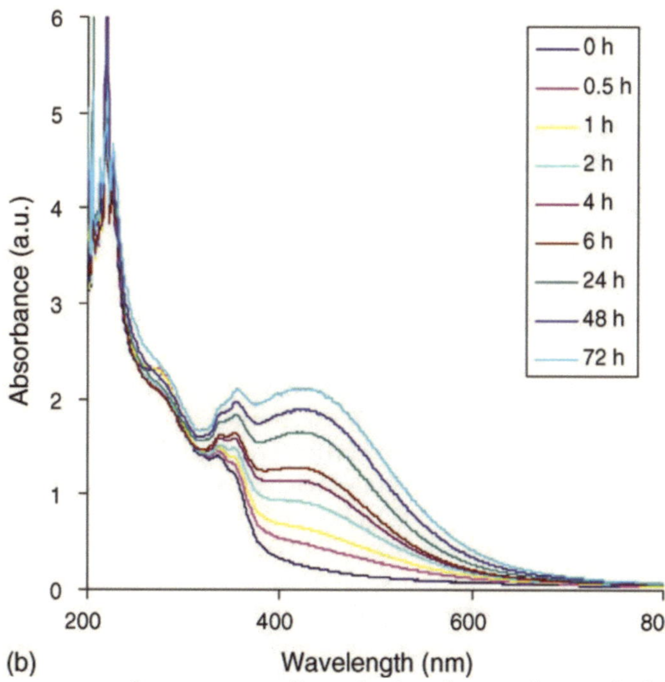

Fig. 7. UV-vis spectrum of aqueous medium during the synthesis of silver nanoparticles (from Sadowski et al. 2008A)

The electrokinetic measurements indicated that zeta potential of silver nanoparticles was negative value (Sadowski et al. 2008B).

The extracellular synthesis of silver and gold-silver nanoparticles by fungus *Fusarium oxysporum* biomass had a contribution on the formation of nanoparticles (Ahmad et al. 2003). The reduction of silver ions by *Fusarium oxysporum* strains has been attributed to a nitrate-dependent reductase and a shuttle quinine extracellular process. In a typical biosynthesis, 10g of fungal biomass was taken in Erlenmeyer flask containing 10 ml of distilled water. A corresponding quantity of $AgNO_3$ was added to Erlenmeyer flask to yield the concentration of Ag^+ ions equal 10^{-3} M. The reaction was carried out in the dark. Periodically, 5 mL of the reaction solution was removed and subjected to UV-vis spectroscopic measurements. Independently, it was observed that the biomass suspension has a yellow color before reaction with the silver ions and brown color on completion of the reaction.

The extracellular biosynthesis of silver nanoparticles using the filamentous fungus *Aspergillus fumigatus* has been investigated (Bhainsa and D'Souza 2006). This study included kinetics of synthesis, spectroscopic and microscopic characterization of the silver nanoparticles. The fungus *Aspergillus fumigatus* (NCIM 902) was grown aerobically in a liquid media containing (g/L) KH_2PO_4, 7.0; K_2HPO_4, 2.0; $MgSO_4$ $7H_2O$, 0.1; $(NH_4)_2SO_4$, 1.0; yeast extract, 0.6; and glucose, 10.0. The biomass was harvested after 72 h of growth, then it was extensive washing with distilled water. 20g of fresh biomass was contacted with 200 mL of deionized water for 72 h and agitated in the same condition as first sample. After incubation, the suspension was filtered using Whatman filter paper No. 1.

The mechanism of leading to formation of silver nanoparticles is not definitely understood at the moment. One hypothesis supports that a first step involve trapping of the Ag^+ ions onto the surface of the fungal cells. It can be realized by electrostatic interaction between Ag+ and a negative charged carboxylate groups on the cell surface. The reduction of metal ions occurs on the surface by the enzymes presented in the cell wall (Mukherjee et al. 2001). The extracellular enzymes such as naphthoquinons and anthraquinones shoved an excellent redox properties, they can act as electron shuttle in silver ions reduction.

It was presented (Duran et al. 2005) that enzyme hydrogenase is present in a filtrate broth obtained from *Fusarium oxysporum* growth. The silver nanoparticles production capacity has been depended on the reductase/electron shuttle relationships.

Next paper presents the extracellular synthesis of stable silver nanoparticles using the fungus *Penicillium brevicompactum* WA 2315 (Shaligram et al. 2009). The analysis of data obtained from transmission electron microscope showed the average size of nanoparticles to be 58.35 ± 17.88 nm. Figure 8 shows the FTIR spectrum of the freeze-dried powder of silver nanoparticles formed after 72 h of incubation with the fungus supernatant. The band seen at 3356 cm^{-1} and 2922 cm^{-1} were assigned to the stretching vibration of primary and secondary amines, respectively. The bands at 1622 cm^{-1} and 1527 cm^{-1} correspond to the stretch molecule vibration. The two bands existing at 1412 cm^{-1} and 1029 cm^{-1} can be assigned to the C-N stretching vibrations of aromatic and aliphatic amines. This FTIR spectrum supports the presence of proteins in the synthesis of silver nanoparticles.

The use of fungus *Fusarium semitectum* for the extracellular synthesis of silver nanoparticles has been reported by Basavaraja and co-workers (Basavaraja et al. 2008). The formation and stability of the reduced silver nanoparticles in colloidal solution was monitored by using UV-vis spectral analysis. It was observed from spectra that the silver surface plasmon resonance band occurred at 420nm and this absorption steadily increased in intensity as a function of time of reaction. IR spectroscopic study has confirmed that amino acid and peptides have formed a coat covering the silver nanoparticles to prevent agglomeration.

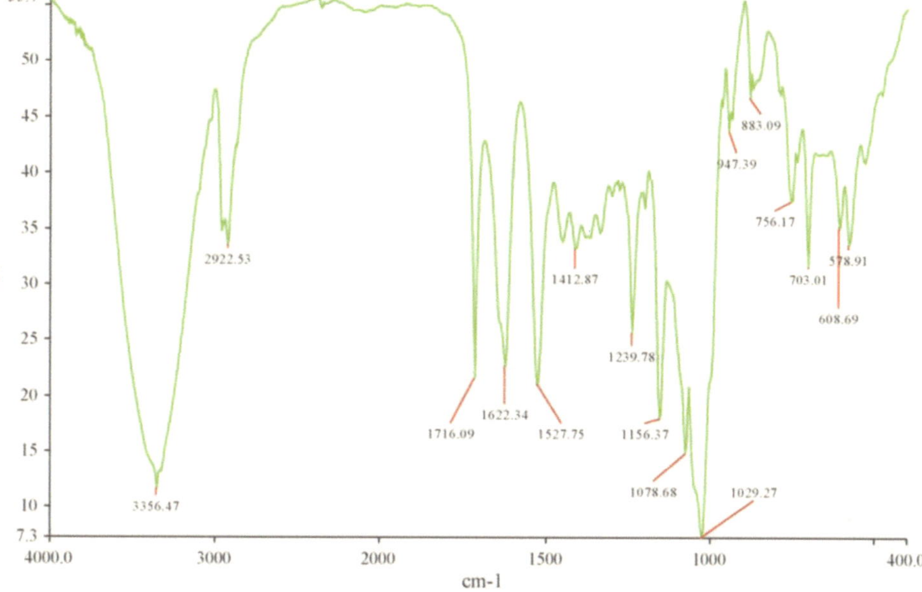

Fig. 8. FTIR spectrum of silver nanoparticles formed after 72 h of incubation of the supernatant of *P. brevicompactum WA 2315*, (from Shaligram et al. 2009)

The mean particles diameter of silver nanoparticles was calculated from the XRD pattern using Scherrer equation. The calculated average particles size of silver nanoparticles was found to be 35 nm.

The extracellular synthesis of silver nanoparticles by a marine fungus *Penicillium fellutanum* has been described by Kathiresan and coworkers (Kathiresan et al. 2009). The fungus *P. fellutanum* was isolated from a costal mangrove sediments. The procedure of biosynthesis of silver nanopaticles was analogous with the procedure has early described. For synthesis of silver nanoparticles $AgNO_3$ 1mM solution was mixed with 50 mL of cell filtrate and agitated in dark. The present of silver nanoparticles in reacting mixture was confirmed by absorption peak at 430 nm. The obtained silver nanoparticles were spherical in shape with size ranging from 5 to 25 nm. The TEM micrograph of silver nanoparticles synthesized by *P. fellutanum* is presented in Fig.9.

Cladosporium cladosporioides is a commonly available fungus found in marshland regions, This fungus was employed to biosynthesis of silver nanoparticles (Balaji et al. 2009). To prepare biomass for the synthesis, the fungus was grown aerobically in liquid medium containing (g/L): K_2HPO_4, 2.5; KNO_3, 5.0; $MgSO_4$ 7 H_2O, 1.00; $MnSO_4$ H_2O 0.001; $CuSO_4$ $5H_2O$, 0.003; $ZnSO_4$ $7H_2O$, 0.01; Na_2MoO_3 $2H_2O$ 0.0015; $FeCl_3$ 0.02 and glucose, 40.0.The fungus was grown for 1 week then the broth was filtered and washed with distilled water. 10 mL of pure solution were brought in contact with 100 mL of double distilled water containing 0.01 mL Ag^+ metal ion solution. The mixture was agitated and kept on a shaker at 27^0C for 78 h. Size and morphology of obtained nanoparticles were analyzed by employing TEM and Fourier transform infrared spectroscopy (FT-IR).

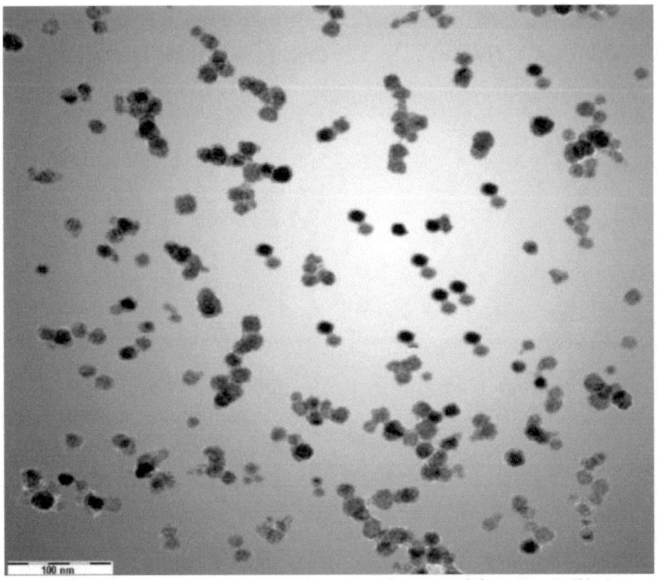

Fig. 9. TEM micrograph of silver nanoparticles synthesized by *Penicilluium fellutanum* (from Kathiresan et al. 2009)

The extra- and intracellular biosynthesis of gold nanoparticles by fungus *Trichothecium* sp. was reported by Absar and coworkers (Absar et al. 2005). It was observed that when the gold ions reacted with the *Trichothecium* sp. fungal biomass under stationary conditions results in the rapid extracellular formation of gold nanoparticles of spherical rod-like and triangular morphology whereas reaction of the biomass under shaking conditions resulted in intracellular growth of the gold nanoparticles.

The biosynthesis of gold nanoparticles using marine alga *Sargassum wightii* has been investigated (Singaravelu et al. 2007). The stable gold nanoparticles were obtained by reduction of aqueous $AuCl_4^-$ ions by extract of marine alga. Seaweed were collected from Mandapad Camp south east coast of Tamil Nadu, India. Collected seaweed were cleaned and dried for 3-5 days. Dried material was ground to powder in glass mortar. Synthesis of gold nanoparticles was carried out by taking 1g of seaweed powder in 500mL Erlenmeyer flask with 100 mL of 10^{-3} M aqueous $HAuCl_4$ solution. Aliquots of the reaction solution were investigated, the absorption of solution was measured.

The 95 % of the gold recovery occurred after 12 h of reaction. UV-vis spectra were recorded from the aqueous chloroauric acid and algae reaction medium. The bends at 527 nm corresponds to the surface plasmon resonance and showed the formation of nanoparticles. These particles were illustrated by TEM micrographs. The diameter of gold nanoparticles was ranging from 8 to 12 nm.

The synthesis of gold nanoparticles by the reduction of gold nanoparticles by the reduction of gold ions using a kind of Chinese herbal extract - *Barbated Skullcup* has been reported (Wang, et al. 2009)

The study on edible mushroom as reducing and protective agent for both silver and gold nanoparticles has been carried out by Philip (Philip 2009). Edible mushroom *Volvariella volvacea* was used for the metal nanoparticles synthesis. 68g of finely cut mushroom was boiled for 2 min in 300 mL water. Then the solution was filtered. This filtrate was cooled to room temperature and used as reducing agent. 30 mL aqueous solution of $HAuCl_4$ $3H_2O$ and 6 mL of mushroom extract were mixed together. Slow reduction taken place and was total in 2.5 h period of time. Gold nanoparticles colloid had a stable purple color. It was showed that the size and shape of Au nanoparticles can be controlled by varying the temperature and relative concentration of the extract with respect to the metal ion. In the case of silver nanoparticles biosynthesis, 35mg of $AgNO_3$ dissolved in 250 mL of water was contacted with a various volume (from 6 to 25 mL) of mushroom extract. The bioreduction was complete in 6 h. Ag-Au bimetallic nanoparticles were prepared by the simultaneous reduction of Au^{3+} and Ag^+ ions using excess of mushroom extract.

The extremophilic actinomycete, *Thermomonos* when exposed to gold ions reduced the metal ions extracellulaly, yielding gold nanopaticles (Sastry et al. 2003). For the synthesis gold nanoparticles, the actinomycete was grown in 250 mL Erlenmeyer flasks containing 50 mL of agar slants. Sodium carbonate was used for pH adjusted. *Thermomonos* sp. had an optimum growth at pH=9 and temperature 50^0C

3. Metal nanoparticles synthesis using plant extracts

An important branch of biosynthesis of nanoparticles is the application of plant extract to the biosynthesis reaction. Fig. 10 shows some popular plants using to the extract preparation.

A rapid reduction of the silver ions was observed when the silver nitrate solution was contacted with geranium (*Pelargonium graveolens)* leaf extract (Shiv Shankar et al. 2003). The extract used for reduction of Ag^+ ions to Ag^0 was prepared by taking 20g of thoroughly washed and finely cut geranium leaves in a 500 mL Erlenmeyer flask with 100 mL of

Fig. 10. Plants used for biosynthesis of metal nanoparticles

distilled water. The suspension was boiling for 1 min. 5 mL of pure broth was added to 100 mL of 10^{-3} M aqueous solution of $AgNO_3$. The bioreduction of the Ag^+ ions was monitored by measuring the UV-vis spectra of the solution.

Fig. 11 shows the UV-vis spectra recorded from the aqueous silver nitrate-geranium leaf extract reaction medium as a function of the reaction time.

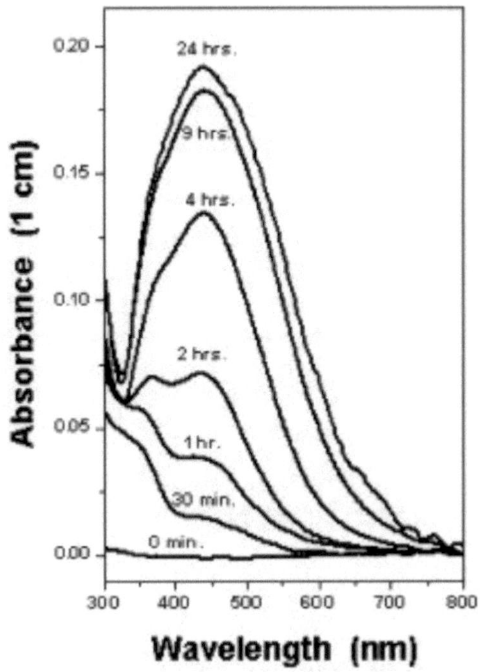

Fig. 11. UV-vis spectra as a function of reaction time of silver ions with *P. gravelens* leaf broth (from Shiv Shankar et al., 2003)

In the case of Neen leaf extract a competition reduction of Au^{3+} and Ag^{+} ions presented simultaneously in solution was observed. It has lead to the synthesis of bimetallic Au core-Ag shell nanoparticles in solution (Shiv Shankar et al. 2004).

Silver nanoparticles ranging from 55 to 80 nm in side and triangular or spherical gold nanoparticles were fabricated using the novel sundried biomass of *Cinnammum canphora* leaf (Huang et al. 2007). It was found that formation of gold nanotrangles by *C. camphore* leaf at ambient temperature strongly depended on the amount of dried biomass. This biomass offered sufficient protective biomolecules.

A simple procedure applying *Aloe vera* leaf extract has been used for gold nanotriangle and spherical silver nanoparticles synthesis (Chandran et al., 2006). The kinetics of gold nanoparticles formation was monitored by UV-vis absorption spectroscopy and transmission electron microscopy (TEM). The effect of the amount of leaf extract on the synthesis of gold nanotriangles was investigated by observation of product formed. Addition of *Aloe vera* extract to 10^{-3} M aqueous solution of $HAuCl_4$ led to the appearance of a red color in solution after bout 5 h of reaction. An analysis of the percentage of triangles formed in the reaction medium as a function of varying amounts of the *Aloe vera* extract showed that more spherical particles were formed with increasing amount of added extract.

Eclipta (known as Bhingraj) belongs to the family Asteraceae. It is a common weed growing mostly in a shade area (Jha et al., 2008). Extract from *Eclipta* leaf has used as medically important herb. The plant is rich in flavonoids, belonging to the group of phenolic compounds. The sample of 5 g of freshly collected leaves of *Eclipta* was washed for 10 min and ringed briefly in distilled water. Prepared biomaterial was taken in 250 mL capacity beaker having 200mL of 50% Et-OH and was placed on boiling steam bath for 15 to 20 min.

till color of the solvent changes to dark green. The cold extract was treated with 20 ml of 0.025 (M) $AgNO_3$ solution and warmed again on the steam bath for 10 min until the color of solution changes. The formation of silver nanoparticles was monitored by UV-vis spectrophotometry and X-ray diffractometer.

Biosynthesis of silver nanoparticles was also conducted using *Cycas* leaf extract. *Cycas* belongs to the *Cycadaceos* family. It is a common gymnospermic plant and is a commercial source of sago (Jha and Prasad 2009). This plant is rich in flavonoids broadly belonging to the lass of phenolic compounds. The procedure of *Cycas* leaf broth preparation was like to made *Eclipta* extract . The *Cycas* extract solution was treated with 20 mL of 0.25 M $AgNO_3$ solution and wormed on the steam bath for 20 min until the color of solution changes to brown.

The particle size histogram obtained silver nanoparticles shows broad distribution of particle size. The size particle ranged from 2-6 nm and the average particle size comes out to be 3.29 ± 0.22 nm. The X-ray diffraction patter obtained for silver nanoparticles synthesized by *Cycas* leaf broth shows that the silver nanopaticles are crystalline in nature.

Silver nanoparticles were successfully synthesized using the latex of *Jatropha curcas* (Bar et al. 2009). The plant, *Jatropha curcas* is commercially important one as bio diesel is extracted from it seeds on industrial scale. Crude latex was obtained by cutting the green stems of *J. curcas* plants. Milky white latex was stored in the refrigeration. For biosynthesis, 20 mL of 3% of latex solution was mixed with 20 mL of 5 10^{-3} M aqueous silver nitrate solution. The reacting mixture was heated at 85°C with constant stirring for 4h.

The bark powder and water extract from *Cynnamn zeylanicum* tree were used for silver synthesis (Sathishkumar et al. 2009). The bark was cut into small pieces powdered in a mixer and then sieved using a 20-mesh sieve. The final sieved fraction of powder was used for further experiments. For the production of extract, 2.5 g of powder was added to a 500 mL Erlenmeyer flask with 100 mL distilled water and then boiled for 5 min. For the silver nanoparticles synthesis 100, 500 and 1000 mg of powder were added to 50 mL of 1mM aqueous $AgNO_3$ solution in a 250 mL Erlenmeyer flask. The flasks were then incubated in the dark at 25° C. In the second way 1, 2.5 and 5 mL of extract were used for the biosynthesis of Ag nanoparticles from 50 mL of 1 mM aqueous $AgNO_3$ solution.

The first report on the formation of gold nanoparticles by living plants was presented by Gardea-Torresdey and coworkers (Gardea-Torresdey et al. 2002). The alfalfa seeds were soaked to avoid fungal contamination in 3 % formaldehyde for 15 min. and washed three times with deionised water. Approximately 100 seeds were transferred to a mason jar and autoclaved for a sterile conditions. The nutrient solution had a composition: $Ca(NO_3)_2\ 4H_2O$ (3.57 10^{-4} M); H_3BO_3 (2.31 10^{-5} M); $CaCl_2\ 2\ H_2O$ (2.14 10^{-3} M); KH_2PO_4 (9.68 10^{-4} M); KNO_3 (2.55 10^{-4} M); $MgClO_4$ (1.04 10^{-3} M); $FeCl_3$ (6.83 10^{-5} M); $MnSO_4\ H_2O$ (7.69 10^{-6} M); MoO_3 1.0 10^{-5} M) $ZnSO_4\ H_2O$ (7.69 10^{-5} M); $CuSO_4\ 5H_2O$ (1.6 10^{-6} M), and agar-agar 1g per 200 mL. Gold(III) ions from potassium tetrachlorourate was used at concentrations of 0, 5, 10, 20, 40, 80, 160, and 320 ppm. Also, the alfalfa plants were harvested after two weeks of growth.

Leaf extracts of two plants *Magnolia kobus* and *Diopyros kaki* were used for extracellular synthesis of gold nanoparticles (Song et al. 2009). The gold nanoparticles were formed by treating an aqueous $HAuCl_4$ solution by the plant extract. Only a few minutes were required for > 90% recovery of gold nanoparticles at a reaction temperature of 90°C.

It has also been published that living alfalfa plants have the capability to take up silver from liquid media (Gardea-Torresday et al. 2003). The alfalfa plant samples grown in silver ions

rich media were embedded in a synthetic resins and dried in an oven at 65^0 C for 24 h. TEM analysis suggested that silver atoms accumulated inside the alfalfa plant tissue under nucleation and nanoparticles formation as a correlated processes.
The extract from Black Tea has been employed as a reducing agent for the synthesis of Au and Ag nanoparticles (Begum at al. 2009). Three different extracts were prepared from Black Tea: (i) tea leaf broth, (ii) ethyl acetate extract and (iii) CH_2Cl_2 extract. Metal nanoparticles were synthesized by adding aqueous solution of $AgNO_3$ or $HAuCl_4$ to any of the three extracts. The formation and growth of the nanoparticles was monitored with the help of absorption spectroscopy and transmission electron microscopy.

4. Application of silver and gold nanoparticles

Gold and silver nanoparticles synthesized by various technique have received special attention because they have found potential application in many fields such as catalysis, sensors, drags delivery system. Additionally, silver nanaparticles possess an excellent biocompatibility and low toxicity.
Nanocatalysis has recently been a rapidly growing field which involves the use of nanoparticles as catalysts. The catalysis properties of gold and silver nanoparticles varied from their sizes and synthesis method. It is well-know that metals such as Au, Ag and Pt and metal ions can catalyzed the decomposition of H_2O_2 to oxygen. In addition, these metal ions can catalyzed luminal-H_2O_2 systems. It was observed, when the Ag colloid was injected, chemiluminoscence emission from the luminal-H_2O_2 system was greatly enhanced (Guo, et al. 2008) . Silver is also the most popular catalyst for the oxidation of ethylene to ethylene oxide and methanol to formaldehyde.
When Au nanoparticles less than 5 nm in size are supported on base metal oxide or carbon, very active catalysts are produced. Understanding the interaction between Au nanoparticles and their support material is a key issue (Hvolbeck et al., 2007). Au nanoparticle catalysts are highly active for the oxidation many compounds, particularly CO and trimethylamine. Gas sensors based on Au nanoparticles have been developed for detecting a number of gases, including CO and NO_x (Thompson 2007).
The most catalytical active material has a Au core (submonolayer Pd shell) nanostructure. Pd-coated silver nanoparticles are very effective catalyst for remediation of trichloroethene (TCE) and common organic pollutant in ground water (Nutt et al., 2005).
One of the potential advantage that Au catalysts offer compared with other precious metal catalysts is lower cost and greater price stability, Au being substantially cheaper and considerable more plentiful then Pt.
The extraordinary optical properties of noble metal nanoparticles have been taken advantage of optical biosensors and chemosensors. One of research subject focused on the measurement of biological binding signal between antigen and antibody using the triangular Ag-nanoparticles (Zhu et al., 2009).
It is well-know that the polymer-gold nanoparticles composites possess the interesting electrical properties (Gou and Wang, 2007).The nanocomposites composed of Au and biopolymer are employed as a novel biosensor. This biosensor exhibited a fast amperometric response and wide linear range of concentrations from $5.0\ 10^{-6}$ M to $4.01\ 10^{-7}$ M.
For biological applications, nanoparticles and quantum dots are conjugated with biospecific molecules such as antibodies, DNA, or enzymes. The binding event is detected by

monitoring nanoparticles property change. Most of these applications are based on the specific optical properties of gold or silver (Huo 2007).
Silver nanoaparticles have important applications in the field of biology such as antibacterial agents and DNA sequencing. Silver has been known to exhibit strong toxicity to wide range of microorganisms (antibacterial applications). Antibacterial property of silver nanoparticles against *Staphyloccocus aureus, Pseudomonas aeruginosa* and *Escherichia coli* has been investigated (Rai et al. 2009). Silver nanoparticles were found to be cytotoxic to *E. coli* It was showed that the antibacterial activity of silver nanoparticles was size dependent. Silver nanoparticles mainly in the range of 1 -10 nm attach to the surface of cell membrane and drastically disturb its proper function like respiration and permeability (Morones et al., 2005).
The fluorescent bacteria were used to investigate the antibacterial properties of silver nanoparticles (Gogoi et al. 2006). The green fluorescent proteins (GFPs) were adapted to these studies. The general understanding is that silver nanoparticles get attached to sulfur-containing proteins of bacteria cell causes the death of the bacteria. The fluorescent measurements of the cell-free supernatant reflected the effect of silver on recombination of bacteria.
The high synergistic activity of silver nanoparticles and antibiotics was observed with erythromycin against *Staphyloccocus aureus* (Shahverdi et al., 2007b). The antibacterial properties of the biosynthesized silver nanoparticles when incorporated on textile fabric were investigated (Kong and Jang 2008). The silver nanoparticles were also used for impregnation of polymeric medical devices to increase their antibacterial activity. Silver impregnated medical devices like surgical masks and implantable devices showed significant antimicrobial efficiency (Furno et al. 2004).
The current investigation that use of silver ion or metallic silver as well as silver nanoparticles can be exploited in medicine for burn treatment, dental materials, coating stainless steel materials, textile fabrics, water treatment, sunscreen lotions, etc. (Duran et al, 2007).
Gold nanoparticles are excellent labels for biosensors because they can be detected by numerous technique, such as optic absorption fluorescence and electric conductivity. Au nanoparticles have been primarily used for labeling application. Gold nanoparticles are a very attractive contrast agent. The interaction of gold nanoparticles with light can be used for the visualization of particles (Sperling et. 2008). In this way, the therapeutic application of metallic nanoparticles is also possible. The metallic structures can be used for hyperthermia therapy. Absorbed light by gold naopaticles leads to heating of these particles and upon transport subsequently to heating of the particle environments. The resulting localized heating causes irreversible thermal cellular destruction (Pissuwan et al. 2006). The plasmonic photothermal therapy is a minimally-invasive oncological treatment strategy (Dickerson et al. 2008).
Generally, gold nanoparticles provide non-toxic routes to drug and gene delivery application. Gold nanoparticles are capable of delivering large biomolecules (peptides, proteins, or nucleic acids like DNA or RNA) (Ghosh at al. 2008) . The gold nanoparticle protected by a thin layer of fluorinated amphiphilic thiols is presented in Fig. 12. It is an example of water soluble gold nanoparticles protected by polymer layer.
Gold nanoparticles can be applied to amplify the biorecognition of the anticancer drug (Shen et al. 2008). Dacarbazine [5-(3, 3-dimethy-1-triazenyl) imidazole-4-carboxamide; DTIC] is a commonly used anticancer drug. Gold nanoparticles were stabilized by PPh_3 with negative charge. The oxidized DTIC is positive charged. Thus, DTIC could be easily

assembled onto the surface of gold nanoparticles. The specific interactions between anticancer drug DTIC and DNA or DNA bases were facilitated by gold nanoparticles.

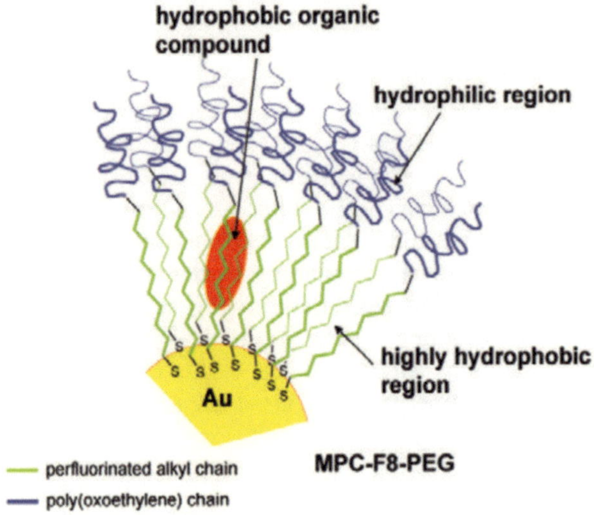

Fig. 12. Schematic of water soluble gold particles (from Agbenyega, 2008)

5. Conclusion

A green chemistry synthetic route has been used for both silver and gold nanoparticles synthesis. The reaction occurred at ambient temperature. Among the nanoparticles biological organism, some microorganisms such as bacteria, fungi, and yeast have been exploited for nanoparticles synthesis. Several plant biomass or plant extracts have been successfully used for extracellular biosynthesis of silver and gold nanoparticles. Analytical techniques, such as ultraviolet-visible spectroscopy (UV-vis), X-ray powder diffraction (XPD), transmission electron microscopy (TEM) and zeta potential measurements were applied to characterize the nanoparticles morphology.

Silver and gold nanoparticles have a number of applications from electronics and catalysis to biology, pharmaceutical and medical diagnosis and therapy. The antibacterial activity of silver ions is well known, however, the antibacterial activity of elementary silver, in the form of nanoparticles has been developed. The antimicrobial activity of Ag nanoparticles was investigated against yeast, *Escherichia coli,* and *Staphylococcus aureus.*

Gold nanoparticles may be supported by a matrix to act as a perfect catalyst. Nanoscale drug delivery systems have the ability to improve a distribution of medicines. The gold nanoparticles were utilized to facilitate the specific interactions between anticancer drugs and DNA. This may created a valuable application of metal nanoparticles in the relative biomedical area.

6. Acknowledgments

This work was supported by MNiSzW grant # N N204 290134

7. References

Absar, A.; Satyajyoti, S.; Khan, M.I.; Rajiv, K.; Sastry, M. (2005). Extra-/intracellular biosynthesis of gold nanoparticles by an alkalotolerant fungus, *Trichothecium* sp., *J. Biomedical Nanotechnnology*, 1, 1, 47-53

Agbenyega, J. (2008). Water soluble gold, *Materials Today*, 11, 12, 13

Ahmad, A.; Mukherjee, P.; Senapati, S.; Mandal, D.; Khan, M.I.; Kumar,R.; Sastry, M. (2003). Extracellular biosynthesis of silver nanoparticles using the fungus *Fusarium oxysporum*, *Colloids Surfaces, B: Biointerfaces*, 27, 313-318

Balaji, S.D.; Basavaraja, S.; Deshpande, R.; Mahesh, B.D.; Prabhakar, K.B.; Venkataraman, A. (2009). Extracellular biosynthesis of functionalized silver nanoparticles by strains of *Cladosporium cladosporioides* fungus, *Colloids Surfaces B: Biointerfaces*, 68, 88-92

Bar, H.; Bhui, Kr.D.; Sahoo, P.G.; Sarkar, P.; De, P.S.; Misra, A. (2009). Green synthesis of silver nanoparticles using latex of *Jatrapha curcas*, *Colloids Surfaces, A. Physicochem. Eng. Aspects*, 339, 134-139

Basavaraja, S.; Balaji, D.S.; Lagashetty, A.; Rajasab, H.A.; Venkataraman, A.; (2008). Extracellular biosynthesis of silver nanoparticles using the fungus *Fusarium semitectum*, *Materials Research Bulletin*, 43, 1164-1170

Bazylinski, A.B.; Frankel, B.R. (2004). Magnetosome formation in Procayotes, *Nature reviews - Microbiology*, 2, 213-230.

Begum, A.N.; Mondal, S.; Basu, S.; Laskar, A.R.; Mandal, D. (2009). Biogenic synthesis of Au and Ag nanoparticles using aqueous solutions of Black Tea leaf extracts, *Colloids Surfaces B. Biointerfaces*, 71, 113-118

Bhainsa, C.K.; D'Souza, F.S. (2006). Extracellular biosynthesis of silver nanoparticles using the fungus *Aspergillus funigatus*, *Colloids Surfaces B; Biointerfaces*, 47, 160-164

Boisselier, E.; Astruc, D. (2009) Gold nanoparticles in nanomedicine: preparation, imaging, diagnostics, therapies and toxicity, *Chem. Soc. Rev.*, 38, 1759-1782

Chan, W.C.W. (2006). Bionanotechnology progress and advances, *Biology Blood Marrow Transplantation*, 12, 87-91

Chandra, P.S.; Chaudhary, M.; Pasricha, R.; Ahmad, A.; Sastry, M. (2006). Synthesis of gold nanotriangles and silver nanoparticles using *Aloe vera* plant extract, *Biotechnology Prog.*, 22, 577-583

Dickerson, B.E.; Dreaden, C.E.; Huang, X., El−Sayed, H. I.; Chu, H.; Pushpanketh, S.; McDonald, H.I., El-Sayed, A.M. (2008). Gold nanorod assisted near-infrared plasmonic photothermal therapy (PPPT) of aquamous cell carcinoma in mice, *Cancer Letters*, 269, 57-66

Duran, N.; Marcato, D.P.; Alves, L.O.; De Souza, G.; Esposito, E. (2005). Mechanical aspect of biosynthesis of silver nanoparticles by several *Fusarium oxysporum* strains. *J. Nanobiotechnology*, 3, 8-15

Duran, N.; Marcato, D.P.; De Souza, H.I.; Alves, L.O.; Espsito, E. (2007). Antibacterial effect of silver nanoparticles produced by fungal process on textile fabrics and their effluent treatment, *J. Biomedical Nanotechnology*, 3, 203-208

Furno, F.; Morley, K.S.; Wong, B.; Sharp, B.L.; Howdle, S.M. (2004). Silver nanoparticles and polymeric medical devices: a new approach to prevention of infection, *J. Antimicrob. Chemother.*, 54, 1019-1024

Gardea-Torresday, L.J.; Gomez, E.; Peratta-Videa, R.J.; Persons, G.J.; Troiani, H.; Jose-Yacaman, M. (2003) Alfalfa sprouts: A natural source for synthesis of silver nanoparticles, *Langmuir*, 19, 1357-1361

Gardea-Torresday, L.J.; Persons, G.J.; Gomez, E.; Peralta-Videa, J.; Troiani, E.H.; Santiago, P.; Yacaman, J.M. (2002). Formation and growth of Au nanoparticles inside live Alfalfa plants, *Nano Letters*, 2, 4, 397-401

Ghosh, P.; Han, G.; De, M.; Kim, K.C.; Rotello, M.V. (2008). Gold nanoparticles in delivery applications, *Advanced Drag Delivery Reviews*, 60, 1307-1315

Glomm, R.W. (2005). Functionalized nanoparticles for application in biotechnology, *J.Dispersion Sci. Technology*, 26, 389-314

Gogoi, K.S.; Gopina, P.; Paul, A.; Ramesh, A.; Ghosh, S.S.; Chattopadhyay, A. (2006). Green fluorescent protein expressing *Escherichia coli* as a model system for investigating the antimicrobial activities of silver nanoparticles, *Langmuir*, 22, 9322-9328

Gourley, L.P. (2005). Brief overview of bio-micronano technologies, *Biotechnol. Prog.*, 21, 2-10

Guo, J-Z.; Cui, H.; Zhou, W.; Wang, W. (2008). Ag nanoparticle-catalyzed chemiluminescent reaction between luminal and hydrogen peroxide, *J. Photochem. Photobiol., A Chemistry*, 193, 89-96.

Guo, S.; Wang, E.(2007). Synthesis and electrochemical application of gold nanoparticles, *Analytica Chimica Acta*, 598, 181-192

He, S.; Guo, Z.; Zhang, Y.; Zhang, S.; Wang, J.; Gu, N.; (2007). Biosynthesis of gold nanoparticles using the bacteria *Rhodopseudomonas capsulate, Materials Latter*, 61, 3984-3987

Huang, J.; Li, Q.; Sun, D.; Lu, Y.; Su, Y.; Yang, X.; Wanh, H.; Wang, Y.; Shao, W.; He, N.; Hong, J.; Chen, C. (2007). Biosynthesis of silver and gold nanoparticles by novel sundried *Cinnamomum canphora* leaf, *Nanotechnology*, 18, 1-11

Huo, Q. (2007). A perspective on bioconjugated nanoparticles and quantum dots, *Colloids Surfaces B: Biointerfaces*, 59, 1 -10.

Husseiny I.M.; El-Aziz, A.M.; Badr, Y.; Mahmoud, A.M. (2007). Biosynthesis of gold nanoparticles using *Pseudomonas aeruginosa, Spectrochimica Acta Part A*, 67, 1003-1006

Hvolbek, B.; Janssens, W.V.T.; Clausen, S.B.; Falsig, H.; Christensen, H.C.; Norskov, K.J. (2007). Catalytic activity of Au nanoparticles, *Nano Today*, 2,4, 14-18

Jha, K.A,; Prasad,; Kumar, V.; Prasad, K. (2008). Biosynthesis of silver nanoparticles using *Eclipta* leaf, *Biotechnology Progress*, in press.

Jha, K.A.; Prasad, K. (2009). Biosynthesis of silver nanoparticles using *Cycas* leaf broth. *Bioprocess Biosystems Eng.*, in press.

Kalimuthu, K.; Babu, S.R.; Venkataraman, D.; Bilal, M.; Gurunathan, S. (2008). Biosynthesis of silver nanoparticles by *Bacillus licheniformis, Colloids Surfaces B: Biointerfaces*, 65, 150-153

Kalishwaralal, K.; Deepak, V.; Ram Kumar Pandian, S.; Gurunathan, S. (2009). Biosynthesis of gold nanocubes from *Bacillus lichemiformis, Bioresource Technology*, 100, 5356-5358

Kalishwaralal, K.; Deepak, V.; Ramkumarpndian, S.; Nellaiah, H.; Sangiliyandi, G. (2008). Extracellular biosynthesis of silver nanoparticles by the culture supernatant of *Bacillus licheniformis. Materials Letters*, 62, 4411-4413

Kasthur, J.; Veerapandian, S.; Rajendiran, N. (2009). Biological synthesis of silver and gold nanoparticles using apin as reducing agent, *Colloids Surfaces B: Biointerface*, 68, 55-60

Kathiresan, K.; Manivannan, S.; Nabeel, A.M.; Dhivya, B. (2009). Studies on silver nanoparticles synthesized by a marine fungus *Penicillum fellutanum* isolated from coastal mangrove sediment, *Colloids Surfaces B: Biointerfaces*, 71, 133-137

Kong, H.; Jang, J. (2008). Antibacterial properties of novel poly(methyl methacrylate) nanofiber containing silver nanparticles, *Langmuir*, 24, 2051-2056

Konishi, Y.; Tsukiyama, T.; Ohno, K.; Saitoh, N.; Nomura, T.; Nagamine, S. (2006). Intracellular recovery of gold by microbial reduction of $AuCl_4^-$ ions using the anaerobic bacterium *Shewanella algae*. *Hydrometallurgy*, 81, 24-29

Lengke, F.M.; Fleet, E.M.; Southam, G. (2007). Biosynthesis of silver nanoparticles by filamentous cyanobacteria a from a silver(I) nitrate complex, *Langmuir*, 23, 2694-2699.

Minaeian, S.; Shahverdi, R.A.; Nohi, S.A.; Shahverdi, R.H. (2008). Extracellular biosynthesis of silver nanoparticles by some bacteria, *J. Sci. I.A.*, 17, 66, 1-4.

Mokhtari, M.; Deneshpojouh, S.; Seyedbagheri, S.; Atashdehghan, R.; Abdi, K.; Sarkar, S.; Minaian, S.; Shahverdi, R.H.: Shahverdi, R.A. (2009). Biological synthesis of very small silver nanoparticles by culture suspernatant of *Klebsiella pneumonia*. The effects of visible-light irradiation and the liquid mixing process, *Materials Research Bull.*, 44, 1415-1421

Morones, J.R.; Elechiguerra, L.J.; Camacho, A.; Holt, K.; Kouri, B.J.; Ramirez, T.J.; Yocaman, J.M. (2005). The bactericidal effect of silver nanoparticles, *Nanotechnology*, 16, 2346-2353

Mukherjee, P.; Ahmad, A.; Mandal, D.; Senapati, S.; Sainkar, R.S.; Khan, I.M; Parishcha, R.; Ajaykumar, V.P.; Alam, M.; Kumar, R.; Sastry, M. (2001). Fungus-mediated synthesis of silver nanoparticles and their immobilization in the mycelia matrix. A novel biological approach to nanoparticle synthesis, *Nano Letters*, 1, 10, 515-519

Nutt, O.M.; Hughes, B.J.; Wong, S.M. (2005). Designing Pd-on-Au bimetallic nanoparticles catalysts for trichloethene hydrodechlorination, *Envir. Sci. Technol.*, 39, 1346-1353

Philip, D. (2009) Biosynthesis of Au, Ag and Au-Ag nanoparticles using edible mushroom extract, *Spectrochemia Acta Part A*, 73, 374-381

Pissuwan, D.; Valenzuela, M.S.; Cortier, B.M. (2006). Therapeutic possibilities of plasmonically heated gold nanoparticles, *Trends Biotechnology*, 24, 2, 62-67

Rai, M.; Yadav, A.; Gade, A. (2009). Silver nanoparticles as a new generation of antimicrobials, *Biotechnology Advances*, 27, 76-83

Sadowski, Z.; Maliszewska, H.I.; Grochowalska, B.; Polowczyk, I.; Kozlecki, T. (2008 B). Synthesis of silver nanoparticles using microorganisms, *Materials Science-Poland*, 26, 2, 419-424

Sadowski, Z.; Maliszewska, I.; Polowczyk, I.; Kozlecki, T.; Grochowalska, B. (2008 A). Biosynthesis of colloidal-silver particles using microorganisms, *Polish J. Chem.*, 82, 377-382

Saifuddin, N.; Wang, W.C.; Nur Yasumira, A.A. (2009). Rapid biosynthesis of silver nanoparticles using culture supernatant of bacteria with microwave irradiation, *E-Journal Chem.*, 6, 2, 61-70

Sastry, M.; Ahmad, A.; Khan, I.M.; Kumar, R. (2003). Biosynthesis of metal nanoparticles using fungi and actinomycete, *Current Sci.*, 85, 2, 162-170

Sathishkumar, M.; Sneha, K.: Won, W.S.; Cho, C-W., Kim, S.; Yun, Y-S. (2009). *Cynamon zeylanicum* bark extract and powder mediated green synthesis of nanocrystalline silver particles and its bactericidal activity, *Colloids Surfaces, B:Biointerfaces*, 73, 332-338.

Shahverdi, R.A., Fakhimi, A.; Shahverdi, H.R.; Minaian, S. (2007b). Synthesis and effect of silver nanoparticles on the antibacterial activity of different antibiotics against *Staphyloccocus aureus* and *Escherichia coli*, *Nanomed: Nanotechnol. Biol. Med.*, 3, 2, 168-171

Shahverdi, R.A.; Minaeian, S.; Shahverdi, R.H.; Jamalifar, H.; Nohi, A-A. (2007a). Rapid synthesis of silver nanoparticles using culture supernatants of *Enterobacteria*: A novel biological approach. *Process Biochemistry*, 42, 919-923

Shaligram, S.N.; Bule, M.; Bhambure, R.; Singhal, S.R.; Singh, K.S.; Szakacs, G.; Pandey, A.; (2009). Biosynthesis of silver nanoparticles using aqueous extract from the compactin producing fungal, *Process Biochemistry*, 44, 939-943

Shen, Q.; Wang, X.; Fu, D. (2008). The amplification effect of functionalized gold nanoparticles on the binding of cancer drug decarbazine to DNA and DNA bases, *Applied Surface Sci.*, 255, 577-580

Shiv Shankar, S.; Ahmad, A.; Sastry, M. (2003). Geranium leaf assisted biosynthesis of silver nanoparticles, *Biotechno. Prog.*, 19, 1627-1631.

Shiv Shankar, S.; Rai, A.; Ahmad, A.; Sastry, M. (2004). Rapid synthesis of Au, Ag and bimetallic Au core-Ag shell nanoparticles using Neem (*Azadirachta indica)* leaf broth. *J. Colloid Inter. Sci.*, 275, 496-502

Singaravelu,G.; Arockiamary, S.J.; Kumar,G.V.; Govindaraju,K. (2007). A novel extracellular synthesis of monodisperse gold nanoparticles using marine alga *Sargassum wightii* Greville, *Colloids Surfaces, B. Biointerfaces*, 57, 97-101

Song, Y.J.; Jang, H-K.; Kim, S.B. (2009). Biological synthesis of gold nanopaticles using *Magnolia kobus* and *Diopyros kaki* leaf extract, *Process Biochemistry*, 44, 1133-1138

Sperling, A.R.; Gil, R.P.; Zhang, F.; Zanella, M.; Parak, J.W. (2008). Biological application of gold nanoparticles,*Chem. Soc. Rev.* 37. 1896-1908

Thompson, T.D. (2007). Using gold nanoparticles for catalysis, *Nano Today*, 2, 4, 40-43

Wang, Y.; He, X,; Wang, K.; Zhang, X.; Tan, W. (2009). *Barbated Skullcup* herb extract-mediated biosynthesis of gold nanoparticles and its primary application in electrochemistry, *Colloids Surfaces B: Biointerfaces*, 73. 75-79

Zhu, S.; Du, CL.; Fu, Y. (2009). Fabrication and characterization of rhombic silver nanoparticles for biosensing, *Optical Materials*, 31, 769-774

14

On-paper Synthesis of Silver Nanoparticles for Antibacterial Applications

Hirotaka Koga and Takuya Kitaoka
Kyushu University
Japan

1. Introduction

Recent years have seen remarkable progress in research and development of metal nanoparticles (NPs) that takes advantage of their unique optical, magnetic, electronic, catalytic and other physicochemical properties, in a wide range of practical and potential applications such as energy, environmental, biomedical and chemical engineering (Feldheim & Foss, 2002; Ma et al., 2006; Jena & Raj, 2008; Zotti et al., 2008). For example, gold (Au) is a chemically inert metallic element in bulk form; however, AuNPs possess colorful plasmon resonances useful for bio-sensing (Shukla et al., 2005; Yokota et al., 2008), and can catalyze chemical reactions (Bond et al., 2006; Ishida & Haruta, 2007), due to their quantum size effects. NPs of inexpensive base metals, such as copper (CuNPs) have recently attracted much attention as innovative nanomaterials for applications such as highly-active gas reforming catalysts for hydrogen production (Gadhe & Gupta, 2007) and effective marine antifouling coatings (Anyaogu et al., 2008). In general, however, practical utilization of nanosized materials involves considerable difficulties since metal NPs are hard to handle directly, and easily aggregate to minimize their surface area. The inevitable aggregation of metal NPs often nullifies their unique functionalities, and eventually yields ordinary bulk metals. For that reason, an area of ongoing research has focused on effective immobilization of metal NPs on easily handled supports such as porous membranes (Dotzauer et al., 2006) and nanostructured inorganic sheets (Wang et al., 2008).

Of the various metal NPs, silver NPs (AgNPs) are of increasing interest because of their high conductivity (Li et al., 2005) and tunable optical responsiveness (McFarland & Van Duyne, 2003). Moreover, it is well known that Ag exhibits potent antibacterial properties with low toxicity for humans and animals by comparison with other heavy metals (Alt et al., 2004; Shah et al., 2008). Ag and Ag-compounded materials are effective for both Gram-negative and Gram-positive bacteria, whereas the efficacy of conventional antibiotics varies with the species of bacteria (Shah et al., 2008). Many researchers have recently reported that AgNPs demonstrate excellent antibacterial activity (Sondi & Salopek-Sondi, 2004; Gogoi et al., 2006; Pal et al., 2007; Navarro et al., 2008). However, effective methods for immobilization of AgNPs for practical use are insufficiently advanced, and incorporation of AgNPs into various matrices is being actively investigated. For instance, it was reported that AgNPs were synthesized on poly(ethylene glycol)-polyurethane-titanium dioxide (TiO_2) films *via*

TiO_2-mediated photocatalysis under UV light irradiation, which facilitated photoreduction of silver nitrate ($AgNO_3$) to form AgNPs on the polymer-inorganic hybrid matrix (Shah et al., 2008). Ag-nanocoated cotton fabrics were successfully developed by an ion-exchange method (Lee et al., 2007). Such types of Ag-organic polymer complexes are generally sensitive to external stimuli (*e.g.* heat, light and pH), resulting in poor stability and durability. Hence, Ag-doped antibacterial inorganics, such as Ag-hydrogen titanate nanobelt sheets (Wang et al., 2008) and thin silica films prepared from ionic Ag-incorporated tetraethyl orthosilicate (Jeon et al., 2003), have been developed using various approaches, including sol-gel processing, ion-exchange and surface modifications. Despite these efforts, however, there is still a need to find more facile ways of making and using AgNPs in practical antibacterial applications, without compromising their excellent bioactive functionality. Thus, one of the present challenges is to develop a novel method for immobilization of highly active AgNPs on multipurpose, convenient support materials.

In this chapter, we present a facile technique for *in situ* synthesis of bioactive AgNPs on an easily handled inorganic 'paper' matrix. Paper is one of the most familiar materials in our daily life, and has the potential for broad applications due to its flexibility and availability. The outline of our strategy is highlighted in Fig. 1. Novel 'on-paper' synthesis of AgNPs is accomplished using a paper matrix that consists of ceramic fibers as the main framework and zinc oxide (ZnO) whiskers as a selective support for AgNPs through a simple soaking treatment with an aqueous solution of $AgNO_3$. These as-prepared AgNPs@ZnO paper composites with paper-like utility demonstrate excellent antibacterial performance. Further promising developments for on-paper synthesis of other metal NPs are also introduced.

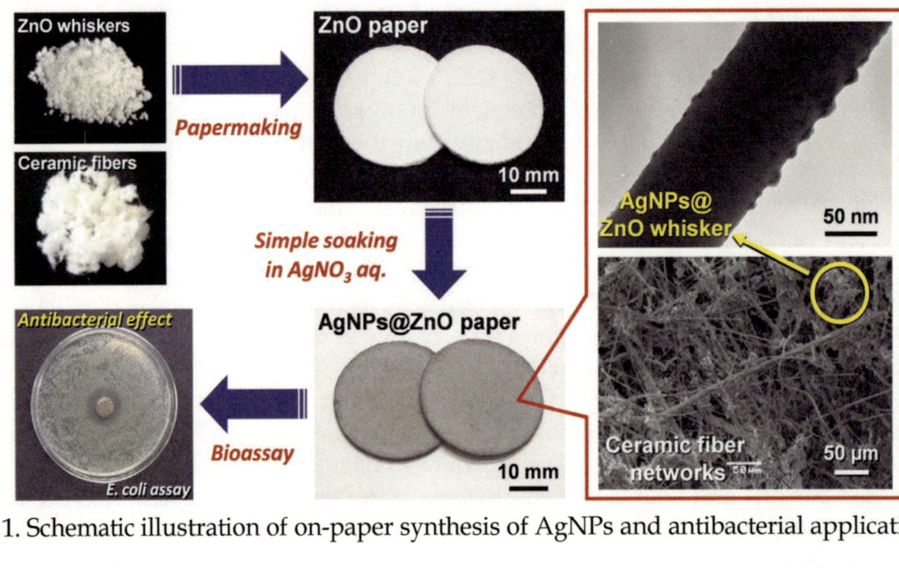

Fig. 1. Schematic illustration of on-paper synthesis of AgNPs and antibacterial application

2. Paper Composites

Paper is a versatile, ubiquitous material, commonly used for writing, printing, wrapping and packaging applications (Roberts, 1996). Paper materials have good practical utility (*e.g.* lightweight, flexible and easy handling), and many researchers and papermakers have developed new types of paper-like composites with various functions such as heat storage capacity and bio-interface properties (Huang et al, 2006; Ichiura et al., 2008; Egusa et al., 2009). Paper has a porous, layered fiber network microstructure, and is consequently useful as a catalyst support. Basic technology of paper composite materials, an outline of the papermaking process, and our paper-like materials that have been developed for catalytic applications are described in the following sections.

2.1 Papermaking

The general papermaking procedure is carried out in an aqueous system, and can be summarized briefly as follows. A well-stirred fiber/water suspension is drained with a wire mesh so that a swollen mat of randomly entangled fibers is laid down on the wire. Water is then removed from the fiber mat by pressing and drying to produce a paper material. Most papers are made from organic fibers such as wood pulp and cotton: inorganic fibers such as ceramic and glass fibers are also used for further functional applications.

Papermakers often use powdery inorganic additives such as calcium carbonate, clay, talc, TiO_2, zeolite and aluminum hydroxide for various purposes during the aqueous paper-forming process (Casey, 1981; Neimo, 1999). These inorganic particles, several micrometers in diameter, are much smaller than the mesh sizes of papermaking wires, and most of the particles are inevitably dropped off through the wire. Hence, various types of charged flocculants are used as aids for improving the retention of fine particles during paper formation. However, most inorganic materials including fibers and powders have weak negative charges in an aqueous papermaking system, and thus insufficient flocculation sometimes occurs. An effective solution to this problem is a dual polymer retention system (Scott, 1996; Ichiura et al.; 2001). The flocculation system can be summarised as follows. First, low-molecular-weight cationic polyelectrolytes with high charge density, such as poly(diallyldimethylammonium chloride) (PDADMAC), are overdosed to a water suspension containing negatively charged inorganics. The polymer molecules strongly adsorb onto the surface of these materials and form a patch of positive charges. Second, high-molecular-weight anionic polyelectrolytes with low charge density, such as anionic polyacrylamide (A-PAM), are added to the water suspension. The anionic polymers immediately connect the positively charged patches, and flocculating bridges are built up between the inorganic particles and fibers. Eventually, the resulting aggregates are effectively retained in the paper composites.

2.2 Catalytic Applications

Functional inorganic particles have been developed for a wide range of uses such as catalysts, adsorbents, electrodes, fillers and others. TiO_2, a typical photocatalyst (Fujishima & Honda, 1972), is being commercialized for self-cleaning glasses, solar panels and air-purifying pigments. Cu/ZnO catalysts used in the methanol reforming process to produce hydrogen for fuel cell applications have been extensively upgraded by tailoring the elemental composition and nanometer-scale morphology of the catalyst surface, to improve their catalytic activity (Matter et al., 2004). However, such particulate materials are generally in the form of fine powders, and are difficult to handle in practical usage.

As described above, a dual polymer retention system enables effective retention of small inorganic particles in an easy to handle paper matrix. We have reported that various catalyst powders, *e.g.* zeolite (Ichiura et al., 2003), TiO_2 (Fukahori et al., 2003, 2007; Iguchi et al., 2003),

Cu/ZnO (Fukahori et al., 2006a, 2006b; Koga et al., 2006) and platinum/aluminum oxide (Pt/Al_2O_3) (Ishihara et al., 2010), can be incorporated into an inorganic paper matrix by a papermaking technique. The papermaking procedure is summarized as follows. A water suspension of ceramic fibers and catalyst powders is mixed with cationic polyelectrolyte PDADMAC (0.5 wt% of total solids), an alumina sol binder and anionic polyelectrolyte A-PAM (0.5 wt% of total solids), in that order. The mixture is added to pulp fiber suspension, and solidified by dewatering using a 200 mesh wire. The wet handsheets are pressed at 350 kPa for 3 min, then dried in an oven at 105°C for 1 h. The paper composites thus obtained are calcined to remove organic pulp fibers and to improve the physical strength by sintering with the alumina sol binder. As-prepared catalyst powder/inorganic fiber composites have the appearance of paperboard, and have good practical utility (Fig. 2).

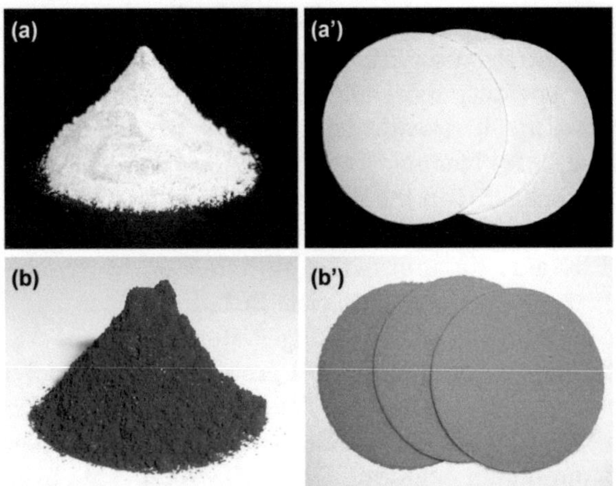

Fig. 2. Optical images of TiO_2 powders (a), Cu/ZnO powders (b), and respective paper composites containing each powder (a′ and b′). The size of each paper composite is 2×10^4 mm^2

Compared to the original powders, these paper composites are far superior in terms of convenience of handling. As shown in Fig. 3, the retention of catalyst powders can be more than 95% with the dual polymer retention system (Fukahori et al., 2006a), and such polymer systems are applicable to large-scale production.

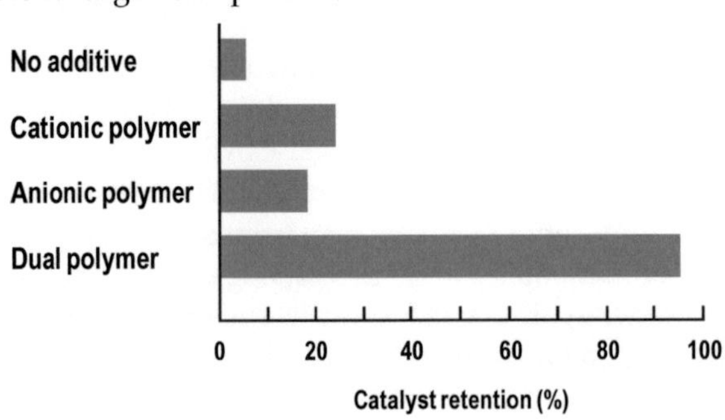

Fig. 3. Representative retention behavior of catalyst powders in paper composites by using polyelectrolyte-assisted retention systems

In addition, catalyst papers are highly effective for a wide range of catalytic processes such as the photodecomposition of bisphenol A in water (Fukahori et al., 2003), methanol reforming for hydrogen production (Fukahori et al., 2006a, 2008; Koga et al., 2006, 2008a) and removal of nitrogen oxides for exhaust gas purification (Ishihara et al., 2010). Moreover, our established papermaking technique allows the compounding of various functional fibers such as silicon carbide fibers (Fukahori et al., 2006b) and carbon fibers (Koga et al., 2009b) with high thermal conductivity, leading to further improvement of the catalytic performance. The results of these investigations suggest that paper-like fiber/catalyst composites prepared by a papermaking technique are promising catalytic materials for practical use. The combinational variation of fibers and functional particles is highly versatile. Further functional development of paper composites will be a significant challenge for exploring a wide range of applications.

2.3 Metal NPs Immobilization

Immobilization of metal NPs on a paper matrix has become of major interest. Immobilization of photoactive TiO_2 NPs on cellulose fibers has been investigated for preparation of self-cleaning paper (Pelton et al., 2006), which has a variety of potential applications such as food packaging. A simple method for immobilizing metal NPs on fibers by using charged polymers has been proposed (Dubas et al., 2006; Pelton et al., 2006). However, these approaches sometimes bring about fatal aggregation of metal NPs: excessive aggregation drastically decreases their specific surface area, resulting in insufficient functionality. There is, therefore, a need to establish more effective approaches for immobilization of metal NPs on paper materials. In the following sections, we describe a new concept whereby scaffold materials for metal NPs are prefabricated in paper composites, followed by *in situ* synthesis of highly-active metal NPs, especially AgNPs, on the paper matrices.

3. *In Situ* Synthesis of AgNPs on a Paper Matrix

This section provides a novel technique for immobilization of bioactive AgNPs on a paper matrix. Direct *in situ* synthesis of AgNPs is successfully achieved on a paper matrix that consists of ceramic fibers as the main framework and ZnO whiskers as a selective support for AgNPs.

3.1 AgNPs Synthesis on ZnO Whiskers

ZnO is an inexpensive, relatively-abundant and non-toxic material, and thus has been widely used as an inorganic additive for various products including rubbers, paints, pigments, foods and adhesives. Recently, a biomimetic membrane for immobilization of biomolecules has been prepared from ZnO due to its high biocompatibility (Kumar & Chen, 2008). Furthermore, ZnO-mediated photocatalysis has become the center of attention in relation to optical (Yang et al., 2002) and photoelectronic (An et al., 2007) applications as semiconductors, sensors, solar cells, etc. The band gap of crystalline ZnO is 3.37 eV (Yang et al., 2002), similar to that of the anatase form of TiO_2 which can reduce ionic Ag species to metallic AgNPs.

We propose a facile new method for using ZnO as a scaffold for the synthesis and immobilization of AgNPs, in which ionic Ag species are adsorbed on a ZnO support by ion exchange and are simultaneously reduced by the photocatalytic function of the ZnO (Koga et al., 2009a). Fig. 4 shows the ZnO whiskers with tetrapod-like nanostructure used in this method.

Fig. 4. Optical (a) and TEM (b) images of original ZnO whiskers

AgNPs synthesis on ZnO whiskers is achieved by the following simple procedure. ZnO whiskers are suspended in an aqueous solution of $AgNO_3$ with continuous stirring for 6 h, then filtered, washed with deionized water and air-dried. AgNPs with size 5–20 nm are formed on the ZnO whiskers through selective ion-exchange between Ag and Zn species, and simultaneous ZnO-mediated photoreduction under natural light irradiation (Fig. 5).

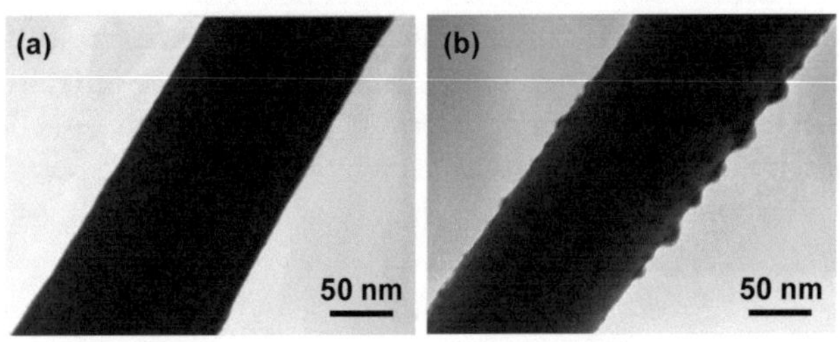

Fig. 5. TEM images of original ZnO whiskers (a) and $AgNO_3$-treated ZnO whiskers (b)

X-Ray photoelectron spectroscopy (XPS) and X-ray diffractometry (XRD) reveal that AgNPs with crystallite size *ca.* 16 nm spontaneously form on ZnO whiskers *via* the simple soaking treatment with $AgNO_3$ solution (Fig. 6).

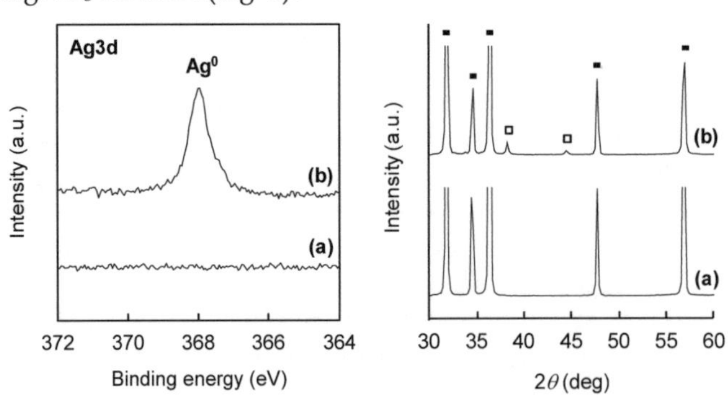

Fig. 6. XPS profiles (left) and XRD patterns (right) of original ZnO whiskers (a) and AgNPs@ZnO whiskers (b). □: Ag, ■: ZnO

In general, immobilization of metal NPs on support materials has required complicated, multi-step processes. For example, AgNPs-modified hydrogen titanate ($H_2Ti_3O_7$) nanobelts have been prepared according to the following procedure (Wang et al., 2008). AgNPs are prepared in advance and capped with 3-aminopropyl triethoxysilane through coordination between Ag and amino group. The modified AgNPs are then covalently fixed on the surfaces of the $H_2Ti_3O_7$ nanobelts, forming AgNPs-modified $H_2Ti_3O_7$ nanobelts. The synthesis and immobilization of PtNPs have been accomplished on a carbon fiber pre-treated with nitric acid to activate the surface (Koga et al., 2009b). Surface-activated carbon fibers are immersed in an aqueous solution of hexachloroplatinic acid, followed by evaporation to complete dryness and reduction by reducing agents such as hydrogen and sodium borohydride. By contrast, our technique using ZnO whiskers allows one-pot, one-step immobilization of AgNPs in the absence of reducing agents, and this straightforward technique is expected to be a promising approach for practical AgNPs immobilization.

3.2 On-paper Synthesis of AgNPs

ZnO whiskers have the unique configuration of the tetrapod-like nanofibers shown in Fig. 4; however, as-prepared AgNPs supported on ZnO whiskers (AgNPs@ZnO whiskers) are still difficult to handle since the ZnO whiskers are fine fibers. Hence, a modified method of fabrication of AgNPs@ZnO whiskers is required for practical use, and paper fabrication is one of the promising methods. However, paper forming of AgNPs@ZnO whiskers inevitably causes functional deterioration of AgNPs, since AgNPs@ZnO whiskers are aggregated by flocculants, and the surfaces of AgNPs are partially covered with the binder components, which are necessarily added to enhance the physical strength of paper composites. To solve that problem, we have investigated direct, *in situ* synthesis of AgNPs on a ZnO whisker-containing paper composite (ZnO paper), that we refer to as on-paper synthesis (Koga et al., 2009a).

The on-paper synthesis of AgNPs is performed by first incorporating ZnO whiskers into a paper composite, then using the embedded whiskers as a scaffold for AgNPs. ZnO whiskers and ceramic fibers are fabricated into a paper composite by a papermaking technique. AgNPs-free ZnO paper composite can be easily prepared by substituting ZnO whiskers for catalyst powders in the aforementioned papermaking procedure (see section 2.2). ZnO whiskers are thus effectively retained inside the paper composite; virtually 100% retention of the inorganic materials is achieved. The paper composite obtained (2×10^4 mm^2) consists of ceramic fibers (5.0 g), ZnO whiskers (3.1 g) and alumina sol (0.50 g). Subsequently, on-paper synthesis of AgNPs is carried out. As-prepared ZnO paper is cut into disk-shaped pieces (8×10^2 mm^2) and immersed in an aqueous solution of $AgNO_3$ (1.3×10^2 mM, 100 mL) for 6 h. The disks are removed from the $AgNO_3$ solution using tweezers, thoroughly washed with deionized water, then dried at 105°C for 2 h. Optical images of the original ZnO paper and the ZnO paper treated with $AgNO_3$ solution are shown in Fig. 7a and b. These paper composites with the appearance of flexible cardboard are lightweight and easy to handle in practical use. The XPS and XRD profiles of $AgNO_3$-treated ZnO paper (data not shown) were similar to those of AgNPs@ZnO whiskers shown in Fig. 6. $AgNO_3$-treated paper without ZnO whiskers was prepared in a similar manner, in which case the amount of Ag adsorbed on the paper composite was less than 2.8% of the amount of Ag in $AgNO_3$-treated ZnO paper, suggesting that AgNPs are synthesized selectively on the ZnO whiskers embedded in the paper composite.

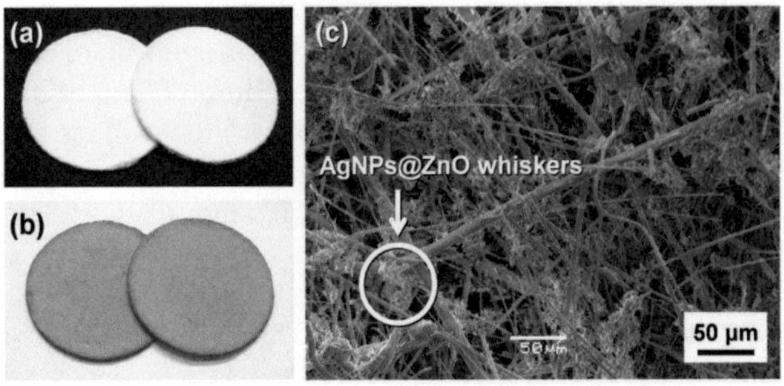

Fig. 7. Optical images of original ZnO paper (a), $AgNO_3$-treated ZnO paper (b), and SEM image of the surface of AgNPs@ZnO paper (c). ZnO whiskers are well scattered throughout the ceramic fiber networks. The size of each paper composite is 8×10^2 mm^2

Fig. 7c displays a scanning electron microscopic (SEM) image of the surface of ZnO paper after the synthesis of AgNPs. AgNP@ZnO paper has a characteristic porous fiber-network microstructure composed of ceramic fibers (average pore size *ca.* 16 µm, porosity *ca.* 50%), in which AgNPs@ZnO whiskers are well dispersed. Thus direct synthesis of AgNPs on an inorganic paper matrix is successfully achieved by using pre-compounded ZnO whiskers as a selective scaffold for AgNPs *via* a simple soaking treatment in $AgNO_3$ solution. To our knowledge, such on-paper synthesis of metal NPs is a hitherto unused approach for one-pot, one-step immobilization of metal NPs on easy to handle supports, *i.e.* paper.

4. Antibacterial Performance of AgNPs@ZnO Paper

In recent years, there has been a major upsurge of interest in AgNPs with unique properties for use in a wide range of fields such as optical (Ohko et al., 2003), electronic (Zhang et al., 2008), catalytic (Mao et al., 2009) and biomedical applications (Alt et al., 2004; Sondi & Salopek-Sondi, 2004; Morones et al., 2005; Lok et al., 2006). In this section, we focus on the antibacterial properties of AgNPs, and describe the practical performance of AgNPs@ZnO paper. The antibacterial activity of AgNPs@ZnO paper is compared with that of Ag-free ZnO paper, $AgNO_3$-impregnated paper and Ag powder-containing paper. $AgNO_3$-impregnated paper, *i.e.* Ag^+ ion-containing ZnO-free paper, was prepared as follows. Disk-shaped ZnO whisker-free paper composite was immersed in aqueous $AgNO_3$, followed by evaporation to dryness at 105°C for 30 min to force precipitation of Ag components on the paper composite. Ag powder-containing paper composites were prepared by the papermaking procedure mentioned previously (see section 2.2). Antibacterial tests of paper samples were carried out by the standard disk diffusion assay previously reported (Lee et al., 2007; Shah et al., 2008; Wang et al., 2008). In summary, Gram-negative *Escherichia coli* (*E. coli*) bacterial suspension (Luria-Bertani (LB) medium, 100 µL, 3.5×10^5 colony forming units per mL) was uniformly inoculated on the solidified LB agar gel (20 mL). Disk-shaped pieces (10 mm in diameter and 1 mm in thickness) of Ag-free ZnO paper, AgNPs@ZnO paper, $AgNO_3$-impregnated paper or Ag powder-containing paper were placed on the LB agar plate, followed by incubation at 37°C for 24 h. The antibacterial activities were compared by the diameter of the zone of inhibition around each paper disk. Fig. 8 displays optical images of the zone of inhibition against Gram-negative *E. coli* for each paper composite.

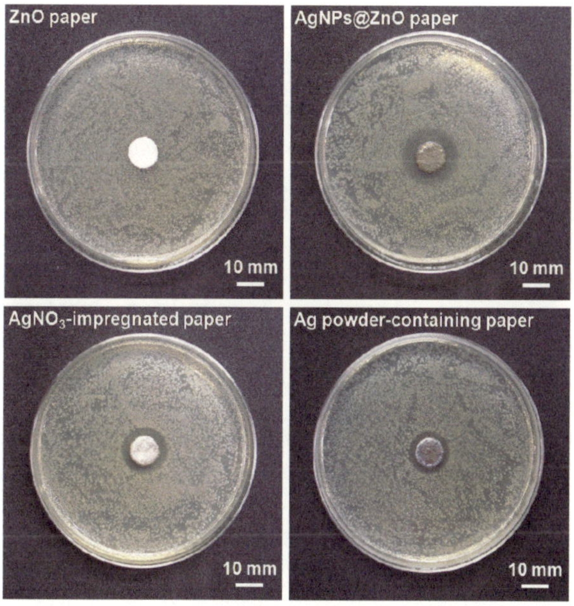

Fig. 8. Optical images of the zone of inhibition against *E. coli* for Ag-free ZnO paper, AgNPs@ZnO paper, $AgNO_3$-impregnated ZnO-free paper, and Ag powder-containing paper. Each paper composite is 10 mm in diameter. Ag content of the latter three samples is 2.0 mg. Incubation condition: 37°C, 24 h

ZnO paper exhibited no antibacterial activity, indicating that ZnO whiskers had no toxic influence on the growth of bacteria in these conditions. By contrast, the three paper composites containing Ag species showed clear zones of inhibition around each paper disk. AgNPs@ZnO paper demonstrated the largest zone of inhibition, *i.e.* highest antibacterial activity compared to paper composites containing either ionic Ag components ($AgNO_3$) or fine Ag powders (particle size: *ca.* 1 μm), although the same amounts of Ag components were used (2.0 mg per sample). In addition, AgNPs@ZnO paper exhibited the highest antibacterial activity even against Gram-positive *Bacillus subtilis* (*B. subtilis*) (Fig. 9), suggesting that AgNPs@ZnO paper is effective for a diverse array of bacteria.

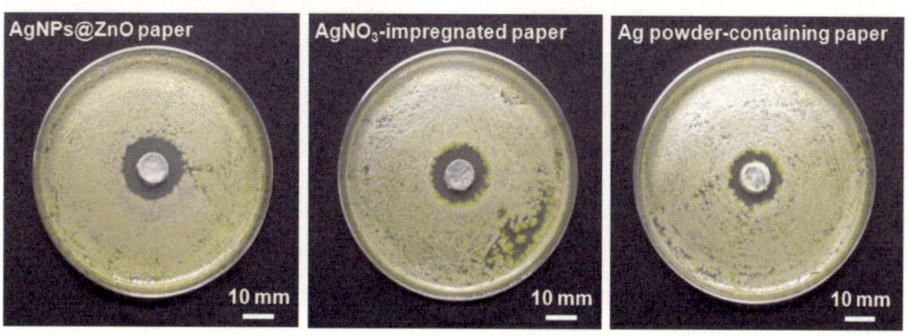

Fig. 9. Optical images of the zone of inhibition against *B. subtilis* for AgNPs@ZnO paper, $AgNO_3$-impregnated ZnO-free paper, and Ag powder-containing paper. Each paper composite is 10 mm in diameter. Ag content: 2.0 mg. Incubation condition: 37°C, 24 h

The antibacterial mechanism of Ag species has been a matter of debate for decades; credible rationales for the antibacterial activity are as follows. (1) Ag^+ ions interact with phosphorous moieties in DNA, resulting in deactivation of DNA replication, and/or (2) they react with sulfur-containing proteins, leading to inhibition of enzyme functions (Gupta et al., 1998; Matsumura et al., 2003). Albeit the mechanism of antibacterial action of AgNPs is still insufficiently understood, many researchers have reported that AgNPs can gradually release Ag^+ ions which play a key role in antibacterial effects (Jeong et al., 2005; Morones et al., 2005). On the other hand, Pal *et al.* have reported that spherical AgNPs had greater antibacterial activity against *E. coli* than Ag^+ ions in the form of $AgNO_3$, and proposed that the nanometer size and the presence of Ag(111) crystal faces synergistically promoted the antibacterial effect of AgNPs (Pal et al., 2007). In that case the elution of Ag^+ ions from and the nanomorphology of AgNPs would synergistically contribute to the excellent antibacterial activity of AgNPs@ZnO paper.
Fig. 10 shows antibacterial behavior for repeated use. In all cases the diameter of the zone of inhibition gradually decreased in successive test cycles.

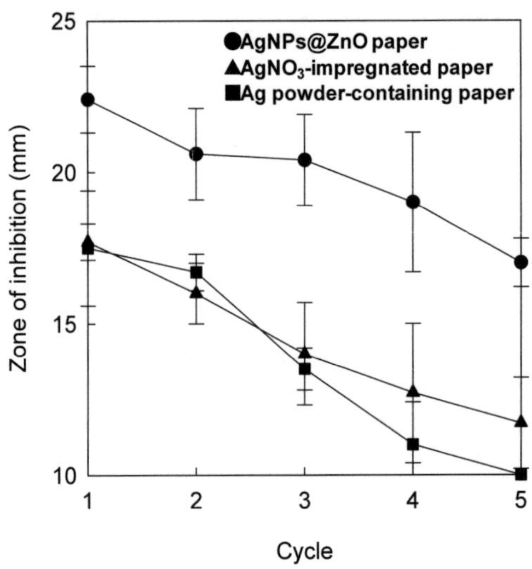

Fig. 10. Antibacterial durability against *E. coli* for repeated use: AgNPs@ZnO paper (circles), $AgNO_3$-impregnated ZnO-free paper (triangles), and Ag powder-containing paper (squares)

The Ag content also decreased due to gradual release from each paper composite during the antibacterial tests, in accordance with the decrease in antibacterial effect. However, although the final Ag contents after the five-cycle test were similar, namely 0.2, 0.2 and 0.3 mg for AgNPs@ZnO paper, $AgNO_3$-impregnated paper and Ag powder-containing paper, respectively, AgNPs@ZnO paper maintained much higher antibacterial activity than the other samples. The good durability of AgNPs@ZnO paper is possibly attributable both to the nanometer size and to the exposure of an active crystal face of AgNPs, leading to efficient antibacterial activity. Thus the AgNPs@ZnO paper with paper-like flexibility and convenience in handling is expected to be a promising bioactive material.
In summary, we have established a novel technique for the efficient immobilization of AgNPs on an easy to handle paper-like support. AgNPs with excellent bioactivity can be

successfully synthesized *in situ* on ZnO whiskers as a selective support, which are pre-incorporated into a ceramic paper matrix, *via* ion-exchange between Ag and Zn species and simultaneous ZnO-mediated photoreduction. The easily fabricated AgNPs@ZnO paper demonstrates excellent antibacterial activity and durability. Furthermore, the use of AgNPs has been growing in various fields such as optical (Ohko et al., 2003), electronic (Zhang et al., 2008), and catalytic applications (Mao et al., 2009). Hence the inorganic AgNPs@ZnO paper composites with paper-like porous structure and practical utility are expected to be one of the advanced materials for a wide range of potential applications of AgNPs.

5. On-paper Synthesis of Metal NPs for Catalytic Applications

Recent years have seen considerable progress in the research and development of metal NPs for use as new catalytic materials with large surface area to volume ratio (Toshima, 2003). In most cases, the electronic properties of metal NPs are significantly different from those of corresponding bulk metals, leading to large enhancement of catalytic activity (Ishida & Haruta, 2007; Park et al., 2008).

Our proposed technique has great potential for wide applications of other metal NPs, and here we describe the on-paper synthesis of CuNPs and AuNPs, and their performance as catalysts.

5.1 CuNPs@ZnO Paper for Hydrogen Production

Catalytic reforming of methanol has widely been used to produce hydrogen for new power generation systems such as polymer electrolyte fuel cells (Palo et al., 2007). In particular, autothermal reforming (ATR), which is the combined process of endothermic steam reforming and exothermic partial oxidation, has recently been investigated as an efficient energy conversion process. The methanol ATR reaction is represented by Eq. (1) (Horny et al., 2004):

$$4CH_3OH + 3H_2O + \frac{1}{2}O_2 \rightarrow 11H_2 + 4CO_2,\ \Delta_r H^0_{573} \approx 0\ \text{kJ mol}^{-1} \quad (1)$$

Methanol reforming has been performed mostly over Cu/ZnO catalysts (Agrell et al., 2002). However, the random packing of catalyst powders into the reactor inevitably causes a high-pressure drop and fluid bypassing, resulting in critical inefficiency of the catalytic reforming process. For that reason, one area of increasing attention has been the development of structured catalysts with micrometer-scale pores that have effective diffusion of heat and reactants, particularly in flow-type reactors (Kiwi-Minsker & Renken, 2005).

As previously described, an inorganic paper matrix prepared using our established papermaking technique has a unique porous microstructure derived from ceramic fiber networks (Fig. 7c). Such a paper matrix is an easy to handle support material, and is an effective microstructured support for catalytic applications. We have shown that a ceramic paper matrix containing Cu/ZnO catalyst powders, called paper-structured catalyst, demonstrates excellent catalytic performance in the methanol ATR process compared with the original catalyst powders and commercial catalyst pellets (Koga et al., 2006, 2008a). The fiber-network microstructure with connected pore spaces inside the paper-structured catalyst provides a suitable catalytic reaction environment by promoting a high degree of desirable gas accessibility to the catalyst surfaces (Koga et al., 2009d).

The on-paper synthesis of highly-active CuNPs using ZnO whiskers is a rational approach toward the further improvement of catalytic performance of paper-structured catalyst.

CuNPs with size 20–50 nm can be successfully synthesized from an aqueous solution of the metal salt precursor copper nitrate ($Cu(NO_3)_2$), due to ion exchange between Cu and Zn species (Fig. 11) (Koga et al., 2008b).

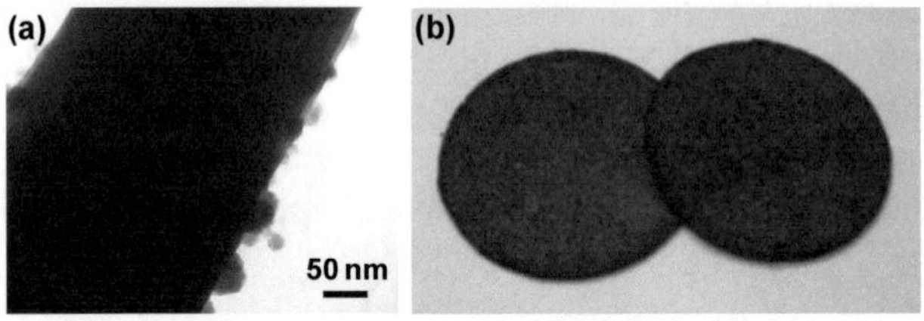

Fig. 11. TEM image of CuNPs@ZnO whiskers (a), and optical image of CuNPs@ZnO paper (b). The size of paper composite is 8×10^2 mm^2

As shown in Fig. 12, CuNPs@ZnO paper offers exceptional catalytic performance in the methanol reforming process for hydrogen production.

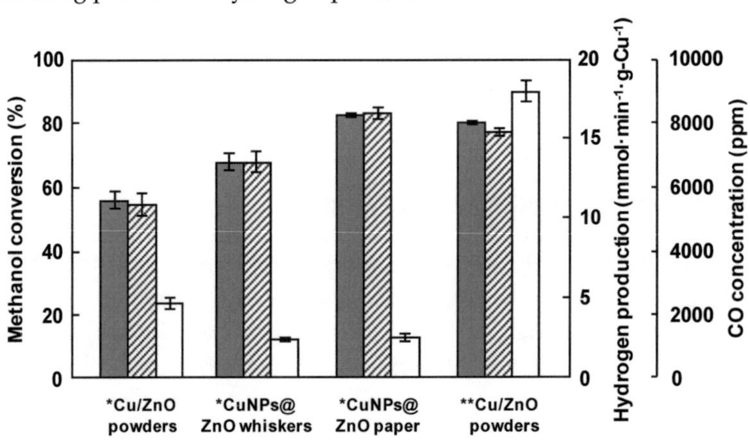

Fig. 12. ATR performance of Cu/ZnO powders, CuNPs@ZnO whiskers and CuNPs@ZnO paper; methanol conversion (filled bars), hydrogen production (striped bars) and CO concentration (open bars). Reaction temperature: 250°C* or 310°C**. Cu:ZnO=1.0:5.2 by weight. Cu content: 0.2 g/8×10^3 mm^3

For 80% conversion of methanol, suppression of by-product carbon monoxide (CO), which acts as a catalytic poison for Pt anode electrocatalysts in fuel cells, to *ca.* 14% is achieved at a 60 K lower reaction temperature than with conventional Cu/ZnO powders. Surprisingly, the catalytic performance of CuNPs@ZnO paper is superior to that of CuNPs@ZnO whiskers. These results suggest that the combination of CuNPs with high catalytic activity and a paper-structured support with a fiber-network microstructure is particularly effective for the ATR reaction, and CuNPs@ZnO paper is an innovative catalytic material for production of pure hydrogen for fuel cell applications.

5.2 AuNPs@ZnO Paper for Low-temperature CO Oxidation

Au has traditionally been considered chemically inert and regarded as catalytically inactive. However, nanosized Au has recently received increasing attention as an innovative nanomaterial with unique properties in various fields such as electronics (Jena & Raj, 2008) and biomedical applications (Skrabalak et al., 2007). In particular, when Au in the form of NPs with diameter less than 10 nm is well dispersed on some metal oxides, it acts as a highly active catalyst for many chemical reactions (Ishida & Haruta, 2007). A great deal of effort has been directed toward low-temperature CO oxidation (represented by Eq. (2)) over Au supported metal oxide catalysts (Bond et al., 2006).

$$CO + \frac{1}{2}O_2 \rightarrow CO_2,\ \Delta_r H^0_{298} = -283\ \text{kJ mol}^{-1} \quad (2)$$

On-paper synthesis of AuNPs is accomplished by using the metal complex precursor tetrachloroauric acid ($HAuCl_4$). AuNPs with particle size <10 nm are spontaneously formed on ZnO whiskers *via* a simple soaking treatment with $HAuCl_4$ solution (Fig. 13), possibly due to electron transfer from Zn(II) in ZnO whiskers to Au(III) species through Zn–O–Au bonds (Koga et al., 2009c).

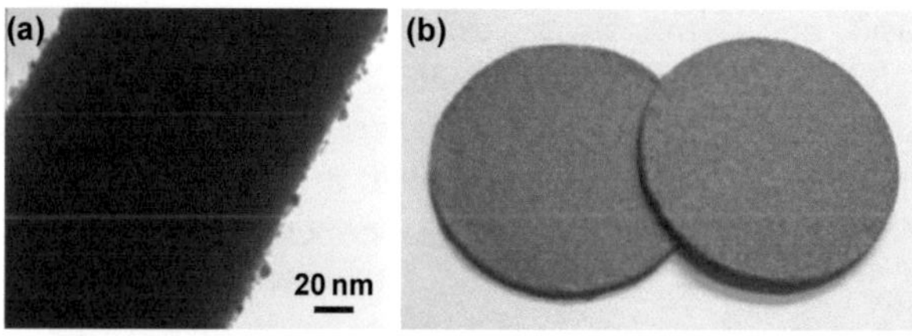

Fig. 13. TEM image of AuNPs@ZnO whiskers (a), and optical image of AuNPs@ZnO paper (b). The size of paper composite is 8×10^2 mm^2

As-prepared AuNPs@ZnO paper has excellent catalytic performance in low-temperature CO oxidation. Complete conversion of CO to CO_2 is achieved at 20°C, which is 140 K lower than the reaction temperature for conventional Au/ZnO catalyst powders (Fig. 14).

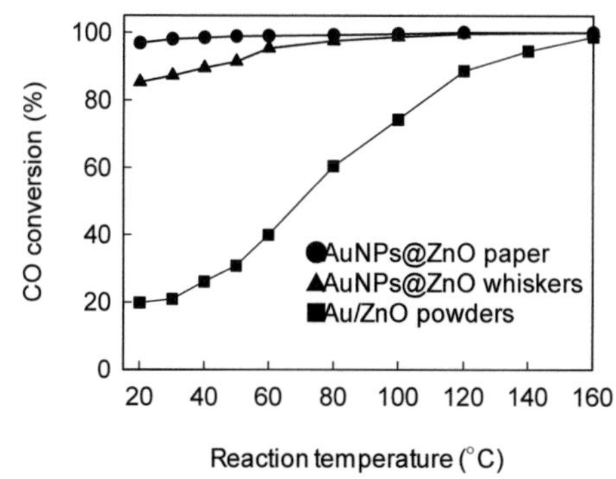

Fig. 14. CO conversion to CO_2 as a function of reaction temperature: AuNPs@ZnO paper (circles), AuNPs@ZnO whiskers (triangles), and Au/ZnO powders (squares). Au content: 10 mg/8×10^3 mm^3

The synthesis of AuNPs described here, which is achieved by use of a metal complex precursor, is clearly different from the formation of CuNPs and AgNPs through simple ion exchange. This strongly suggests that the on-paper synthesis of metal NPs can be performed using various metal precursors, and has great potential for future applications for a diverse array of metal species. Catalytic metal NPs@ZnO paper composites with paper-like flexibility are able to fit various reactor configurations and are thus expected to perform as advanced catalytic materials for improving the practical utility and catalytic performance of these systems, for a wide range of industrial chemical processes.

6. Conclusion

This chapter provides a new concept and strategic approach to optimising use of AgNPs in practical applications without compromising their excellent functionalities. Direct, *in situ* synthesis of AgNPs, called on-paper synthesis, is successfully achieved by using a paper matrix composed of ceramic fibers as a main framework and ZnO whiskers as a selective support for AgNPs *via* a simple soaking treatment with $AgNO_3$ solution. As-prepared AgNPs@ZnO paper with the appearance of flexible cardboard is convenient for a wide range of practical uses, and demonstrates excellent antibacterial performance against both Gram-negative *E. coli* and Gram-positive *B. subtilis*, with much greater efficacy than paper composites containing either ionic Ag or commercial crystalline Ag microparticles. Thus, this method provides highly-functional AgNPs in a form that can be easily handled, and will open up a new avenue for promoting the effective use of AgNPs in practical applications.

On-paper synthesis has broad applications and can be extended to a diverse array of metal NPs. CuNPs are successfully synthesized from an aqueous solution of $Cu(NO_3)_2$, and CuNPs@ZnO paper offers exceptional catalytic performance in the methanol reforming process for hydrogen production. AuNPs@ZnO paper, which is prepared from $HAuCl_4$ complex as a precursor, can mediate the complete conversion of CO to CO_2 at room temperature. Thus the novel on-paper synthesis approach allows us to design a variety of nanomaterials containing metal NPs that maintain their original excellent functionality, and metal NPs@paper composites with convenience in handling are expected to be promising materials in a wide range of future applications. Such a new concept for practical immobilization of metal NPs will break new ground in metal NPs engineering fields.

7. References

Agrell, J.; Birgersson, H. & Boutonnet, M. (2002). Steam reforming of methanol over a $Cu/ZnO/Al_2O_3$ catalyst: a kinetic analysis and strategies for suppression of CO formation. *Journal of Power Sources*, 106, 1-2, 249-257, ISSN 0378-7753

Alt, V.; Bechert, T.; Steinrücke, P.; Wagener, M.; Seidel, P.; Dingeldein, E.; Domann, E. & Schnettler, R. (2004). An in vitro assessment of the antibacterial properties and cytotoxicity of nanoparticulate silver bone cement. *Biomaterials*, 25, 18, 4383-4391, ISSN 0142-9612

An, X.Q.; Cao, C.B. & Zhu, H.S. (2007). Bio-inspired fabrication of ZnO nanorod arrays and their optical and photoresponse properties. *Journal of Crystal Growth*, 308, 2, 340-347, ISSN 0022-0248

Anyaogu, K.C.; Fedorov, A.V. & Neckers, D.C. (2008). Synthesis, characterization, and antifouling potential of functionalized copper nanoparticles. *Langmuir*, 24, 8, 4340-4346, ISSN 0743-7463

Bond, G.C.; Louis C. & Thompson, D.T. (2006). *Catalysis by Gold*, Imperial College Press, ISBN 1-86094-658-5, London

Casey, J.P. (1981). *Pulp and Paper: Chemistry and Chemical Technology Volume III*, John Wiley & Sons, Inc., ISBN 978-0-471-03177-2, New York

Dotzauer, D.M.; Dai, J.; Sun, L. & Bruening, M.L. (2006). Catalytic membranes prepared using layer-by-layer adsorption of polyelectrolyte/metal nanoparticle films in porous supports. *Nano Letters*, 6, 10, 2268-2272, ISSN 1530-6984

Dubas, S.T.; Kumlangdudsana, P. & Potiyaraj, P. (2006). Layer-by-layer deposition of antimicrobial silver nanoparticles on textile fibers. *Colloids and Surfaces A: Physicochemical and Engineering Aspects*, 289, 1-3, 105-109, ISSN 0927-7757

Egusa, S.; Yokota, S.; Tanaka, K.; Esaki, K.; Okutani, Y.; Ogawa, Y.; Kitaoka, T.; Goto, M. & Wariishi, H. (2009). Surface modification of a solid-state cellulose matrix with lactose by a surfactant-enveloped enzyme in a nonaqueous medium. *Journal of Materials Chemistry*, 19, 13, 1836–1842, ISSN 0959-9428

Feldheim, D.L. & Foss, C.A. (2002). *Metal nanoparticles: synthesis, characterization, and applications*. Marcel Dekker, Inc., ISBN 0-8247-0604-8, New York

Fujishima, A. & Honda, K. (1972). Electrochemical photolysis of water at a semiconductor electrode. *Nature*, 238, 5358, 37-38, ISSN 0028-0836

Fukahori, S.; Ichiura, H.; Kitaoka, T. & Tanaka, H. (2003). Photocatalytic decomposition of bisphenol A in water using composite TiO_2-zeolite sheets prepared by a papermaking technique. *Environmental Science and Technology*, 37, 5, 1048-1051, ISSN 0013-936X

Fukahori, S.; Kitaoka, T.; Tomoda, A.; Suzuki, R. & Wariishi, H. (2006a). Methanol steam reforming over paper-like composites of Cu/ZnO catalyst and ceramic fiber. *Applied Catalysis A: General*, 300, 2, 155-161, ISSN 0926-860X

Fukahori, S.; Koga, H.; Kitaoka, T.; Tomoda, A.; Suzuki, R. & Wariishi, H. (2006b). Hydrogen production from methanol using a SiC fiber-containing paper composite impregnated with Cu/ZnO catalyst. *Applied Catalysis A: General*, 310, 138-144, ISSN 0926-860X

Fukahori, S.; Iguchi, Y.; Ichiura, H.; Kitaoka, T.; Tanaka, H. & Wariishi, H. (2007). Effect of void structure of photocatalyst paper on VOC decomposition. *Chemosphere*, 66, 11, 2136-2141, ISSN 0045-6535

Fukahori, S.; Koga, H.; Kitaoka, T.; Nakamura, M. & Wariishi, H. (2008). Steam reforming behavior of methanol using paper-structured catalysts: Experimental and computational fluid dynamic analysis. *International Journal of Hydrogen Energy*, 33, 6, 1661-1670, ISSN 0360-3199

Gadhe, J.B. & Gupta, R.B. (2007). Hydrogen production by methanol reforming in supercritical water: Catalysis by in-situ-generated copper nanoparticles. *International Journal of Hydrogen Energy*, 32, 13, 2374-2381, ISSN 0360-3199

Gogoi, S.K.; Gopinath, P.; Paul, A.; Ramesh, A.; Ghosh S.S. & Chattopadhyay, A. (2006). Green fluorescent protein-expressing *Escherichia coli* as a model system for investigating the antimicrobial activities of silver nanoparticles. *Langmuir*, 22, 22, 9322-9328, ISSN 0743-7463

Gupta, A.; Maynes, M. & Silver, S. (1998). Effects of halides on plasmid-mediated silver resistance in *Escherichia coli*. *Applied and Environmental Microbiology*, 64, 12, 5042-5045, ISSN 1098-5336

Horny, C.; Kiwi-Minsker, L. & Renken, A. (2004). Micro-structured string-reactor for autothermal production of hydrogen. *Chemical Engineering Journal*, 101, 1-3, 3-9, ISSN 1385-8947

Huang, J.; Ichinose, I. & Kunitake, T. (2006). Biomolecular modification of hierarchical cellulose fibers through titania nanocoating. *Angewandte Chemie-International Edition*, 45, 18, 2883–2886, ISSN 1433-7851

Ichiura, H.; Kubota, Y.; Wu, Z. & Tanaka, H. (2001). Preparation of zeolite sheets using a papermaking technique - Part I - Dual polymer system for high retention of stock components. *Journal of Materials Science*, 36, 4, 913–917, ISSN 0022-2461

Ichiura, H.; Nozaki, M.; Kitaoka, T. & Tanaka, H. (2003). Influence of uniformity of zeolite sheets prepared using a papermaking technique on VOC adsorptivity. *Advances in Environmental Research*, 7, 4, 975-979, ISSN 1093-0191

Ichiura, H.; Ohi, T.; Oyama, H.; Yokota, H.; Kunitake, T.; Ohashi, S. & Morikawa, M. (2008). Paper–paraffin composites prepared by interfacial polymerization reaction on paper surface and its function of thermal energy storage. *Journal of Materials Science*, 43, 4, 1486–1491, ISSN 0022-2461

Iguchi, Y.; Ichiura, H.; Kitaoka, T. & Tanaka, H. (2003). Preparation and characteristics of high performance paper containing titanium dioxide photocatalyst supported on inorganic fiber matrix. *Chemosphere*, 53, 10, 1193-1199, ISSN 0045-6535

Ishida, T. & Haruta, M. (2007). Gold catalysts: Towards sustainable chemistry. *Angewandte Chemie-International Edition*, 46, 38, 7154-7156, ISSN 1433-7851

Ishihara, H.; Koga, H.; Kitaoka, T.; Wariishi, H.; Tomoda, A. & Suzuki, R. (2010). Paper-structured catalyst for catalytic NOx removal from combustion exhaust gas. *Chemical Engineering Science*, 65, 1, 208-213, ISSN 0009-2509

Jena, B.K. & Raj, C.R. (2008). Morphology dependent electrocatalytic activity of Au nanoparticles. *Electrochemistry Communications*, 10, 6, 951–954, ISSN 1388-2481

Jeon, H.-J.; Yi, S.-C. & Oh, S.-G. (2003). Preparation and antibacterial effects of Ag-SiO_2 thin films by sol-gel method. *Biomaterials*, 24, 27, 4921-4928, ISSN 0142-9612

Jeong, S.H.; Hwnag Y.H. & Yi, S.C. (2005). Antibacterial properties of padded PP/PE nonwovens incorporating nano-sized silver colloids. *Journal of Materials Science*, 40, 20, 5413-5418, ISSN 0022-2461

Kiwi-Minsker, L. & Renken, A. (2005). Microstructured reactors for catalytic reactions. *Catalysis Today*, 110, 1-2, 2-14, ISSN 0920-5861

Koga, H.; Fukahori, S.; Kitaoka, T.; Tomoda, A.; Suzuki, R. & Wariishi, H. (2006). Autothermal reforming of methanol using paper-like Cu/ZnO catalyst composites prepared by a papermaking technique. *Applied Catalysis A: General*, 309, 2, 263-269, ISSN 0926-860X

Koga, H.; Fukahori, S.; Kitaoka, T.; Nakamura, M. & Wariishi, H. (2008a). Paper-structured catalyst with porous fiber-network microstructure for autothermal hydrogen production. *Chemical Engineering Journal*, 139, 2, 408-415, ISSN 1385-8947

Koga, H.; Kitaoka, T. & Wariishi, H. (2008b). *In situ* synthesis of Cu nanocatalysts on ZnO whiskers embedded in a microstructured paper composite for autothermal hydrogen production. *Chemical Communications*, 43, 5616-5618, ISSN 1359-7345

Koga, H.; Kitaoka, T. & Wariishi, H. (2009a). *In situ* synthesis of silver nanoparticles on zinc oxide whiskers incorporated in a paper matrix for antibacterial applications. *Journal of Materials Chemistry*, 19, 15, 2135-2140, ISSN 0959-9428

Koga, H.; Umemura, Y.; Ishihara, H.; Kitaoka, T.; Tomoda, A.; Suzuki, R. & Wariishi, H. (2009b). Paper-structured fiber composites impregnated with platinum nanoparticles synthesized on a carbon fiber matrix for catalytic reduction of nitrogen oxides. *Applied Catalysis B: Environmental*, 90, 3-4, 699-704, ISSN 0926-3373

Koga, H.; Kitaoka, T. & Wariishi, H. (2009c). On-paper synthesis of Au nanocatalysts from Au(III) complex ions for low-temperature CO oxidation. *Journal of Materials Chemistry*, 19, 29, 5244-5249, ISSN 0959-9428

Koga, H.; Kitaoka, T.; Nakamura, M. & Wariishi, H. (2009d). Influence of a fiber-network microstructure of paper-structured catalyst on methanol reforming behavior. *Journal of Materials Science*, 44, 21, 5836-5841, ISSN 0022-2461

Kumar, S.A.; Chen, S.-M. (2008). Nanostructured zinc oxide particles in chemically modified electrodes for biosensor applications. *Analytical Letters*, 41, 2, 141-158, ISSN 0003-2719

Lee, H.Y.; Park, H.K.; Lee, Y.M.; Kim, K. & Park, S.B. (2007). A practical procedure for producing silver nanocoated fabric and its antibacterial evaluation for biomedical applications. *Chemical Communications*, 28, 2959-2961, ISSN 1359-7345

Li, Y.; Wu, Y. & Ong, B.S. (2005). Facile synthesis of silver nanoparticles useful for fabrication of high-conductivity elements for printed electronics. *Journal of the American Chemical Society*, 127, 10, 3266-3267, ISSN 0002-7863

Lok, C.-N.; Ho, C.-M.; Chen, R.; He, Q.-Y.; Yu, W.-Y.; Sun, H.; Tam, P.K.-H.; Chiu, J.-F. & Che, C.-M. (2006). Proteomic analysis of the mode of antibacterial action of silver nanoparticles. *Journal of Proteome Research*, 5, 4, 916-924, ISSN 1535-3893

Ma, D.; Guan, J.; Normandin, F.; Dénommée, S.; Enright, G.; Veres, T. & Simard, B. (2006). Multifunctional nano-architecture for biomedical applications. *Chemistry of Materials*, 18, 7, 1920-1927, ISSN 0897-4756

Mao, J.P.; Deng, M.M.; Xue, Q.S.; Chen, L. & Lu, Y. (2009). Thin-sheet Ag/Ni-fiber catalyst for gas-phase selective oxidation of benzyl alcohol with molecular oxygen. *Catalysis Communications*, 10, 10, 1376-1379, ISSN 1566-7367

Matsumura, Y.; Yoshikata, K.; Kunisaki, S. & Tsuchido, T. (2003). Mode of bactericidal action of silver zeolite and its comparison with that of silver nitrate. *Applied and Environmental Microbiology*, 69, 7, 4278-4281, ISSN 1098-5336

Matter, P.H.; Braden, D.J. & Ozkan, U.S. (2004). Steam reforming of methanol to H_2 over nonreduced Zr-containing CuO/ZnO catalysts. *Journal of Catalysis*, 223, 2, 340-351, ISSN 0021-9517

McFarland, A.D. & Van Duyne, R.P. (2003). Single silver nanoparticles as real-time optical sensors with zeptomole sensitivity. *Nano Letters*, 3, 8, 1057-1062, ISSN 1530-6984

Morones, J.R.; Elechiguerra, J.L.; Camacho, A.; Holt, K.; Kouri, J.B.; Ramirez, J.T. & Yacaman M.J. (2005). The bactericidal effect of silver nanoparticles. *Nanotechnology*, 16, 10, 2346-2353, ISSN 0957-4484

Navarro, E.; Piccapietra, F.; Wagner, B.; Marconi, F.; Kaegi, R.; Odzak, K.; Sigg, L. & Behra, R. (2008). Toxicity of silver nanoparticles to *Chlamydomonas reinhardtii*. *Environmental Science and Technology*, 42, 23, 8959-8964, ISSN 0013-936X

Neimo, L. (1999). *Papermaking Chemistry*, Fapet Oy, ISBN 952-5216-04-7, Helsinki

Ohko, Y.; Tatsuma, T.; Fujii, T.; Naoi, K.; Niwa, C.; Kubota, Y. & Fujishima, A. (2003). Multicolour photochromism of TiO2 films loaded with silver nanoparticles. *Nature Materials*, 2, 1, 29-31, ISSN 1476-1122

Pal, S.; Tak, Y.K. & Song, J.M. (2007). Does the antibacterial activity of silver nanoparticles depend on the shape of the nanoparticle? A study of the Gram-negative bacterium *Escherichia coli*. *Applied and Environmental Microbiology*, 73, 6, 1712-1720, ISSN 1098-5336

Palo, D.R.; Dagle, R.A. & Holladay, J.D. (2007). Methanol Steam Reforming for Hydrogen Production. *Chemical Reviews*, 107, 10, 3992–4021, ISSN 1520-6890

Park, J.Y.; Zhang, Y.; Grass, M.; Zhang, T. &, G.A. (2008). Tuning of catalytic CO oxidation by changing composition of Rh−Pt bimetallic nanoparticles. *Nano Letters*, 8, 2, 673-677, ISSN 1530-6984

Pelton, R.; Geng, X.L. & Brook, M. (2006). Photocatalytic paper from colloidal TiO_2 - fact or fantasy. *Advances in Colloid and Interface Science*, 127, 1, 43-53, ISSN 0001-8686

Roberts, J.C. (1996). *Paper Chemistry*, Blackie Academic & Professional, ISBN 0-7514-0236-2, London

Scott, W.E. (1996). *Principles of Wet End Chemistry*, TAPPI PRESS, ISBN 0-89852-286-2, Atlanta

Shah, M.S.A.S.; Nag, M.; Kalagara, T.; Singh, S. & Manorama, S.V. (2008). Silver on PEG-PU-TiO_2 polymer nanocomposite films: An excellent system for antibacterial applications. *Chemistry of Materials*, 20, 7, 2455-2460, ISSN 0897-4756

Shukla, R.; Bansal, V.; Chaudhary, M.; Basu, A.; Bhonde R.R. & Sastry, M. (2005). Biocompatibility of gold nanoparticles and their endocytotic fate inside the cellular compartment: a microscopic overview, *Langmuir*, 21, 23, 10644-10654, ISSN 0743-7463

Skrabalak, S.E.; Chen, J.; Au, L.; Lu, X.; Li, X. & Xia, Y. (2007). Gold nanocages for biomedical applications. *Advanced Materials*, 19, 20, 3177-3184, ISSN 0935-9648

Sondi, I. & Salopek-Sondi, B. (2004). Silver nanoparticles as antimicrobial agent: a case study on *E-coli* as a model for Gram-negative bacteria. *Journal of Colloid and Interface Science*, 275, 1, 177-182, ISSN 0021-9797

Toshima, N. (2003). Metal nanoparticles for catalysis, In: *Nanoscale Materials*, Liz-Marzan, L.M. & Kamat, P.V. (Ed.), 79-96, Kluwer Academic Publishers, ISBN 978-1-4020-7366-3, Boston

Wang, Y.; Du, G.; Liu, H.; Liu, D.; Qin, S.; Wang, N.; Hu, C.; Tao, X.; Jiao, J.; Wang, J. & Wang, Z.L. (2008). Nanostructured sheets of Ti-O nanobelts for gas sensing and antibacterial applications. *Advanced Functional Materials*, 18, 7, 1131-1137, ISSN 1616-301X

Yang, P.D.; Yan, H.Q.; Mao, S.; Russo, R.; Johnson, J.; Saykally, R.; Morris, N.; Pham, J.; He, R.R. & Choi, H.J. (2002). Controlled growth of ZnO nanowires and their optical properties. *Advanced Functional Materials*, 12, 5, 323-331, ISSN 1616-301X

Yokota, S.; Kitaoka, T.; Opietnik, M.; Rosenau, T. & Wariishi, H. (2008). Synthesis of gold nanoparticles for in situ conjugation with structural carbohydrates. *Angewandte Chemie-International Edition*, 47, 51, 9866–9869, ISSN 1433-7851

Zhang, H.; Wang, G.; Chen, D.; Lv, X.J. & Li, J.H. (2008). Tuning photoelectrochemical performances of Ag-TiO_2 nanocomposites via reduction/oxidation of Ag. *Chemistry of Materials*, 20, 20, 6543-6549, ISSN 0897-4756

Zotti, G.; Vercelli, B. & Berlin, A. (2008). Gold nanoparticles linked by pyrrole- and thiophene-based thiols. Electrochemical, optical, and conductive properties. *Chemistry of Materials*, 20, 2, 397-412, ISSN 0897-4756

15

The Silver Nanoparticle (Nano-Ag): a New Model for Antifungal Agents

Juneyoung Lee, Keuk-Jun Kim, Woo Sang Sung,
Jong Guk Kim and Dong Gun Lee
School of Life Sciences and Biotechnology, College of Natural Sciences, Kyungpook National University
Republic of Korea

1. Introduction

In recent years, a rapid increase in microbes that are resistant to conventional antibiotics has been observed (Goffeau, 2008). Especially, the frequency of infections provoked by opportunistic fungal strains has increased dramatically. Even though the majority of invasive fungal infections are still due to the *Aspergillus* or *Candida* species, the spectrum of fungal pathogens has changed and diversified (Denning, 1991; Ellis et al., 2000; Odds et al., 2003). Azoles that inhibit sterol formation and polyenes that bind to mature membrane sterols have been the mainstays regarding antifungal therapy for several decades (Kullberg & de Pauw, 1999; Sheehan et al., 1999). However, not only the emergence of fluconazole resistance among different pathogenic strains but also the high toxicity of amphotericin B (Alexander & Perfect, 1997; Mukherjee et al., 2003) has prompted research on new antifungal agents (Kontoyiannis et al., 2003).

Bio-nanotechnology has emerged as an integration between biotechnology and nanotechnology for developing biosynthesis and environmental-friendly technology for synthesis of nanomaterials. We specifically regarded nanoparticles as clusters of atoms in the size of 1-100 nm. 'Nano' is a Greek word synonymous to dwarf meaning extremely small. The use of nanoparticles is gaining impetus in the present century as they posses defined chemical, optical and mechanical properties. Among them, the metallic nanoparticles are most promising as they contain remarkable antibacterial properties due to their large surface area to volume ratio, which is of interest to researchers due to the growing microbial resistance against metal ions, antibiotics, and the development of resistant strains (Rai et al., 2009; Gong et al., 2007).

Different types of nanomaterials like copper, zinc, titanium (Schabes-Retchkiman et al., 2006), magnesium, gold (Gu et al., 2003), alginate (Ahmad et al., 2005) and silver have been developed but silver nanoparticles (Nano-Ag) have proved to be most effective as they exhibit potent antimicrobial efficacy against bacteria, viruses and eukaryotic micro-organisms. However, Nano-Ag used as a disinfectant drug also has some risks as the exposure to silver can cause argyrosis or argyria; it can be toxic to mammalian cells (Gong et al., 2007).

The current investigation supports the theory that the use of silver ions or metallic silver as well as Nano-Ag can be exploited in medicine for burn treatment, dental materials, coating stainless steel materials, textile fabrics, water treatment, sunscreen lotions, etc. and posses low toxicity to human cells, high thermal stability and low volatility (Duran et al., 2007).
Many studies have shown the biological effects of Nano-Ag, however, its effects against fungal pathogens have not yet been fully studied. In this study, therefore, the antifungal properties and mechanism of actions of Nano-Ag against human pathogenic fungal strains were investigated. Furthermore, the therapeutic potential of Nano-Ag for treating fungal diseases in humans was suggested.

2. Materials and methods

2.1. Materials

Amphotericin B, carbonyl cyanide *m*-chlorophenylhydrazone (CCCP), trehalase and RNase A were purchased from the Sigma Chemical Co.. Stock solutions of amphotericin B were prepared in dimethyl sulfoxide (DMSO), and stored at -20 °C. For all the experiments, a final concentration of 2% DMSO was used as the solvent carrier.

2.2. Preparation of Nano-Ag

One hundred grams of solid silver were dissolved in 100 ml of 100% nitric acid at 90 °C, and then 1 l of distilled water was added. By adding sodium chloride to the silver solution, the Ag ions were precipitated and then clustered together to form monodispersed nanoparticles in the aqueous medium. The sizes and morphology of Nano-Ag were examined by using a transmission electron microscope (TEM) (H-7600, HITACHI, LTD). The result showed that the Nano-Ag was in a spherical form and had an average size of 3 nm (Fig. 1).

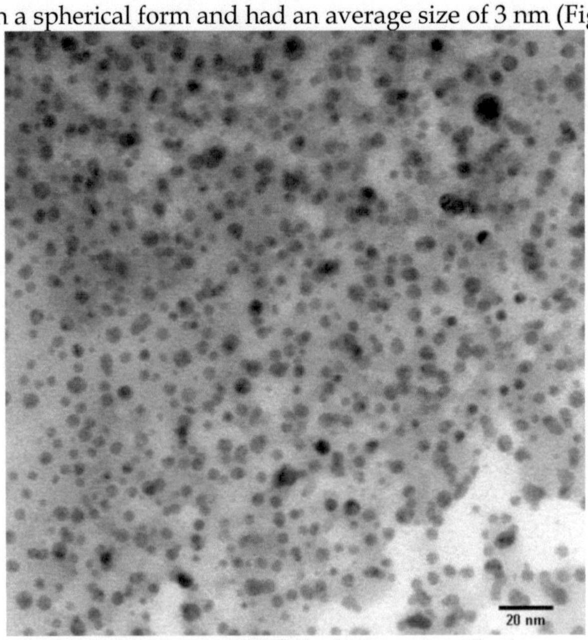

Fig. 1. Transmission electron micrograph (TEM) of Nano-Ag. The bar corresponds to 20 nm.

As the final concentration of colloidal silver was 60,000 ppm, this solution was diluted, and then samples of different concentrations were used to investigate the antifungal effects of Nano-Ag.

2.3. Fungal strains and culture conditions

A total of 44 strains of 2 fungal species were used in this study. *Candida albicans* (ATCC 90028), *Candida glabrata* (ATCC 90030), *Candida parapsilosis* (ATCC 22019), and *Candida krusei* (ATCC 6258) were obtained from the American Type Culture Collection (ATCC) (Manassas, VA, USA). Clinical isolates of *Candida* spp. were obtained from the Department of Laboratory Medicine, Chonnam National University Medical School (Gwangju, Korea), and clinical isolates of *Trichophyton mentagrophytes* were obtained from the Institute of Medical Mycology, Catholic Skin Clinic (Daegu, Korea). *Candida* spp. and *Trichophyton mentagrophytes* were cultures in a Sabouraud dextrose agar (SDA) and a potato dextrose agar (PDA) at 35 °C, respectively.

2.4. Antifungal susceptibility testing

The minimum inhibitory concentration (MIC) for *Candida* spp. and *T. mentagrophytes* was determined by a broth microdilution method based on the National Committee for Clinical Laboratory Standards (NCCLS; now renamed as Clinical and Laboratory Standards Institute, CLSI, 2000) method outlined in documents M-27A and M-38P, respectively. An RPMI 1640 medium buffered to pH 7.0 with 3-(*N*-morpholino) propanesulfonic acid (MOPS) was used as the culture medium, and the inoculum size of *Candida* spp. was 0.5 x 10^3 to 2.5 x 10^3 cells/ml, and that of *T. mentagrophytes* was 0.4 x 10^4 to 5 x 10^4 cells/ml. The microdilution plates inoculated with fungi were incubated at 35 °C, and the turbidity of the growth control wells was observed every 24 hrs. The 80% inhibitory concentration (IC_{80}) was defined as the lowest concentration that inhibited 80% of the growth as determined by a comparison with the growth in the control wells. The growth was assayed with a microplate reader (Bio-Tek Instruments, Winooski, VT, USA) by monitoring absorption at 405 nm. In the current study, amphotericin B and fluconazole were used as a positive control toward fungi; amphotericin B is a fungicidal agent widely used in treating serious systemic infections (Hartsel & Bolard, 1996), and fluconazole is used in the treatment of superficial skin infections caused by dermatophytes and *Candida* species (Amichai & Grunwald, 1998).

2.5. The effect on the dimorphic transition of *C. albicans*

C. albicans cells were maintained by periodic subculturing in a liquid yeast extract/peptone/dextrose (YPD) medium. Cultures of yeast cells (blastoconidia) were maintained in a liquid YPD medium at 37 °C. To induce mycelial formation, cultures were directly supplemented with 20% of fetal bovine serum (FBS). The dimorphic transition in *C. albicans* was investigated from cultures containing 2 mg/ml of Nano-Ag (at the IC_{80}), which were incubated for 48 hrs at 37 °C (Jung et al., 2007; Sung et al., 2007). The dimorphic transition to mycelial forms was detected by phase contrast light microscopy (NIKON, ECLIPSE TE300, Japan).

2.6. Hemolytic activity assay

The hemolytic effect of Nano-Ag was examined by measuring the release of hemoglobin from a 4% suspension of fresh human red blood cells (hRBCs). The hRBCs were washed with phosphate-buffered saline (PBS: 35 mM phosphate buffer/150 mM NaCl, pH 7.4). One–hundred microliters of hRBC suspension was added to the 96-well microtiter plates, and then 100 μl of the compound solution in PBS was mixed into each well. After incubating the mixtures for 1 hr at 37 ˚C, the mixtures were centrifuged at 1,500 rpm for 10 min, and the aliquots were transferred to new 96-well microtiter plates. The absorbance of the aliquots was measured at 414 nm by using a microtiter ELISA Reader. Hemolytic rates of 0 and 100% were determined in PBS and 0.1% Triton X-100, respectively (Park et al., 2003). The percentage of hemolysis was calculated by employing the following equation:

$$\text{Percentage hemolysis} = [(\text{Abs}_{414\,nm} \text{ in the compound solution} - \text{Abs}_{414\,nm} \text{ in PBS})/(\text{Abs}_{414\,nm} \text{ in 0.1\% Triton X-100} - \text{Abs}_{414\,nm} \text{ in PBS})] \times 100.$$

2.7. Flow cytometric analysis for plasma membrane potential

For analysis of the membrane integrity after the treatment of Nano-Ag, log-phased cells of *C. albicans* (1 x 10^8 cells), cultured in a YPD medium, were harvested and resuspended with 1 ml of a fresh YPD medium, containing 30 μg/ml of Nano-Ag (at 15 times the MIC) or 10 μM of CCCP, used as a positive control. After incubation for 3 hrs, the cells were washed three times with PBS. To detect any depolarization of the cell membrane, 1 ml of PBS, containing 50 μg of bis-(1,3-dibutylbarbituric acid) trimethine oxonol [$DiBAC_4(3)$], was added and the samples were incubated for 1 hr at 4°C in the dark (Liao et al., 1999). Flow cytometric analysis was performed using a FACSCalibur flow cytometer.

2.8. Measurement of plasma membrane fluorescence intensity

The fluorescence intensity from exponential *C. albicans* cells labeled by 1,6-diphenyl-1,3,5-hexatriene (DPH) was used to monitor changes in membrane dynamics. The cells (1 x 10^8 cells in a YPD medium), containing 20, 40, 60 and 80 μg/ml of Nano-Ag or amphotericin B, were incubated at a physiological temperature of 28°C on a rotary shaker at 140 rpm for 2 hrs. The control cells were incubated without a compound. The cells were fixed with formaldehyde (0.37%, v/v) for 45 min and they were collected by centrifugation at 3,000 rpm, then washed several times with PBS buffer (pH 7.4), and the pellets were frozen in liquid nitrogen. For DPH labeling, the pellets were resuspended in PBS buffer and incubated at 28°C for 45 min in the presence of 0.6 mM of DPH, followed by several washings in PBS buffer. The steady-state fluorescence intensity was measured by using a SHIMADZU RF-5301PC spectrofluorophotometer at 350 nm excitation and 425 nm emission wavelengths (Fernandes et al., 2000). The results represent the average of the triplicate measurements from three independent assays.

2.9. Determining released glucose and trehalose

Fungal strains were grown at 28°C in a YPD medium. *C. albicans* cells were washed three times with PBS, and then 1 ml of the *C. albicans* cell suspension (1 x 10^8 cells), containing 20 μg/ml of Nano-Ag (at 20 times the MIC), was incubated for 2 hrs at 28°C in PBS. The negative control was incubated without Nano-Ag, and a positive control was incubated with 100 μg/ml of amphotericin B (at 20 times the MIC). The fungal cells were settled by centrifugation (12,000 rpm for 20 min). The pellets were dried to calculate their dry weight and supernatants were transferred to a new tube. Released glucose and trehalose-containing supernatants were added to 0.05 units of trehalase. After 1 hr of enzymatic reaction at 37°C, the reaction suspension was mixed with water and 16% DNS reagent (3,5-dinitrosalicylic acid 1%, NaOH 2%, sodium potassium tartrate 20%) was added. For the reaction of glucose with the DNS reagent, the mixture was boiled for 5 min and cooled. Color formations were measured at 525 nm. The results represent the average of the measurements conducted in triplicate of three independent assays.

2.10. Transmission electron microscopy (TEM)

Log-phased cells of *C. albicans* (1 x 10^8 cells) cultured in a YPD medium, were harvested and incubated in the presence of several different amounts of Nano-Ag for 24 hrs at 28°C. TEM was used as a complementary technique to examine sections of the treated cells, using standard procedures for fixing and embedding sensitive biological samples, which are described elsewhere (Osumi, 1998; Mares et al., 1998).

2.11. Flow cytometric analysis for a fungal cell cycle

Log-phased cells of *C. albicans* (1 x 10^8 cells) cultured in a YPD medium, were harvested and treated with 40 μg/ml of Nano-Ag (at 20 times the MIC). After incubation for 8 hrs, the cells were washed with PBS and fixed with 70% ethanol overnight at 4°C. The cells were treated with 200 μg/ml of RNase A and the mixture was left to react for 2 hrs at 37°C. For DNA staining, 50 μg/ml of propidium iodide (PI) were added and incubated for 1 hr at 4°C in the dark (Green et al., 1999). Flow cytometric analysis was performed by a FACSCalibur flow cytometer. The values represent the average of the measurements conducted in triplicate of three independent assays.

3. Results and discussion

3.1. Antifungal activity of Nano-Ag

In the past, silver has been in use for the treatment of burns or chronic wounds. In the 1940s, however, after penicillin was introduced, the use of silver for the treatment of bacterial infections decreased (Hugo & Russell, 1982; Demling & DeSanti, 2001; Chopra, 2007).
Silver again came in picture in the 1960s when Moyer introduced the use of 0.5% silver nitrate for the treatment of burns. He proposed that this solution does not interfere with epidermal proliferation and possess antibacterial properties against *Staphylococcus aureus, Pseudomonas aeruginosa and Escherichia coli* (Moyer et al., 1965; Bellinger & Conway, 1970). In 1968, silver nitrate was combined with sulfonamide to form silver sulfadazine cream, which served as a broad-spectrum antibacterial agent and was used for the treatment of burns. Silver sulfadazine is effective against various bacterial strains like *E. coli, S. aureus, Klebsiella* sp., and *Pseudomonas* sp. (Rai et al., 2009). It also possesses some antifungal and antiviral activities (Fox & Modak, 1974). Recently, due to the emergence of antiobiotic-resistant bacteria and limitations of the use of antibiotics clinicians have returned to using silver wound dressings, containing varying level of silver (Gemmell et al., 2006; Chopra, 2007). For

these reasons, the antifungal activity and its mechanism of silver, Nano-Ag specifically, was investigated.

Nano-Ag, in an IC_{80} range of 1-25 μg/ml, showed significant antifungal activity against *T. mentagrophytes* and *Candida* species. Toward all fungal strains, Nano-Ag exhibited similar activity with amphotericin B, showing IC_{80} values of 1-5 μg/ml, but more potent activity than fluconazole, showing IC_{80} values of 10-30 μg/ml. However, this compound exhibited less potent activity than amphotericin B, showing IC_{80} values of 2-4 μg/ml for *C. parapsilosis* and *C. krusei* (Table 1).

Fungal strains	IC_{80} (μg/ml)		
(no. of strains)	Nano-Ag	Amphotericin B	Fluconazole
C. albicans (4)	2-4	5	10-16
C. tropicalis (2)	7	2-4	13
C. glabrata (4)	1-7	2	10-16
C. parapsilosis (3)	4-25	2	13
C. krusei (1)	13	4	13
T. mentagrophytes (30)	1-4	1-2	20-30

Table 1. Antifungal activity of Nano-Ag.

In order to elucidate the antifungal activity of Nano-Ag, the dimorphic transition of *C. albicans*, induced by Nano-Ag, was investigated. The dimorphic transition of *C. albicans* from yeast form to mycelial form is responsible for pathogenicity, with mycelial shapes being predominantly found during the invasion of host tissue. A mycelial form can be induced by temperature, pH, and serum (McLain et al., 2000). As shown in Fig. 2, the serum-induced mycelia were significantly inhibited from extending and forming in the presence of Nano-Ag (Fig. 2C), but the mycelia formed was normal in the absence of Nano-Ag (Fig. 2B).

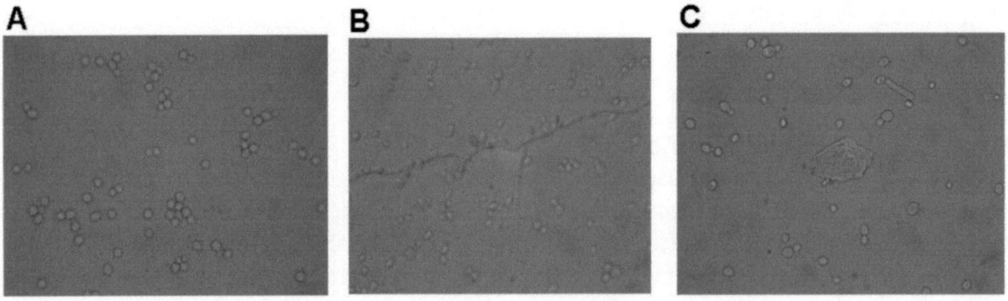

Fig. 2. The effect of Nano-Ag on the dimorphic transition in *C. albicans*. (A) Yeast control without 20% FBS and Nano-Ag, (B) Cells treated with only 20% FBS, (C) Cells treated with 2 μg/ml of Nano-Ag.

These results confirmed the antifungal activity of Nano-Ag and suggested that Nano-Ag can be a potential compound in the treatment of fungal infectious diseases.

3.2. Hemolytic activity of Nano-Ag against human erythrocytes

Many antimicrobial agents are limited regarding clinical applications, as they can induce cytolysis of human cells. The hemolytic activity of Nano-Ag was investigated as an indicator of its cytotoxicity to mammalian cells. The hemolytic activity was evaluated by the percentage of hemolysis in the concentration range of 1.25-10 μg/ml. The result showed that Nano-Ag caused 6% lysis of erythrocytes at a concentration of 10 μg/ml, whereas amphotericin B induced 10% lysis at the same level (Table 2).

Compound	% Hemolysis (μg/ml)			
	10	5	2.5	1.25
Nano-Ag	6	0	0	0
Amphotericin B	10	0	0	0

Table 2. Hemolytic activity of Nano-Ag against human erythrocytes.

This result suggested that Nano-Ag could be applied to therapeutic agents regarding human fungal diseases with low cytotoxicity.

3.3. Changes of plasma membrane potential induced by Nano-Ag

To assess whether Nano-Ag can affect the function of a fungal plasma membrane, the dissipation of the fungal plasma membrane potential was investigated. *C. albicans* cells were cultured in the presence of Nano-Ag or CCCP used as a positive control, and the amounts of accumulated DiBAC$_4$(3) in the cells were measured *via* flow cytometry by staining with DiBAC$_4$(3). CCCP is an H^+ ionophore which dissipates the H^+ gradient and thus uncouples electron transport from ATP synthesis (Ghoul et al., 1989). DiBAC$_4$(3) has a high voltage sensitivity and it enters depolarized cells, where it binds to lipid-rich intracellular components (Liao et al., 1999). Therefore, the fluorescence of DiBAC$_4$(3) increases upon membrane depolarization. The addition of Nano-Ag to *C. albicans* cells caused an increase in fluorescence intensity, similar to the increase observed in the presence of CCCP, which is indicative of membrane depolarization (Fig. 3). This result indicated that Nano-Ag affected yeast cells by attacking their membranes, thus disrupting membrane potential.

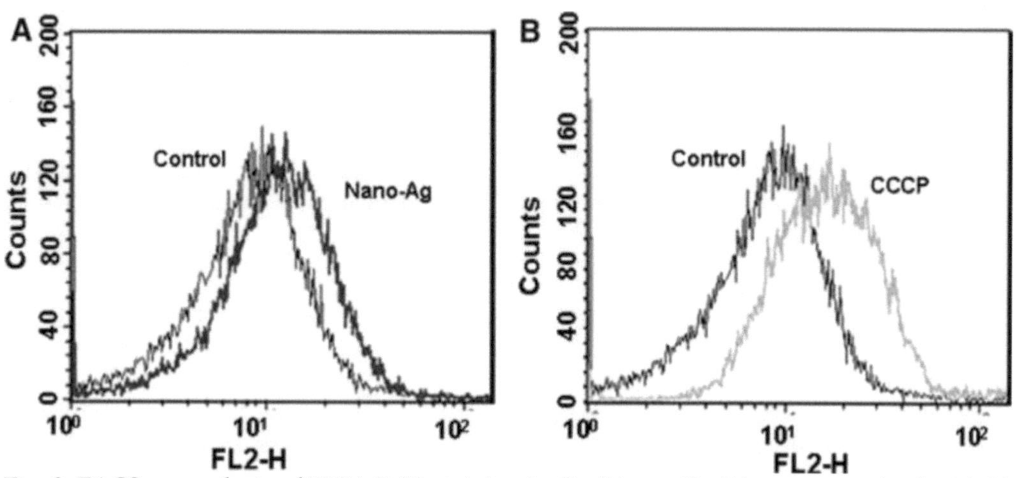

Fig. 3. FACScan analysis of DiBAC$_4$(3) staining in *C. albicans*. *C. albicans* were mixed with 30 μg/ml of Nano-Ag and 10 μM of CCCP and incubated at 28 °C for 3 hrs under constant shaking. This figure shows the fluorescence intensity of stained *C. albicans* treated with the

compounds. FL2-H indicates the log fluorescent intensity of $DiBAC_4(3)$, and y-axis indicates cell numbers (events).

3.4. Changes in the plasma membrane dynamics of fungal cells induced by Nano-Ag

The effect of Nano-Ag on the fungal plasma membrane was further investigated by using DPH as a membrane probe. If the antifungal activities exerted by Nano-Ag on *C. albicans* are at the level of the cell plasma membrane, DPH, which can interact with an acyl group of the plasma membrane lipid bilayer, can not be inserted into the membrane. As shown in Fig. 4, the plasma membrane DPH fluorescence anisotropy was significantly decreased by increasing the concentrations of Nano-Ag or amphotericin B. This is consistent with the disruption of the plasma membrane by Nano-Ag as well as by a positive control, amphotericin B.

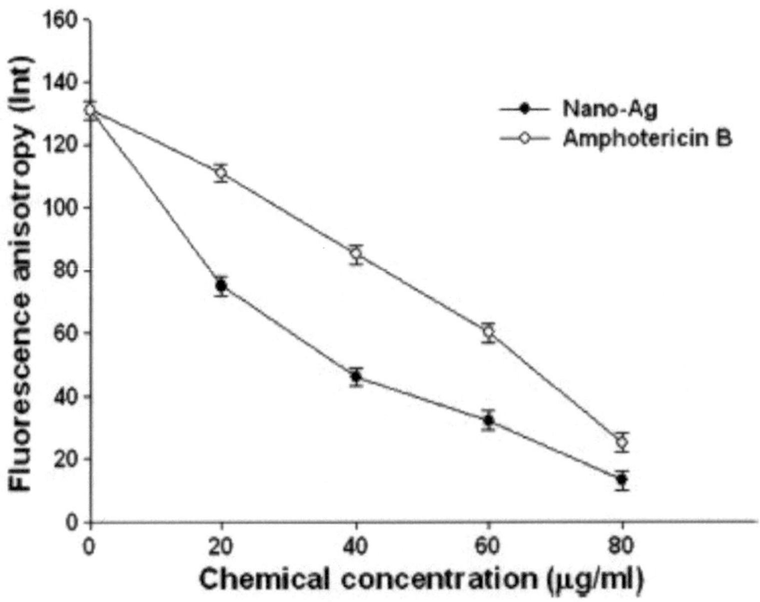

Fig. 4. DPH fluorescence anisotropy after the addition of Nano-Ag or amphotericin B. The error bars represent the standard deviation (SD) values for three independent experiments, performed in triplicate.

3.5. Intracellular glucose and trehalose release induced by Nano-Ag

The ability of Nano-Ag to disturb the integrity of the plasma membrane of fungal cells was also assessed by measuring the glucose and trehalose released in cell suspensions which were exposed to this compound. Trehalose can protect proteins and biological membranes from inactivation or denaturation caused by a variety of stress conditions, including desiccation, dehydration, heat, cold, oxidation, and toxic agents in yeast (Alvarez-Peral et al., 2002; Elbein et al., 2003). The result showed that Nano-Ag or amphotericin B-treated cells both accumulated more intracellular glucose and trehalose than the compound-untreated cells. In addition, these cells also increased extracellular glucose and trehalose than the compound-untreated cells. The extracellular glucose and trehalose, induced by amphotericin B, were measured as being 27.4 μg per fungal dry weight of 1 mg. The extracellular glucose and trehalose amounts, however, induced by Nano-Ag were measured as being 30.3 μg per fungal dry weight of 1 mg. This rate was significantly higher than that induced in the compound-untreated cells (Table 3). The result suggests that it may be one of several intracellular components released during membrane disruption by Nano-Ag. As for the mechanism by which Nano-Ag breaks down the membrane permeability barrier, it is possible that Nano-Ag perturbs the membrane lipid bilayers, causing the leakage of ions and other materials as well as forming pores and dissipating the electrical potential of the membrane.

	Amounts of trehalose and glucose concentrations (μg/mg)	
	Intracellular glucose and trehalose	Released glucose and trehalose
Control	7.2	6.8
Nano-Ag	16.1	30.3
Amphotericin B	20.5	27.4

Table 3. The concentrations of trehalose and glucose from *C. albicans* by Nano-Ag and amphotericin B.

3.6. Transmission electron microscopic (TEM) analysis

TEM was used to evaluate the ability of Nano-Ag to disrupt the fungal envelope structure. The results showed that the Nano-Ag-treated fungal cells showed significant damage, which was characterized by the formation of a "pit" in their cell walls and pores in their plasma membrane (Fig. 5).

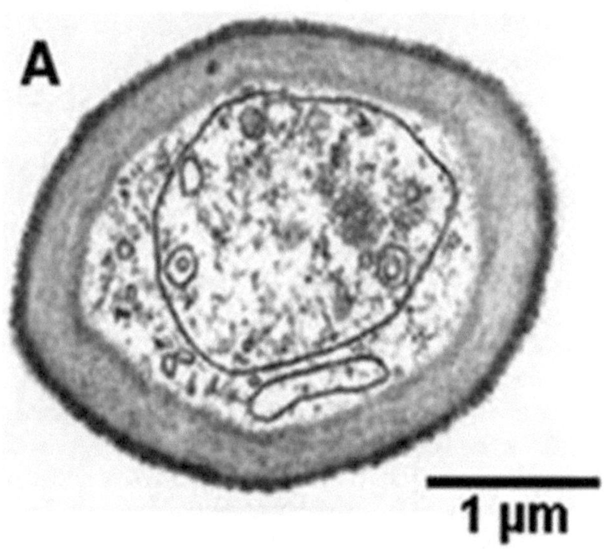

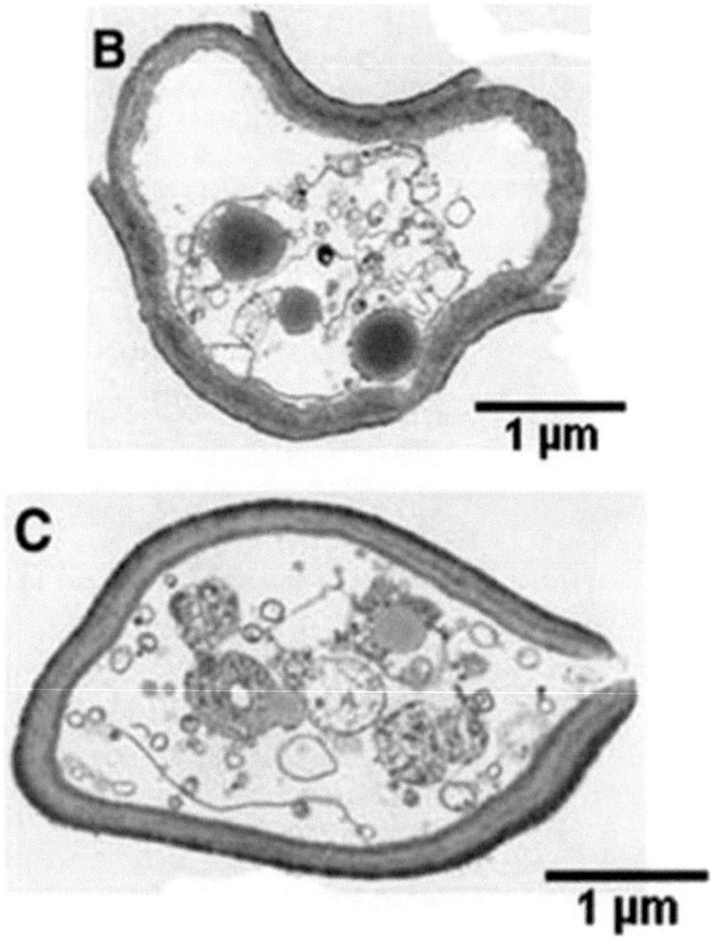

Fig. 5. Transmission electron micrograph (TEM) of *C. albicans* cells. *C. albicans* were incubated in the presence of different amounts of Nano-Ag for 24 hrs at 28 °C. The bar marker represents 1 µm. (A) Control, (B) Cells treated with 170 µg/ml of Nano-Ag, (C) Cells treated with 400 µg/ml of Nano-Ag.

3.7. The arrest of fungal cell cycle induced by Nano-Ag

In order to elucidate the physiological changes of the fungal cells induced by Nano-Ag, a flow cytometric analysis of the cell cycle was performed. The cells were cultured in the presence or absence of Nano-Ag and their DNA content was determined *via* flow cytometry by staining with propidium iodide (PI). PI is a DNA-staining dye that intercalates between the bases of DNA or RNA molecules (Tas & Westerneng, 1981). As shown in Fig. 6, the percentage of cells in the G_2/M phase increased by 15%, while that in the G_1 phase significantly decreased by about 20% in the presence of Nano-Ag. This data suggests that Nano-Ag inhibited some cellular processes which are involved in normal bud growth. Endo et al. have reported that the inhibition of bud growth correlates with membrane damage

(Endo et al., 1997). Therefore, this report suggests that Nano-Ag inhibits the normal budding process, probably through the destruction of membrane integrity.

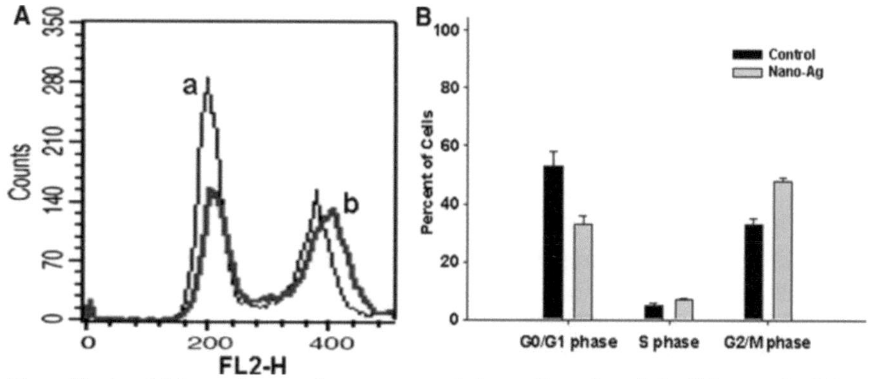

Fig. 6. The effects of Nano-Ag on the process of the cell cycle of *C. albicans*. *C. albicans* were incubated in the presence of Nano-Ag for 8 hrs at 28 ˚C. FL2-H indicates the fluorescent intensity of PI, and y-axis indicates cell numbers (events). (A) FACS diagram of cell cycle; (a) Control without Nano-Ag, (b) Cells treated with 40 μg/ml Nano-Ag, (B) A histogram indicating the percentage of each cell cycle progress.

4. Conclusion

Many studies have shown the antimicrobial effects of Nano-Ag (Klasen, 2000; Russell & Hugo, 1994; Silver, 2003), but the effects of Nano-Ag against fungal pathogens of the skin, including clinical isolates of *T. mentagrophytes* and *Candida* species are mostly unknown. The primary significance of this study is the observation that Nano-Ag can inhibit the growth of dermatophytes, which cause superficial fungal infections (Kim et al., 2008). Nano-Ag also exhibited potent antifungal effects, probably through destruction of membrane integrity (Kim et al., 2009). To the author's knowledge, this is the first study to apply Nano-Ag successfully to dermatophytes and pathogenic fungal strains. Secondly, the fact that preparation method of Nano-Ag described here is cost-effective is also of importance. Therefore, it can be expected that Nano-Ag may have potential as an anti-infective agent for human fungal diseases.

5. References

Ahmad, Z.; Pandey, R.; Sharma, S. & Khuller, G.K. (2006). Alginate nanoparticles as antituberculosis drug carriers: formulation development, pharmacokinetics and therapeutic potential. *Indian journal of chest diseases and allied sciences*, Vol. 48, No. 3, (July-September) 171-176, ISSN 0377-9343

Alexander, B.D. & Perfect, J.R. (1997). Antifungal resistance trends towards the year 2000. Implications for therapy and new approaches. *Drugs*, Vol. 54, No. 5, (November) 657-678, ISSN 0012-6667

Alvarez-Peral, F.J.; Zaragoza, O.; Pedreno, Y. & Argüelles, J.C. (2002). Protective role of trehalose during severe oxidative stress caused by hydrogen peroxide and the adaptive oxidative stress response in *Candida albicans*. *Microbiology*, Vol. 148, No. Pt 8, (August) 2599-2606, ISSN 1350-0872

Amichai, B. & Grunwald, M.H. (1998). Adverse drug reactions of the new oral antifungal agents—terbinafine, fluconazole, and itraconazole. *International journal of dermatology*, Vol. 37, No. 6, (June) 410-415, ISSN 0011-9059

Bellinger, C.G. & Conway, H. (1970). Effects of silver nitrate and sulfamylon on epithelial regeneration. *Plastic and reconstructive surgery*, Vol. 45, No. 6, (June) 582-585, ISSN 0032-1052

Chopra, I. (2007). The increasing use of silver-based products as antimicrobial agents: a useful development or a cause for concern?. *Journal of antimicrobial chemotherapy*, Vol. 59, No. 4, (April) 587-590, ISSN 0305-7453

Demling, R.H. & DeSanti, L. (2001). Effects of silver on wound management. *Wounds*, Vol. 13, No. Suppl A, 5-15, ISSN 1044-7946

Denning, D.W. (1991). Epidemiology and pathogenesis of systemic fungal infections in the immunocompromised host, Journal of antimicrobial chemotherapy, Vol. 28, No. Suppl B, (October) 1-16, ISSN 0305-7453.

Durán, N.; Marcarto, P.D.; De Souza, G.I.H.; Alves, O.L. & Esposito, E. (2007). Antibacterial effect of silver nanoparticles produced by fungal process on textile fabrics and their effluent treatment. *Journal of biomedical nanotechnology*, Vol. 3, No. 2, (June) 203-208, ISSN 1550-7033

Elbein, A.D.; Pan, Y.T.; Pastuszak, I. & Carroll, D. (2003). New insights on trehalose: a multifunctional molecule. *Glycobiology*, Vol. 13, No. 4, (April) 17R-27R, ISSN 0959-6658

Ellis, M.; Richardson, M. & de Pauw, B. (2000). Epidemiology. *Hospital medicine*, Vol. 61, No. 9, (September) 605-609, ISSN 1462-3935

Endo, M.; Takesako, K.; Kato, I. & Yamaguchi, H. (1997). Fungicidal action of aureobasidin A, a cyclic depsipeptide antifungal antibiotic, against *Saccharomyces cerevisiae*. *Antimicrobial agents and chemotherapy*, Vol. 41, No. 3, (March) 672-676, ISSN 0066-4804

Fernandes, A.R.; Prieto, M. & Sá-Correia, I. (2000). Modification of plasma membrane lipid order and H+-ATPase activity as part of the response of Saccharomyces cerevisiae to cultivation under mild and high copper stress. *Archives of microbiology*, Vol. 173, No. 4, (April) 262-268, ISSN 0302-8933

Fox, C.L.Jr. & Modak, S.M. (1974). Mechanism of silver sulfadiazine action on burn wound infections. *Antimicrobial agents and chemotherapy*, Vol. 5, No. 6, (June) 582-588, ISSN 0066-4804

Gemmell, C.G.; Edwards, D.I.; Fraise, A.P.; Gould, F.K.; Ridgway, G.L. & Warren, R.E. (2006). Guidelines for the prophylaxis and treatment of methicillin-resistant *Staphylococcus aureus* (MRSA) infections in the UK. *Journal of antimicrobial chemotherapy*, Vol. 57, No. 4, (April) 589-608, ISSN 0305-7453

Ghoul, M.; Pommepuy, M.; Moillo-Batt, A. & Cormier, M. (1989). Effect of carbonyl cyanide m-chlorophenylhydrazone on *Escherichia coli* halotolerance. *Applied and environmental microbiology*, Vol. 55, No. 4, (April) 1040-1043, ISSN 0099-2240

Goffeau, A. (2008). Drug resistance: the fight against fungi. *Nature*, Vol. 452, No. 7187, (April) 541-542, ISSN 0028-0836

Gong, P.; Li, H.; He, X.; Wang, K.; Hu, J.; Zhang, S. & Yang, X. (2007). Preparation and antibacterial activity of Fe_3O_4@Ag nanoparticles. *Nanotechnology*, Vol. 18, No. 28, (June) 604-611, ISSN 0957-4484

Green, L.J.; Marder, P.; Mann, L.L.; Chio, L.C. & Current, W.L. (1999). LY303366 exhibits rapid and potent fungicidal activity in flow cytometric assays of yeast viability. *Antimicrobial agents and chemotherapy*, Vol. 43, No. 4, (April) 830-835, ISSN 0066-4804

Gu, H.; Ho, P.L.; Tong, E.; Wang, L. & Xu, B. (2003). Presenting vancomycin on nanoparticles to enhance antimicrobial activities. *Nano letters*, Vol. 3, No. 9, (August) 1261-1263, ISSN 1530-6984

Hartsel, S. & Bolard, J. (1996). Amphotericin B: new life for an old drug. *Trends in pharmacological sciences*, Vol. 17, No. 12, (December) 445-449, ISSN 0165-6147

Hugo, W.B. & Russell, A.D. (1982). *Principles and practice of disinfection, preservation and sterilization*, Oxford, ISBN 0632041943, London

Jung, H.J.; Seu, Y.B. & Lee, D.G. (2007). Candicidal action of resveratrol isolated from grapes on human pathogenic yeast *C. albicans*. *Journal of microbiology and biotechnology*, Vol. 17, No. 8, (August) 1324-1329, ISSN 1017-7825

Kim, K.J.; Sung, W.S.; Moon, S.K.; Choi, J.S.; Kim, J.G. & Lee, D.G. (2008). Antifungal effect of silver nanoparticles on dermatophytes. Journal of microbiology and biotechnology, Vol. 18, No. 8, (August) 1482- 1484, ISSN 1017-7825

Kim, K.J.; Sung, W.S.; Suh, B.K.; Moon, S.K.; Choi, J.S.; Kim, J.G. & Lee, D.G. (2009). Antifungal activity and mode of action of silver nano-particles on *Candida albicans*. *Biometals*, Vol. 22, No. 2, (April) 235-242, ISSN 0966-0844

Klasen, H.J. (2000). A historical review of the use of silver in the treatment of burns. II. Renewed interest for silver. *Burns*, Vol. 26, No. 2, (May) 131-138, ISSN 0305-4179

Kontoyiannis, D.P.; Mantadakis, E. & Samonis, G. (2003). Systemic mycoses in the immunocompromised host: an update in antifungal therapy. *Journal of hospital infection*, Vol. 53, No. 4, (April) 243-258, ISSN 0195-6701

Kullberg, B.J. & de Pauw, B.E. (1999). Therapy of invasive fungal infections. *Netherlands journal of medicine*, Vol. 55, No. 3, (September) 118-127, ISSN 0300-2977

Liao, R.S.; Rennie, R.P. & Talbot, J.A. (1999). Assessment of the effect of amphotericin B on the vitality of *Candida albicans*. *Antimicrobial agents and chemotherapy*, Vol. 43, No. 5, (May) 1034-1041, ISSN 0066-4804

Mares, D.; Romagnoli, C.; Sacchetti, G.; Vicentini, C.B. & Bruni, A. (1998). Morphological study of *Trichophyton rubrum*: ultrastructural findings after treatment with 4-amino-3-methyl-1-phenylpyrazolo-(3,4-c)isothiazole. *Medical mycology*, Vol. 36, No. 6, (December) 379-385, ISSN 1369-3786

McLain, N.; Ascanio, R.; Baker, C.; Strohaver, R.A. & Dolan, J.W. (2000). Undecylenic acid inhibits morphogenesis of *Candida albicans*. *Antimicrobial agents and chemotherapy*, Vol. 44, No. 10, (October) 2873-2875, ISSN 0066-4804

Moyer, C.A.; Brentano, L.; Gravens, D.L.; Margraf, H.W. & Monafo, W.W.Jr. (1965). Treatment of large human burns with 0.5 percent silver nitrate solution. *Archives of surgery*, Vol. 90, (June) 812-867, ISSN 0272-5533

Mukherjee, P.K.; Chandra, J.; Kuhn, D.M. & Ghannoum, M.A. (2003). Mechanism of fluconazole resistance in *Candida albicans* biofilms: phase-specific role of efflux pumps and membrane sterols. *Infection and immunity*, Vol. 71, No. 8, (August) 4333-4340, ISSN 0019-9567

Odds, F.C.; Brown, A.J. & Gow, N.A. (2003). Antifungal agents: mechanisms of action. *Trends in microbiology*, Vol. 11, No. 6, (June) 272-279, ISSN 0966-842X

Osumi, M. (1998). The ultrastructure of yeast: cell wall structure and formation. *Micron*, Vol. 29, No. 2-3, (April-June) 207-233, ISSN 0968-4328

Park, Y.; Lee, D.G.; Jang, S.H.; Woo, E.R.; Jeong, H.G.; Choi, C.H. & Hahm, K.S. (2003). A Leu-Lys-rich antimicrobial peptide: activity and mechanism. *Biochimica et biophysica acta*, Vol. 1645, No. 2, (February) 172-182, ISSN 0006-3002

Rai, M.; Yadav, A. & Gade, A. (2009). Silver nanoparticles as a new generation of antimicrobials. *Biotechnology advances*, Vol. 27, No. 1, (January-February) 76-83, ISSN 0734-9750

Russell, A.D. & Hugo, W.B. (1994). Antimicrobial activity and action of silver. *Progress in medicinal chemistry*, Vol. 31, 351-370, ISSN 0079-6468

Schabes-Retchkiman, P.S.; Canizal, G.; Herrera-Becerra, R.; Zorrilla, C.; Liu, H.B. & Ascencio, J.A. (2006). Biosynthesis and characterization of Ti/Ni bimetallic nanoparticles. *Optical materials*, Vol. 29, No. 1, (October) 95-99, ISSN 0925-3467

Sheehan, D.J.; Hitchcock, C.A. & Sibley, C.M. (1999). Current and emerging azole antifungal agents. *Clinical microbiology reviews*, Vol. 12, No. 1, (January) 40-79, ISSN 0893-8512

Silver, S. (2003). Bacterial silver resistance: molecular biology and uses and misuses of silver compounds. *FEMS microbiology reviews*, Vol. 27, No. 2-3, (June) 341-353, ISSN 0168-6445

Sung, W.S.; Lee, I.S. & Lee, D.G. (2007). Damage to the cytoplasmic membrane and cell death caused by lycopene in *Candida albicans*. *Journal of microbiology and biotechnology*, Vol. 17, No. 11, (November) 1797-1804, ISSN 1017-7825

Tas, J. & Westerneng, G. (1981). Fundamental aspects of the interaction of propidium diiodide with nuclei acids studied in a model system of polyacrylamide films. *Journal of histochemistry and cytochemistry*, Vol. 29, No. 8, (August) 929-936, ISSN 0022-1554

16

Silver nanoparticles interactions with the immune system: implications for health and disease

Rebecca Klippstein[1], Rafael Fernandez-Montesinos[1], Paula M. Castillo[2], Ana P. Zaderenko[2] and David Pozo[1]
[1] *CABIMER-Andalusian Center for Molecular Biology & Regenerative Medicine CSIC-University of Seville-UPO-Junta de Andalucia, Seville, Spain*
[2] *Department of Physical, Chemical & Natural Systems, Pablo de Olavide University, Seville, Spain*

1. Introduction

Our immune system constantly interacts with our internal environment, protects us from our external environment and provides the inherent knowledge to sense the difference between friend and foe with important implications in human health and disease (Pozo, 2008). For these reasons, it is important to identify functional alteration of key immune responses as the number of silver nano-enabled products grows while the current data strongly suggest that other related nanomaterials, such as polymer nanoparticles, fullerenes, dendrimers and gold nanoparticles, interact with the immune system.

In the present chapter, we will focus on the effects of engineered silver nanoparticles on the initiation and regulation of the immune response. Particular attention will be paid on the potential clinical usefulness of silver nanoparticles in the context of its effects on the production of key immunological mediators, as well as its significance on bacterial and viral infections. By a critical analysis of the current state of knowledge, the chapter will help to reduce the serious lack of information and controversial issues concerning the biological effects of silver nanoparticles on the immune system.

2. The immune system

Studies on the biological consequences of new nanomaterials suited for biomedical applications are of importance, particularly those related to the immune system. Therefore, the interactions of silver nanoparticles with the immune system and its potential effects and implications are key questions for nanomaterials that are intended for biomedical applications or extensive industrial manufacturing such as those considered in this chapter. For this reason, before explaining the variety of interactions, effects and implications, we will start with a brief introduction about the functions of the immune system and its cellular components.

The immune system is a dynamic network of cells, tissues, and organs that work coordinated to defend the body against attacks by "foreign" invaders and protects against disease by identifying "self" and "non-self" (for example virus, fungus, bacterium) cells and tissues (Christensen & Thomsen, 2009).

To deal with antigens, the system uses specialized cells to recognize infiltrators and eliminate them. Detection is complicated as pathogens can evolve rapidly, producing adaptations that avoid the immune system and allow the pathogens to successfully infect their hosts (Christensen & Thomsen, 2009). For cellular communication there is a category of signaling molecules called cytokines. They are small secreted proteins which are critical to the development and functioning of both the innate and adaptive immune response, although they are not limited just to the immune system. They act by binding to specific membrane receptors, which produce cascades of intracellular signalling to alter cell functions. Cytokines have been classified as, lymphokines (cytokines made by lymphocytes), monokines (cytokines made by monocytes), chemokines (cytokines with chemotactic activities), and interleukins (cytokines made by one leukocyte to act on other leukocytes). The actions of cytokines may be grouped as, autocrine action (the cytokine acts on the cell that secretes it), paracrine action (the target is restricted to nearby cells), and endocrine action (the cytokine diffuses to distant regions of the body).

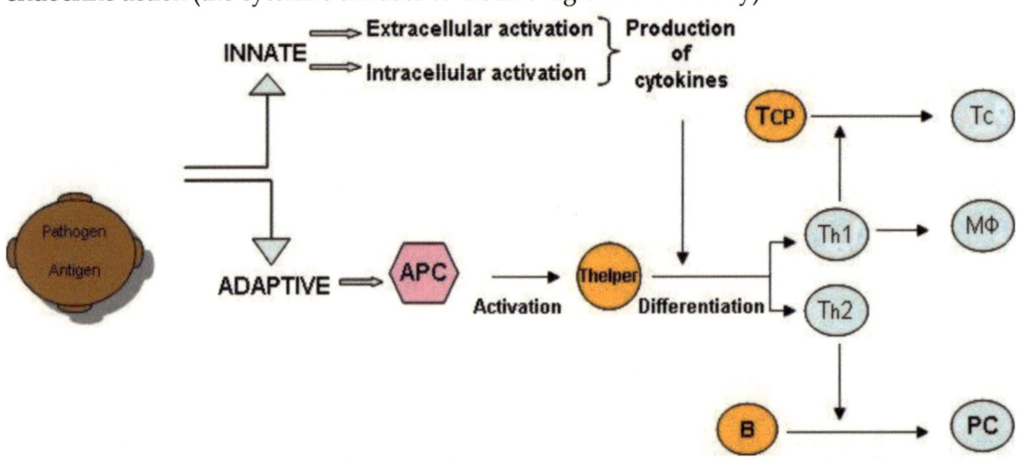

Fig. 1. The dynamic network of cells in the immune system. Recognition of pathogens leads the activation of T helper cells (Th). This process involves adaptive events that occur on the surface of the antigen presenting cells (APC). They display a foreign antigen complex with MHC on its surface recognized by T-cells using their T-cell receptor (TCR). The immune system contains three types of APC macrophages, dendritic cells and B cells. APC collaborate with the innate response by cytokine communication, which steer the differentiation of Th cells into Th1 or Th2 subsets. These T cells activate other cells of the immune system such as cytotoxic T cells (Tc), and macrophages (MΦ). Silver nanoparticles can interact with the immune system and carry out different functions like binding and reacting with cells or proteins for drug delivery, detection, diagnosis and therapy (targeted and non-targeted).

2.1 Adaptive immunity

As part of this complex immune response, the human immune system adapts over time to recognize specific pathogens more efficiently. This adaptation process is referred to as adaptive immunity and creates immunological memory from a primary response to a specific pathogen, provides an enhanced response to secondary encounters with that same, specific pathogen.

2.2 Innate immunity

The innate immune system comprises the cells and mechanisms that defend the host from infection by other organisms, in a non-specific manner (Castellheim *et al.*, 2009). This means that cells of the innate system recognize and respond to pathogens in a generic way, but unlike the adaptive immune system, does not confer long-lasting or protective immunity to the host. Innate immune systems provide immediate defense against infection (Rasmussen *et al.*, 2009).

3. Silver nanoparticles

In nanotechnology, a nanoparticle is defined as a material with dimensions and tolerance limits of 0.1-100nm that behaves as a whole unit in terms of its transport, properties and unique characteristics.

Metallic nanoparticles have unique optical, electrical and biological properties that have attracted significant attention due to their potential use in many applications, such as catalysis, biosensing, drug delivery and nanodevice fabrication. Capped silver nanoparticles (AgNPs) have many biomedical applications due to its excellent biocompatibility and antibacterial properties. It has been reported that silver nanoparticles interact with virus, bacteria, and the immune system, being the reason why in this chapter we will explain how the size, shape and composition of silver nanoparticles can have a significant effect on their efficacy and have to be kept in mind when using bioconjugates.

3.1 Interaction of silver nanoparticles with key signaling pathways and soluble mediators

Activation of the innate immune system is mediated by pattern recognition receptors (PRRs) on particular immunocompetent cells that recognize pathogen-associated molecular patterns. The best characterized signaling PRRs to date are the Toll-like receptors (TLRs) present in plants, invertebrates and vertebrates that represent a primitive host defense mechanism against bacteria, fungi and viruses. Toll-like receptors (TLRs) play an important role in innate immunity. Individual TLRs recognise microbial components that are conserved among pathogens, such recognition initiates necessary inflammatory immune responses and induces subsequent activation of adaptive immunity (Uematsu & Akira, 2006). In conclusion TLRs are a type of pattern recognition receptors (PRRs) which recognize molecules broadly shared by pathogens but distinguishable from host molecules. TLRs can be divided into two groups according to their cellular localization: TLRs 1, 2, 4, 5 and 6 are mainly located on the cell surface and primarily recognize bacterial components, while TLRs 3, 7, 8 and 9 are mostly found in the endocytic compartments and mainly recognize viral products (Akira *et al.*, 2006).

Toll-like receptors (TLR) and their ligands are one of the main players in the initiation of innate immunity which precedes, and is required, for the establishment of adaptive immunity. Manipulating the immune response by using TLR agonists or antagonists might be of therapeutic and/or prophylactic value (Makkouk & Abdelnoor, 2009). Exogenous signals are provided by TLRs mechanisms which affect the initiation, maintenance and progression of inflammatory diseases. Moreover, reagents that enhance TLR signalling pathways can be powerful adjuvants for fighting pathogens or cancer. By contrast TLR antagonists or signalling inhibitors could block the activation of TLRs by neutralizing the ligands, blocking the receptors or preventing signalling, which can also have other beneficial therapeutic effects in autoimmune diseases and sepsis (Zuany-Amorim *et al.*, 2002).

This opens up new avenues in the world of adjuvants and illustrates the basic requirements for the design of NP conjugates that efficiently reach their target.

It has recently been studied whether AgNPs with a narrow size distribution and protected with a monolayer of adsorbed tiopronin (Ag@tiopronin) had any functional impact on specific TLR stimulation of Interleukin-6 (IL-6) secretion in Raw 264.7 macrophages, a murine monocyte/macrophage cell line (Castillo *et al.*, 2008), and the effects of silver nanoparticles on cytokine expression by peripheral blood mononuclear cells (PBMC) (Shin *et al.*, 2007), human mesenchymal stem cells (hMSCs) (Greulich *et al.*, 2009), and J774 A1 macrophages (Yen *et al.*, 2009).

Nanoparticles	Cell type	Concentration	Size (nm)	Cytokines	Ref
Ag@tiopronin	Raw 264.7 macrophages	10 µg/mL	~5	IL-6	Castillo et al (2008)
AgNPs	PBMC	1, 3, 5, 10, 20 µg/mL	~1.5	IL-5 INF- γ TNF-α	Shin et al (2007)
AgNPs	J774 A1 macrophages	1 µg/mL	Small: 2-4 Medium: 5-7 Large: 20-40	IL-1 IL-6 TNF-α	Greulich et al (2009)
AgNPs	hMSCs	0.05, 0.5, 1, 2.5, 5, 50 µg/mL	100	IL-6 IL-8 IL-11	Yen et al (2009)

Table 1. The effect of silver nanoparticles on the production of cytokines. Silver NPs were found to strongly inhibit cytokine production by Raw 264.7 macrophages, PBMC, hMSCs and more weakly by J774 A1 macrophages.

The results demonstrated that Ag@tiopronin do not show proinflammatory effects on macrophages and that interestingly, Ag@tiopronins differentially inhibits the IL-6 secretion mediated by specific TLRs located in the cell surface or in the endocytic compartments (Castillo *et al.*, 2008).

Moreover, the effects of silver nanoparticles on the production of cytokines by PBMC, hMSCs, were found to strongly inhibit cytokine production, for example, of INF- γ, IL-6, IL-8, IL-11, TNF-α and more weakly IL-5. In the case of J774 A1 macrophages the expression levels of IL-1 and IL-6 were similar to controls.
The biological effects of nanoparticles were analyzed by enzyme-linked immunosorbent assay (ELISA) in the case of PBMCs stimulated by phytohaemagglutinin (PHA) in the presence of NPs (Shin *et al.*, 2007) and hMSCs stimulated previously by lipopolysaccaride (LPS) (Greulich *et al.*, 2009). In the case of non-stimulated J774 A1 macrophages, the cytokine mRNA expression levels were analyzed in cells treated with NPs and were compared to the resting state, measured by RT-PCR (Yen *et al.*, 2009).
In contrast there are other nanoparticles such as SiO_2 nanoparticles (Lucarelli *et al.*, 2004), gold nanoparticles (AuNPs)(Yen *et al.*, 2009), or very small size proteoliposomes (VSSP) (Venier *et al.*, 2007), that are dramatically pro-inflamatory.
However, in these studies, nanoparticles had a very heterogeneous size (from 1.5 to 100 nm) and concentration (from 0.05 to 50 μg/mL) (see Table 1), so it is difficult to establish whether the effect is related to the chemical nature, the size or the concentration. In any case, in these different experiments, exposures to concentrations lower than 3 ppm did not produce a significant decrease of cytokine production. These effects can be considered for future medical applications, but we have to keep in mind the characteristics previously mentioned, as others like the easily uptake into cells or tissues and not so easy metabolism.

3.2 Interaction of silver nanoparticles with virus

The interaction of silver nanoparticles with viruses is a largely unexplored field. A virus is a sub-microscopic infectious agent that can only multiply in living cells of animals, plants, or bacteria. Viruses are about 1/100th the size of bacteria and consist of a single- or double-stranded nucleic acid (DNA or RNA) surrounded by a protein shell called a capsid. Some viruses have an outer envelope composed of lipids and proteins. The viral capsid proteins bind to the host cellular surface specific receptors. This attachment can induce the viral envelope protein to undergo changes that results in the fusion of viral and cellular membranes and may lead to an infection. For this reason there is a high interest studying possible mechanisms of binding NPs to the viral capsid and inhibit the later fusion.
Recently, it has been suggested that nanoparticles bind with a viral envelope glycoprotein and inhibit the virus by binding to the disulfide bond regions
of the CD4 binding domain within the gp120 glycoprotein, as demonstrated in vitro (Elechiguerra *et al.*, 2005). Silver nanoparticles undergo a size-dependent interaction with HIV-1, nanoparticles ranging from 1 to 10 nm attached to the virus, and their surface chemistry can modify their interactions with viruses, tested with silver NPs that had three different surface chemistries: foamy carbon, poly (N-vinyl-2-pyrrolidone) (PVP), and bovine serum albumin (BSA). Differences have been observed in HIV-1 inhibition and can be justified because BSA and PVP are directly bounded to the nanoparticle surface and are totally encapsulated, while the foamy carbon silver nanoparticles have fundamentally a free surface area, which exhibit higher inhibitory effect and cytotoxicity as they are able to have stronger interactions (Elechiguerra *et al.*, 2005).

Within this unexplored area there are also other studies that analyse the interaction of silver nanoparticles with hepatitis B virus. The effects of silver NPs on hepatitis B virus (HBV) have been reported using an infection model HepAD38 human hepatoma cell line (Lu *et al.*, 2008). The binding affinity of NPs with different sizes (10 and 50 nm) for HBV DNA and extracellular virions resulted very high and could also inhibit the production of HBV RNA and extracellular virions in vitro, which was determined using a UV-vis absorption titration assay(Lu *et al.*, 2008). It will need further investigation to find whether this binding activity prevents HBV virions from entering into host cells or not. In vivo studies with silver NPs are necessary to design therapeutics and vaccines that can specifically target viruses in order to increase therapeutic benefit and minimize adverse effects.

3.3 Interaction of silver nanoparticles with bacteria

Bacteria are prokaryotic, microscopic, single-celled organisms that lack membrane bound organelle in the cytoplasm. They can inhabit all kinds of environments and exist either as independent (free-living) organisms or as parasites.

Silver has been used for at least six millennia in order to prevent microbial infections. It has been used to treat a wide variety of infections and has been effective against almost all organisms tested. Silver compounds were major weapons against wound infection in World War I until the advent of antibiotics and between 1900 and 1940, tens of thousands of patients consumed colloidal silver, and several million of doses were given intravenously. But it was shown that high doses of silver, when administered intravenously could cause convulsions or even death, and that oral administration of high doses could cause gastrointestinal disturbances (Alexander, 2009). For this reason the biomedical applications of silver can be effective by the use of biologically synthesized NPs, which minimize the factors such as toxicity and cost, and are found to be exceptionally stable and by virtue of extremely small size silver NPs exhibit unusual physicochemical properties and biological activities. Due to the large surface area (generally the diameter is smaller than 100nm and contains 20–15,000 silver atoms) for reaction of the NPs, the dose of silver used in medical applications can be reduced.

The mechanisms of action and binding of silver nanoparticles to microbes remain unclear but it is known that silver binds to the bacterial cell wall and cell membrane and inhibits the respiration process (Klasen, 2000) by which the chemical energy of molecules is released and partially captured in the form of ATP. Silver nanoparticles interact with sulfur-containing proteins of the bacterial membrane as well as with the phosphorus containing compounds like DNA to inhibit replication (Silver *et al.*, 2006). Bactericidal effect of silver has also been attributed to inactivation of the enzyme phosphomannose isomerase (Bhattacharya & Mukherjee, 2008), that catalyzes the conversion of mannose-6-phosphate to fructose-6-phosphate which is an important intermediate of glycolysis, the most common pathway in bacteria for sugar catabolism.

The antimicrobial activity of silver nanoparticles has been investigated against yeast, gram negative and positive bacteria (Kim *et al.*, 2007) (Sondi & Salopek-Sondi, 2004; Yu, 2007). When silver nanoparticles were tested in yeast and *Escherichia coli* (Gram -), bacterial growth was inhibited (Sondi & Salopek-Sondi, 2004), but the inhibitory effect in *Staphylococcus aureus* (Gram +) was mild (Kim *et al.*, 2007). Therefore, this suggests that the antimicrobial effects of silver nanoaparticles can be associated with different characteristics of the membrane structure, in order to the considerable differences between the membrane structures of Gram+ and Gram-. These differences mainly rely on the peptydoglycan layer thickness, the rigidity and extended cross linking that makes the penetration of nanoparticles very difficult.

Recently, due to the emergence of antibiotic-resistant bacteria and the use limitations of antibiotics that can cause serious diseases and is an important public health problem (Furuya & Lowy, 2006), the synergetic effect of silver nanoparticles with antibiotics has been studied combinating silver nanoparticles with different antibiotics like ampicillin, kanamycin, erythromycin and chloramphenicol against gram positive and gram negative bacteria. The antibacterial activities of these antibiotics increase in the presence of silver nanoparticles against gram positive and gram negative bacteria determined by the disk diffusion method. Different diameters of inhibition zones have been shown around the different antibiotic disk with or without AgNPs. The combination effect of nanosilver and ampicillin has more potential compared to the other antibiotics and may be caused by both, the cell wall lysis action of the ampicillin and the DNA binding action of the silver nanoparticles (Fayaz *et al.*, 2009). The antibiotic molecules contain many active groups such as hydroxyl and amido groups, which reacts easily with silver nanoparticles by chelation, for this reason, the synergistic effect may be caused by the bonding reaction with antibiotic and silver nanoparticles.

The bactericidal effects of silver nanoparticles are part of an extensive research field due to its potential translation for biomedical applications such as, wound-healing (Tian *et al.*, 2007; Silver *et al.*, 2006), clothes, coating for medical devices (Roe *et al.*, 2008), antimicrobial gel (Jain *et al.*, 2009), and orthopaedic implants (Nair & Laurencin, 2008).

It is well known that the use of central venous catheters is associated with bactericidal line infections, which is a usual problem (Stevens *et al.*, 2009). Contaminated or infected catheters are a major source of nosocomial infections responsible for > 40% of all episodes of nosocomial sepsis in acute-care hospitals (Samuel & Guggenbichler, 2004). For this reason catheters coated with silver NPs are important to confer antimicrobial activity and play an essential part in the prevention of catheter-related infections.

In vivo studies have been performed to test the antimicrobial activity of catheters coated with silver NPs and it has been reported that the coating process is slowly reversible, yielding sustained release of silver for at least 10 days (Roe *et al.*, 2008). Each animal (C57B1/6J male mice) was implanted with the equivalent of a 28 cm silver coated catheters and showed no sign of toxicity, inflammation or infection at the site of catheter implantation. The released silver is active against microorganisms with no risk of systemic toxicity and safety of use in animals. This suggests that catheters coated with this method could provide local protection against infections (Roe *et al.*, 2008).

However, some questions need to be addressed and more elaborate experimental evidences will be needed to clarify such as the exact mechanism of interaction of silver nanoparticles with the bacterial cells and the killing activity of the nanoparticle surface (Rai *et al.*, 2009).

3.4 Cytotoxicity, biodistribution and clearance.

Biomedical applications of silver nanoparticles need to be cytocompatible and have the capacity to restore natural morphology of the tissue in contact without triggering immunogenecity. Therefore, the evaluations of these effects are of high importance for future medical purposes and prior to in vivo applications, in vitro methods are needed to evaluate cytocompatibility as a prerequisite. Silver nanoparticles studies work as basic tools

for evaluating nanoparticle safety in order to foster the efficient movement of AgNPs products through preclinical and clinical development.
Different assays are used to measure cell viability such as MTT assays, LDH assays and apoptosis/necrosis assays among others. MTT [3-(4,5-dimethylthiazoyl-2-yl)-2,5-diphenyltetrazoliumbromide] assays, used to demonstrate the viability of cells through the reduction of the pale yellow MTT dye to a dark blue formazan product by the activity of succinate dehydrogenase present in the mithocondria of living cells, LDH (lactate dehydrogenase) assays, that consists in measuring the release of this LDH stable cytoplasmic enzyme that is rapidly released into the cell-culture supernantant upon damage of the plasma membrane and apoptosis/necrosis assays detects changes in cell membrane permeability (Weyermann *et al.*, 2005).

As it has been reported, tiopronin silver nanoparticles (Castillo *et al.*, 2008), BSA capped Ag-Pt alloy nanoparticles (Singh *et al.*, 2009) and silver nanoparticles protected with Na^+-poly (γ -glutamic acid) (Yu, 2007) are not cytotoxic, while, by contrast bare silver nanoparticles have been found to be rather toxic (Braydich-Stolle *et al.*, 2005; Kim *et al.*, 2009; Hussain *et al.*, 2005). This supports the idea that the toxicity is associated to the presence of bare metallic nanoparticle surfaces, while particles protected by and organic layer are much less toxic except starch capped nanoparticles which present mithocondrial dysfunction , induction of reactive oxygen species (ROS), DNA damage and cell cycle arrest (AshaRani *et al.*, 2009).
Serious limitations of these studies assessing biological properties of nanoparticles exist, such as the partial characterization of the material used and its size heterogeneity. This is one of the main causes of discrepancies in the literature (See Table 2).

Nanoparticles	Concentration	Size (nm)	Cell type	Toxicity signs	Ref
AgNPs	10 µg/mL	2-40	Murine macrophages	Spread shape	Greulich et al (2009)
Starch-capped-NPs	25-400 µg/mL	6-20	IMR-90 U251	◦Mithocondrial damage ◦ Increase of ROS ◦ Reduced ATP content ◦ DNA damage ◦ Cell cycle arrest	AshaRani et al (2009)
AgNPs	10 µg/mL	15	C18-4	◦ Cell morphology changes ◦ Reduced mithocondrial function ◦Slight LDH leakage	Braydich-Stolle et al (2009)

AgNPs	0.1-10 µg/mL	5-10	HepG2	◦ Reduced mithocondrial function ◦LDH leakage ◦Increase of ROS	Kim et al (2009)
AgNPs	5-50 µg/mL	15, 100	BRL 3A	◦Reduced mithocondrial function ◦LDH leakage ◦ Depletion of GSH level ◦Increase of ROS	Hussain et al (2005)
BSA capped Ag-Pt alloy NPs	50-100 µM	10-15	HGF	None	Singh et al (2009)
Ag@tiopronin	10 µg/mL	~5	Raw 264.7 macrophages	None	Castillo et al (2008)

Table 2. In vitro cytotoxicity studies of silver NPs on different cell lines. Toxicity signs seem to be associated to the presence of bare metallic nanoparticle surfaces, while particles protected by and organic layer seem to be much less toxic.

Toxicity studies with silver NPs have been made in vivo using a zebrafish embryonic model, based on the putative similarity between the zebrafish and human genomes (Bar-Ilan *et al.*, 2009). All sizes of colloidal silver NPs caused toxicity in zebrafish embryos in a size-dependent manner for certain concentrations and time points. Exposure to 250 µM (3, 10, 50 and 100nm) of colloidal silver nanoparticles caused significantly different percentages of mortality, with 80%, 64%, 36%, and 3% respectively, by 24 hours post-fertilization; 250 µM of colloidal silver nanoparticles of all sizes caused almost 100% mortality by 120 hours post-fertilization. Lethality caused by 100 µM is significantly lower than mortality caused by 250 µM at almost all time points, although 100 µM of colloidal silver NPs causes low lethality that increases with time, the induced sublethal toxic effects are represented by high average toxicity values. Overall, there are other many different parameters that could be responsible for the adverse effects that induce nanoparticle toxicity, such as, concentration, stability, chemistry and/or functionalization (Bar-Ilan *et al.*, 2009)

The human body has several semi-open interfaces such as the respiratory tract, or gastrointestinal tract and skin for direct substance exchange with the environment and are also the principle routes of exposure to silver nanoparticles (Chen & Schluesener, 2008). At these sites, silver nanoparticles can carry out different functions like binding and reacting with proteins, deposition, clearance and translocation. If the administration comes from other portals, the nanoparticles will have a direct contact with blood components, with the cardiovascular system and are distributed throughout the body. For this reason, the need for comprehensive Absorption, Distribution, Metabolism and Excretion (ADME) studies for nanomaterials are crucial to provide examples of how physical properties affect the state of agglomeration or aggregation, surface characteristics and stability(Zolnik & Sadrieh, 2009).

However, the need for an open dialogue between industry, academia, research labs and regulatory agencies cannot be overemphasized and as such, active collaboration should be facilitated, so that safe and effective nanotechnology products are developed for clinical use to treat the complex diseases.

4. Preclinical studies and clinical trials

In terms of healing, the elucidation of pro-inflammatory and anti-inflammatoyry pathways is important for the development of strategies to defend regenerative tissue from damage caused by imbalances in cytokines, oxidants, antioxidants within the wound. Recently, information about specific subsets of inflammatory cell lineages and the cytokine network orchestrating inflammation associated with tissue repair has increased (Eming *et al.*, 2007).

It is known that silver NPs can promote wound healing and reduce scar appearance in a dose-dependent manner and that cytokines play an important role in these processes by their capacity to decrease wound inflammation and modulate fibrogenic cytokines (Tian *et al.*, 2007; Wong *et al.*, 2009). It has been shown in vivo that silver NPs act decreasing inflammation through cytokine modulation.

Two animal models have been used for these experiments, twenty-week-old male BALB/C mice were used for all thermal injury experiments and C57BLKs/J-m +ldb, db/db (genetically diabetic) and C57BLKs/J-m (nondiabetic control) male mice were used for the impaired wound-healing animal model.

For thermal injury the dorsum of each mouse was shaved from the base of the tail to the base of the neck, anesthetized and lade on a burn template. For the chronic wound model, the hair on the back of each mouse was shaved, and a piece of full-thickness skin was excised with scissors. The experiments were performed with spherical silver NPs (~14nm) at 1mM. The expression patterns of IL-6, TGF-β1, IL-10, VEGF (vascular endothelial growth factor, a polypeptide that stimulates the growth of new blood vessels), and IFN-γ were investigated by real-time RT-PCR. Expression levels of IL-6 mRNA in the wound areas treated with silver NPs were maintained at statistically significant lower levels during the healing process while TGF-β1 levels were higher in the initial period of healing but decreased at a lower level during the later phase of healing. Furthermore, IL-10, VEGF and IFN-γ expression levels stayed higher in animals treated with silver NPs relative to the no-treated ones. These results can confirm that silver NPs can modulate cytokine expression (Tian *et al.*, 2007).

At the time of injury, the production of pro-inflammatory cytokines and the expression of E-selectin, chemokines and integrin ligands on endothelial cells mediate the selective recruitment of cutaneous lymphocyte-associated antigen positiveT-cells into the wound where they recognise the antigen for which their receptor is specific and become activated. The macrophages act as antigen presenting-cells and also express the co-stimulatory molecules that are essential for the T-cell activation. After antigen binding; T-cells differentiate into subtypes, preferentially into Th1 subsets and secrete interferon-gamma.

Interferon-gamma is the major macrophage activating cytokine and enhances the effective functions of macrophages (Tsirogianni *et al.*, 2006). The functional diversity of cytokines is thought to be important in dictating different phases of immunoinflammatory responses. Th1 cytokines are mostly involved in cell-mediated immunity associated with autoimmune disorders and allograft rejection, whereas Th2 cytokines are mostly involved in mediating

allergic inflammation and chronic fibroproliferative disorders. The initial period of 4–5 weeks of infection is largely driven by the Th1 cytokine. The Th1 response in the early phase of the infection is initiated by higher numbers of IFN-γ- secreting CD4 and CD8 cells in the spleen and lymph nodes (Azouz *et al.*, 2004). For this reason to accomplish successful wound repair and tissue regeneration, the inflammatory response must be tightly regulated by silver NPs. The lack of amplification of the inflammatory cytokine cascade is important in providing a permissive environment for wound repair and tissue regeneration to proceed. These results have given an insight into the actions of silver and provided a novel therapeutic direction for wound treatment in clinical practice. Other experiments need to be focused on the effect of silver NPs on the lymphocyte subset of the Th1/Th2 profile and see how silver NPs affect their particular cytokine production.
Moreover, other studies have been developed in vivo with a rat model of ulcerative colitis treated with nanocrystalline silver (40, 4 and 0.4 mg/kg) which demonstrated to have antimicrobial and anti-inflammatory properties (Bhol & Schechter, 2007). Ulcerative colitis is a form of inflamatory bowel disease which is a chronic inflammatory condition of the gastrointestinal tract. Ulcerative colitis is a disease of the large intestine or colon, which includes characteristic ulcers, or open sores. Nanocrystalline silver (NP 32101) was administered intracolonically or orally and the study revealed that intracolonic treatment of NP 32101 at concentrations of 40 and 4 mg/kg significantly reduced colonic inflammation and that oral treatment with 40mg/kg also improved colonic lesions but was not effective at concentrations of 0.4 and 4 mg/kg. These results suggest that NP 32101 is much more effective when delivered locally to the target organ due to the increased potency with intracolonic treatment (Bhol & Schechter, 2007).

Other experiments with nanocrystalline silver cream have been made using a murine model of allergic contact dermatitis. Dermatitis was induced on the ears of BALB/c mice using dinitrofluorobenzene and later treated with 1% of nanocrystalline cream, tacrolimus ointment (topical drug used for the treatment of eczema) or a high potency steroid, applied once a day for four days. The results showed significant reductions of ear swelling, erythema and histopathological inflammation with no significant difference between treatments. The effect of topical nanocrystalline silver on the induction of apoptosis of inflammatory cells and the role of nanocrystalline silver in suppresion of inflammatory cytokines was examined by measuring the mRNA expression and protein expression of IL-12 and TNF-α. The expression of these inflammatory cytokines were significantly suppressed by nanocrystalline silver, tacrolimus and high potency steroid, but test sites treated with nanocrystalline silver showed more extensive apoptosis of inflammatory cells than the test sites treated with the other treatments (Bhol & Schechter, 2005) .

Other in vivo studies will be needed for further assessment of the NPs in various fields such as medical devices, antimicrobial systems and drug delivery.

At present clinical trials are being performed comparing central venous catheters with silver NPs versus conventional catheters. This study is currently recruiting participants. Catheters coated and/or impregnated with different antimicrobial agents have been proposed to reduce the risk of such infections. However, results obtained so far did not reach enough clinical relevance to consider these medicated catheters as a valid alternative to the conventional ones. (Data was obtained from Clinicaltrials.gov, a service of the U.S Natoinal Institutes of Health)

Other clinical studies are being carried out about the efficacy of silver nanoparticle gel versus a common antibacterial hand gel. This study is also recruiting participants and the specific aims are to compare the immediate antimicrobial efficacy of a one-time application of silver nanoparticle gel, compare the persistent antimicrobial efficacy of a one-time application of silver nanoparticle gel (SilvaSorb®) versus an alcohol-based hand gel (Purell®) over a 10 minute time frame in producing a persistent reduction on transient bacterial counts isolated from hands seeded with S. marcescens and finally to Compare user acceptability of silver nanoparticle gel (SilvaSorb®) versus an alcohol-based hand gel (Purell®) using a self-assessment questionnaire.

5. Future research

The future research is now being shaping by a number of research groups that are actively trying to combine a variety of functions into so-called multifunctional silver nanoparticles. We can expect examples of applications that could include a metallic or semi-metallic core that responds to external energy field or which contains a delivered agent; targeting biomolecules for delivery to specific cellular or disease sites; an image contrast agent for tracking of movement and accumulation of the particles round the body. Such multifunctional particles could also be tailored in size for delivery to different desired sites, tissues or cells. A major effort towards successful nanoparticle-based therapeutics will be to avoid extensive and non specific immunostimulatory or immunosuppressive reactions to the nanomaterials once administered into the body. Also, these developments bring some degree of risk with it, and to bring a product to the clinic entails identifying all possible hazards, characterizing and quantifying the associated risks, including probabilities and severities, given current scientific knowledge, reducing risks to an acceptable level, balancing any remaining risk against benefit to the patient, and communicating effectively and appropriately on the nature of such remaining or "residual" risks. This implies the future development of new or adapted methods appropriate to assess new medicinal products and devices involving silver nanoparticles. Although many questions still require thorough investigation, the available data suggest that silver nanoparticles can be engineered to become part of the next generation of biocompatible drug delivery platforms.

6. Glossary

- MHC: The major histocompatibility complex is expressed on the surface of cells and displays self and non self antigens to T cells, primarily with the goal of eliminating foreign organisms and other invaders that can result in disease.
- APC: The antigen-presenting cell or accessory cell displays foreign antigen complex with MHC on its surface. T-cells may recognize this complex using their T-cell receptor (TCR).
- TLRs: Toll-like receptors recognize molecules that are broadly shared by pathogens but distinguishable from host molecules and activate immune cell responses.
- ROS: Reactive oxygen species are molecules or ions that are highly reactive due to the incomplete one-electron reduction of oxygen. These reactive oxygen intermediates include oxygen ions, superoxides, peroxides and free radicals.

- Gram +: Gram-positive bacteria retain the violet stain used in Gram's method. This is characteristic of bacteria that have a cell wall composed of a thick layer of peptidologlycan.
- Gram -: Gram-negative bacteria do not retain the violet stain (and take the color of the red counterstain) in Gram's method of staining. This is characteristic of bacteria that have a cell wall composed of a thin layer of peptydoglycan.
- RT-PCR: a high sensitive technique used for the detection and quantification of DNA or messenger RNA (mRNA) in a sample. This technique consists of two parts, the synthesis of cDNA (complementary DNA), a reaction applied when the target sequence is RNA by reverse transcription (RT) and the amplification of a specific cDNA by the polymerase chain reaction (PCR).

7. References

Akira, S.; Uematsu, S. & Takeuchi, O. (2006). Pathogen recognition and innate immunity. *Cell*, 124, (4), 783-801

Alexander, J. W. (2009). History of the medical use of silver. *Surg Infect (Larchmt)*, 10, (3), 289-292

AshaRani, P. V.; Low Kah Mun, G.; Hande, M. P. & Valiyaveettil, S. (2009). Cytotoxicity and genotoxicity of silver nanoparticles in human cells. *ACS Nano*, 3, (2), 279-290

Azouz, A.; Razzaque, M. S.; El-Hallak, M. & Taguchi, T. (2004). Immunoinflammatory responses and fibrogenesis. *Med Electron Microsc*, 37, (3), 141-148

Bar-Ilan, O.; Albrecht, R. M.; Fako, V. E. & Furgeson, D. Y. (2009). Toxicity assessments of multisized gold and silver nanoparticles in zebrafish embryos. *Small*, 5, (16), 1897-1910

Bhattacharya, R. & Mukherjee, P. (2008). Biological properties of "naked" metal nanoparticles. *Adv Drug Deliv Rev*, 60, (11), 1289-1306

Bhol, K. C. & Schechter, P. J. (2005). Topical nanocrystalline silver cream suppresses inflammatory cytokines and induces apoptosis of inflammatory cells in a murine model of allergic contact dermatitis. *Br J Dermatol*, 152, (6), 1235-1242

Bhol, K. C. & Schechter, P. J. (2007). Effects of nanocrystalline silver (NPI 32101) in a rat model of ulcerative colitis. *Dig Dis Sci*, 52, (10), 2732-2742

Braydich-Stolle, L.; Hussain, S.; Schlager, J. J. & Hofmann, M. C. (2005). In vitro cytotoxicity of nanoparticles in mammalian germline stem cells. *Toxicol Sci*, 88, (2), 412-419

Castellheim, A.; Brekke, O. L.; Espevik, T.; Harboe, M. & Mollnes, T. E. (2009). Innate immune responses to danger signals in systemic inflammatory response syndrome and sepsis. *Scand J Immunol*, 69, (6), 479-491

Castillo, P. M.; Herrera, J. L.; Fernandez-Montesinos, R.; Caro, C.; Zaderenko, A. P.; Mejias, J. A. & Pozo, D. (2008). Tiopronin monolayer-protected silver nanoparticles modulate IL-6 secretion mediated by Toll-like receptor ligands. *Nanomed*, 3, (5), 627-635

Chen, X. & Schluesener, H. J. (2008). Nanosilver: a nanoproduct in medical application. *Toxicol Lett*, 176, (1), 1-12

Christensen, J. E. & Thomsen, A. R. (2009). Co-ordinating innate and adaptive immunity to viral infection: mobility is the key. *APMIS*, 117, (5-6), 338-355

Elechiguerra, J. L.; Burt, J. L.; Morones, J. R.; Camacho-Bragado, A.; Gao, X.; Lara, H. H. & Yacaman, M. J. (2005). Interaction of silver nanoparticles with HIV-1. *J Nanobiotechnology*, 3, 6

Eming, S. A.; Krieg, T. & Davidson, J. M. (2007). Inflammation in wound repair: molecular and cellular mechanisms. *J Invest Dermatol*, 127, (3), 514-525

Fayaz, M.; Balaji, K.; Girilal, M.; Yadav, R.; Kalaichelvan, P. T. & Venketesan, R. (2009). Biogenic synthesis of silver nanoparticles and its synergetic effect with antibiotics: A study against Gram positive and Gram negative bacteria. *Nanomedicine*,

Furuya, E. Y. & Lowy, F. D. (2006). Antimicrobial-resistant bacteria in the community setting. *Nat Rev Microbiol*, 4, (1), 36-45

Greulich, C.; Kittler, S.; Epple, M.; Muhr, G. & Koller, M. (2009). Studies on the biocompatibility and the interaction of silver nanoparticles with human mesenchymal stem cells (hMSCs). *Langenbecks Arch Surg*, 394, (3), 495-502

Hussain, S. M.; Hess, K. L.; Gearhart, J. M.; Geiss, K. T. & Schlager, J. J. (2005). In vitro toxicity of nanoparticles in BRL 3A rat liver cells. *Toxicol In Vitro*, 19, (7), 975-983

Jain, J.; Arora, S.; Rajwade, J.; Khandelwal, S. & Paknikar, K. M. (2009). Silver nanoparticles in therapeutics: development of an antimicrobial gel formulation for topical use. *Mol Pharm*,

Kim, J. S.; Kuk, E.; Yu, K. N.; Kim, J. H.; Park, S. J.; Lee, H. J.; Kim, S. H.; Park, Y. K.; Park, Y. H.; Hwang, C. Y.; Kim, Y. K.; Lee, Y. S.; Jeong, D. H. & Cho, M. H. (2007). Antimicrobial effects of silver nanoparticles. *Nanomedicine*, 3, (1), 95-101

Kim, S.; Choi, J. E.; Choi, J.; Chung, K. H.; Park, K.; Yi, J. & Ryu, D. Y. (2009). Oxidative stress-dependent toxicity of silver nanoparticles in human hepatoma cells. *Toxicol In Vitro*,

Klasen, H. J. (2000). A historical review of the use of silver in the treatment of burns. II. Renewed interest for silver. *Burns*, 26, (2), 131-138

Lu, L.; Sun, R. W.; Chen, R.; Hui, C. K.; Ho, C. M.; Luk, J. M.; Lau, G. K. & Che, C. M. (2008). Silver nanoparticles inhibit hepatitis B virus replication. *Antivir Ther*, 13, (2), 253-262

Lucarelli, M.; Gatti, A. M.; Savarino, G.; Quattroni, P.; Martinelli, L.; Monari, E. & Boraschi, D. (2004). Innate defence functions of macrophages can be biased by nano-sized ceramic and metallic particles. *Eur Cytokine Netw*, 15, (4), 339-346

Makkouk, A. & Abdelnoor, A. M. (2009). The potential use of toll-like receptor (TLR) agonists and antagonists as prophylactic and/or therapeutic agents. *Immunopharmacol Immunotoxicol*,

Nair, L. S. & Laurencin, C. T. (2008). Nanofibers and nanoparticles for orthopaedic surgery applications. *J Bone Joint Surg Am*, 90 Suppl 1, 128-131

Pozo, D. (2008). Immune-based disorders: the challenges for translational immunology. *J Cell Mol Med*, 12, (4), 1085-1086

Rai, M.; Yadav, A. & Gade, A. (2009). Silver nanoparticles as a new generation of antimicrobials. *Biotechnol Adv*, 27, (1), 76-83

Rasmussen, S. B.; Reinert, L. S. & Paludan, S. R. (2009). Innate recognition of intracellular pathogens: detection and activation of the first line of defense. *APMIS*, 117, (5-6), 323-337

Roe, D.; Karandikar, B.; Bonn-Savage, N.; Gibbins, B. & Roullet, J. B. (2008). Antimicrobial surface functionalization of plastic catheters by silver nanoparticles. *J Antimicrob Chemother*, 61, (4), 869-876

Samuel, U. & Guggenbichler, J. P. (2004). Prevention of catheter-related infections: the potential of a new nano-silver impregnated catheter. *Int J Antimicrob Agents*, 23 Suppl 1, S75-78

Shin, S. H.; Ye, M. K.; Kim, H. S. & Kang, H. S. (2007). The effects of nano-silver on the proliferation and cytokine expression by peripheral blood mononuclear cells. *Int Immunopharmacol*, 7, (13), 1813-1818

Silver, S.; Phung, L. T. & Silver, G. (2006). Silver as biocides in burn and wound dressings and bacterial resistance to silver compounds. *J. Ind. Microbiol. Biotechonol.* ,

Singh, A. V.; Patil, R.; Kasture, M. B.; Gade, W. N. & Prasad, B. L. (2009). Synthesis of Ag-Pt alloy nanoparticles in aqueous bovine serum albumin foam and their cytocompatibility against human gingival fibroblasts. *Colloids Surf B Biointerfaces*, 69, (2), 239-245

Sondi, I. & Salopek-Sondi, B. (2004). Silver nanoparticles as antimicrobial agent: a case study on E. coli as a model for Gram-negative bacteria. *J Colloid Interface Sci*, 275, 177-182

Stevens, K. N.; Crespo-Biel, O.; van den Bosch, E. E.; Dias, A. A.; Knetsch, M. L.; Aldenhoff, Y. B.; van der Veen, F. H.; Maessen, J. G.; Stobberingh, E. E. & Koole, L. H. (2009). The relationship between the antimicrobial effect of catheter coatings containing silver nanoparticles and the coagulation of contacting blood. *Biomaterials*, 30, (22), 3682-3690

Tian, J.; Wong, K. K.; Ho, C. M.; Lok, C. N.; Yu, W. Y.; Che, C. M.; Chiu, J. F. & Tam, P. K. (2007). Topical delivery of silver nanoparticles promotes wound healing. *ChemMedChem*, 2, (1), 129-136

Tsirogianni, A. K.; Moutsopoulos, N. M. & Moutsopoulos, H. M. (2006). Wound healing: immunological aspects. *Injury*, 37 Suppl 1, S5-12

Uematsu, S. & Akira, S. (2006). The role of Toll-like receptors in immune disorders. *Expert Opin Biol Ther*, 6, (3), 203-214

Venier, C.; Guthmann, M. D.; Fernandez, L. E. & Fainboim, L. (2007). Innate-immunity cytokines induced by very small size proteoliposomes, a Neisseria-derived immunological adjuvant. *Clin Exp Immunol*, 147, (2), 379-388

Weyermann, J.; Lochmann, D. & Zimmer, A. (2005). A practical note on the use of cytotoxicity assays. *Int J Pharm*, 288, (2), 369-376

Wong, K. K.; Cheung, S. O.; Huang, L.; Niu, J.; Tao, C.; Ho, C. M.; Che, C. M. & Tam, P. K. (2009). Further evidence of the anti-inflammatory effects of silver nanoparticles. *ChemMedChem*, 4, (7), 1129-1135

Yen, H. J.; Hsu, S. H. & Tsai, C. L. (2009). Cytotoxicity and Immunological Response of Gold and Silver Nanoparticles of Different Sizes. *Small*,

Yu, D. G. (2007). Formation of colloidal silver nanoparticles stabilized by Na+-poly(gamma-glutamic acid)-silver nitrate complex via chemical reduction process. *Colloids Surf B Biointerfaces*, 59, (2), 171-178

Zolnik, B. S. & Sadrieh, N. (2009). Regulatory perspective on the importance of ADME assessment of nanoscale material containing drugs. *Adv Drug Deliv Rev*, 61, (6), 422-427

Zuany-Amorim, C.; Hastewell, J. & Walker, C. (2002). Toll-like receptors as potential therapeutic targets for multiple diseases. *Nat Rev Drug Discov*, 1, (10), 797-807

17

Potential use of silver nanoparticles as an additive in animal feeding

Manuel Fondevila
Instituto Universitario de Experimentación en Ciencias Ambientales, Departamento de Producción Animal y Ciencia de los Alimentos, Universidad de Zaragoza, M. Servet 177, 50013 Zaragoza Spain

1. Abstract

Among other uses, metallic silver and silver salts have currently been applied as antimicrobial agents in many aspects of medical industries, such as coating of catheters, dental resin composites and burn wounds, as well as in homeopathic medicine, with a minimal risk of toxicity in humans. However, their use in animal feeding as prebiotics have remain minimised, mostly because of the low cost antibiotics used as growth promoters in the second half of the XX Century. However, after the ban of this practice in the European Community, silver compounds appear as a potential alternative to other already in use, such as organic acids, oligosaccharides, plant extracts, etc. The major concerns about the safe use of an additive in animal feeding are its effective role as antimicrobial, acting selectively over potential pathogens but not over symbiotic microbial communities; a low toxic effect over the animal and its human consumer; and a low risk of environmental pollution.

Metallic silver nanoparticles (up to 100 nm) allow for a higher antimicrobial effect than silver salts, are more resistant to deactivation by gastric acids and have a low absorption rate through the intestinal mucosa, thus minimising its potential risk of toxicity. Besides, it has been shown that the doses that promote animal physiological and productive effects are very low (20 to 40 ppm), especially compared to the 10 to 100-fold higher concentration used with other metallic compounds such as copper and zinc, thus precluding a harmful environmental effect. This chapter describes the reasons why silver nanoparticles could be applied to animal feeding, and provides with some available data in this regard. In any case, its registration as feed additive is a previous requisite before being applied in practical conditions.

2. Introduction

From the second half of the XX Century, the modern application of technology on animal production has been associated to the intensification of the applied systems, looking for a higher economic profitability by reducing the time and increasing the total magnitude of production. The necessary shortening of the productive cycles and the earlier weaning of animals leads to an increasing of sensitivity of animals, adapted to focus all their physiological resources to high growth performances and consequently making them more sensitive to the environmental conditions and the infection by different diseases, not necessarily of severe gravity, but that in any case produce considerable reductions in productivity. In terms of animal feeding and nutrition, this situation allowed to the transition from the concept of giving nutrients to meet the needs for improving growth as the basic rule to, once this has been assumed, the use of additives to improve productive performances over nutritive standards by reaching an optimum health status of animals.

Any substance is considered as a feed additive when, not having a direct utilisation as nutrient, is included at an optimum concentration in diet or in the drinking water to exert a positive action over the animal health status or the dietary nutrient utilisation. Because of their chemical nature as active principles, are generally included in very small proportions in diet. With the onset of the mentioned productive situation, the use of antibiotics as feed additives – or growth promoters – became predominant over other alternatives, because of their low cost and high and uniform response. It has to be considered that the use of antibiotics as growth promoters, given at sub therapeutical levels to all animals and for prolonged periods of time, is different to their use as therapeutics, administered at higher proportions to sick animals and only until recovery. Briefly, if a small amount of a substance selectively acts against some harmful microbial species occasionally established or transient in the digestive tract, thus controlling the microbial equilibrium of its microbiota, host animals would need to spend less metabolic effort in the immunological control of the situation. Then they would use the extra nutrients for other physiological purposes, thus reaching better productive performances. In this scenario, the magnitude of such growth promoter substances will be highest in young weaned animals, which low immune development and high growth requirements make them more exposed to pathological challenges. It has been reported that using antibiotics as growth promoters in diets increases weight gain and reduces feed to gain ratio (the amount of feed ingested to reach each unit of weight gain) in pigs by 0.16 and 0.07, respectively (Cromwell, 1991). It has to be noticed that, whereas this concept of host health improvement through microbial manipulation is generally applied for monogastric animals (pigs, poultry, rabbits, etc.), it is not totally so for ruminants, where the search of the digestive health interacts with the presence in former sites of the tract of a large fermentation chamber of extreme importance for the ruminant physiology.

As it has been shown that the continuous use of antibiotics as growth promoters provoke the retention in animal tissues and that the human consumption of such animal products would potentially increase processes of antibiotic resistance, movements of social pressure towards food security were claiming for a strict control and against their use in animal feeding, reaching the banning of using antibiotics as growth promoters from 2006 in the European Community (CE 1831/2003). In other way, the use of some trace elements such as zinc and copper, that have been systematically included as growth promoters in diets for weaned piglets because of their beneficial role in pig health status (Hahn & Baker 1993; Smith et al., 1997) have been also restricted to those levels that satisfy the metabolic needs of animals because of both their retention in animal tissues and environmental hazard. The addition of high doses of zinc (from 2500 to 3500 ppm, as zinc oxide) or copper (from 150 to 250 ppm, as copper sulphate) modulates the microbial status of the digestive tract and reduce the incidence of post-weaning diarrhoea (Jensen-Waern et al., 1998; Broom et al., 2006), generally promoting increases in productive performances (Hill et al., 2000; Case &

Carlson, 2002). However, it remains unclear to what extent the response is associated with its role over the digestive microbial ecosystem (Hogberg et al., 2005) or directly over the piglet metabolism (Zhou et al., 1994), by affecting the secretion and activity of pancreatic and intestinal digestive enzymes or the maintenance of the morphology of the intestinal mucosa (Li et al., 2001; Hedemann et al., 2006).

Considerable efforts have been made to look for alternatives to antibiotics growth promoters in animal feeding during the last three decades. Among the most widely used products in pig and poultry production can be cited the organic acids (Partanen & Mroz, 1999; Ravindran & Kornegay, 1993), plant extracts (Cowan 1999; Burt 2004), oligosaccharides (Mull & Perry 2004) or probiotics (Gardiner et al., 2004).

3. Silver as antimicrobial

Silver compounds have been historically used to control microbial proliferation (Wadhera & Fung, 2005). The antifungal and antibacterial effect of silver nanoparticles, even against antibiotic-resistant bacteria (Wright et al., 1994; 1999) has been demonstrated in *in vitro* conditions. Nowadays, silver compounds are routinely applied in a wide array of industrial and sanitary fields, such as coating of catheters and surgery material, the production of synthetic compounds for odontology, treatment of burn injuries, homeopathic medicine or water purification (Spencer, 1999; Klasen, 2000; Wadhera & Fung, 2005; Atiyeh et al., 2007; Hwang et al., 2007).

Traditionally, silver has been used as salts (ionic form), mainly nitrate, sulphate or chloride. However, silver cation is converted into the less effective silver chloride in the stomach or bloodstream, and can form complexes with various ligands. Silver nitrate is unstable, and can be toxic to tissues (Atiyeh et al., 2007). In contrast, metallic silver in form of colloidal solution or as 5 to 100 nm nanoparticles is more stable to hydrochloric acid, is absorbed at a much lower extent by eucharìotic cells and therefore is minimally toxic, and at the same time exert a higher antimicrobial effect (Choi et al., 2008), which explains why its use has been promoted in the last decades (Atiyeh et al., 2007). Lok et al. (2006) showed that, even though silver nanoparticles and silver ions in form of silver nitrate have a similar mechanism of action, their effective concentrations are at nanomolar and micromolar levels, respectively.

Silver exerts its antimicrobial activity through different mechanisms. It has been reported to uncouple the respiratory electron transport from oxidative phosphorylation and to inhibit respiratory chain enzymes (Schreurs & Rosemberg, 1982; Bard & Holt, 2005). Silver also adheres to bacterial surface, thus altering membrane functions, leading to a dissipation of the proton motive force (Percival et al., 2005; Lok et al., 2006), and interacts with nucleic acid bases, inhibiting cell replication (Wright et al., 1994; Yang et al., 2009). Some authors have demonstrated its toxic effect over different serovars of *Escherichia coli* (Zhao & Stevens, 1998; Sondi & Salopek-Sondi, 2004; Jung et al., 2008) and *Streptococcus faecalis* (Zhao & Stevens, 1998), but its observed effect over *Staphylococcus aureus* has been variable (Li et al., 2006; Kim et al., 2007; Jung et al., 2008). Yoon et al. (2007) observed a higher effect of silver nanoparticles on *Bacillus subtilis* than on *Escherichia coli*, suggesting a selective antimicrobial effect, possibly related to the structure of the bacterial membrane, although Singh et al. (2008) assume higher sensitivity of Gram-negative bacteria to treatment with nanoparticles.

The possible effects of metallic silver and silver ions over microorganisms from the digestive tract are scarcely documented. The selective response of silver in such ecosystem, with a wide diversity of species that can exert either symbiotic (positive) or pathogen (negative) effects, deserves further attention.

4. Other effects of silver

Despite its potential effect on digestive microbial biodiversity and function, other effects of metallic silver related with host physiological status, such as the immunological status, the digestive enzymatic activity and intestinal structure can be expected. This can be assumed considering the chemical similarity of silver with other metals such as zinc and copper and the characteristics of their antimicrobial response. The capability of zinc and copper to minimise the negative effect of weaning on the of height of intestinal villi, thus ensuring its absorbing potential (Li et al., 2001) and the enhancement of the metabolic pancreatic activity (Zhou et al., 1994) could also be potentially expected with the use of silver. Besides, studies related with the role of silver nanoparticles on wound treatment show its role on metalloproteinases regulation, reducing inflammation and favouring cellular apoptosis and cicatrisation (Wright et al., 2002; Warriner & Burrell, 2005). Lansdown (2002) indicates that the topic use of silver promotes an increase of zinc and copper concentration over epithelial tissue, thus indirectly stimulating its positive effects.

A cytotoxic effect of silver on the host animal must also be considered. This has been occasionally observed in human medicine when chronic (extended in time) treatments with high doses of silver have been used, often related with the use of silver compounds for wound healing or in dental implants (Abe et al., 2003; Lam et al., 2004). Chronic ingestion of silver compounds may lead to its retention in skin, eyes and other organs such as liver, but it has been generally considered as a cosmetic problem, with minor or nil pathological symptoms (Lansdown, 2006). Wadhera & Fung (2005) state that no physiological alterations or damage of organs of patients with argyria (subcutaneous accumulation of silver associated with silver salts treatment), even with daily intake of 650 mg ionic silver for 10 months (corresponding to a total of 200 g silver intake). The minimal dose causing generalised argyria in humans has been fixed in 4 to 5 g (Brandt et al., 2005). According to Ricketts et al. (1970), the minimal dose of silver nitrate to cause inhibition of cell respiration in tissues is about 25-fold higher to that inhibits growth of *Pseudomonas aeruginosa*, and Gopinath et al. (2008) concluded that a necrotic effect on human cells of silver nanoparticles occur at concentrations above 44 μg/ml (44 ppm). However, no limiting concentration of silver intake has been fixed for humans, although the US Environmental Protecting Agency (EPA) recommends a maximum silver dose in drinking water for chronic or short term (1 to 10 days) intake of 0.05 and 1.14 ppm, respectively (ATSDS 1990).

5. Potential use of silver in animal feeding

In the 50´s, colloidal silver was used as zootechnical additive in poultry diets, but its high cost at that time avoided its possibility to compete with the lower cost of antibiotics. Nowadays, the development of industrial processes of silver nanoparticles allows for its consideration as a potential feed additive, once the banning of the use of antibiotics as growth promoters. However, the availability of results testing metallic silver nanoparticles in animal production experiments is very scarce. It has been observed *in vitro* that the proportion of coliforms in pigs ileal contents was linearly reduced ($P<0.05$), whereas no effect was observed on lactobacilli proportion, when the concentration of colloidal silver in the medium increased from 0 to 25, 50 or 100 ppm (Fondevila et al., 2009). According to these results, metallic silver nanoparticles would reduce the viability of organisms with a potentially harmful effect, such as coliforms, whereas it does not affect lactobacilli, which positively compete against pathogens proliferation and reduce their virulence (Blomberg et al., 1993). A trend (P = 0.07) to a coliform reduction in ileal contents was also observed *in vivo* by Fondevila et al. (2009) when 20 and 40 ppm of metallic silver nanoparticles were given to weaned piglets as metallic silver adsorbed in a sepiolite matrix (ARGENTA, Laboratorios Argenol S.L., Spain) as antimicrobial and growth promoter for weaned pigs during their transition phase (from 5 to 20 kg weight). Besides, although concentration of major bacterial groups in the ileum of pigs were not markedly affected, the concentration of the pathogen *Clostridium perfringens/ Cl. histolyticum* group was reduced with 20 ppm silver (P = 0.012). In the same way, Sawosz et al. (2007) did not observed a major effect of colloidal silver on bacterial concentration in the digestive tract of quails, but only a significant increase in lactic acid bacteria was observed with 25 ppm.

Results on productive performances in several experiments with pigs and poultry carried out by our group were variable (Table 1): a numerical increase in daily growth was generally observed when 20 ppm silver were added compared with the control (no silver), but this effect was not generally significant. As the productive responses to an additive that improves the sanitary status of animals are in general inversely proportional to the environmental quality of the productive site (Cromwell 1995), it is likely that under the stress conditions of commercial farms the concentration of pathogenic bacteria increased and thus the effect of silver would be more manifested. In the same way, a lack of effect of adding zinc oxide had also been sometimes reported (Jensen-Waern et al., 1998; Broom et al., 2006), which would partly explain this lack of significant results. Studies in animals as models for humans have shown that high silver concentrations (between 95 and 300 ppm, corresponding to 2.4 and 7.5-fold the concentrations used in these experiments) in form of silver salts and given as chronic dose (for more than 18 weeks) reduce weight of mice (Rungby & Danscher 1984) and turkeys (Jensen et al., 1974). However, these dosing conditions are considered of much higher toxic potential than low concentration metallic silver given for short periods of time (Wadhera & Fung 2005). In an experiment (E. Gonzalo, M.A. Latorre & M. Fondevila, unpublished) where pigs were given 0, 20 and 40 ppm silver from weaning to slaughter weight (91 kg), the feed to gain ratio (amount of feed per unit of increased weight) was reduced (P= 0.03) by silver addition, indicating a higher growth efficiency and showing a reduction in overall production cost.

Another important aspect to verify when an additive is promoted to use is to what extent it does not challenge the health of the potential consumer. Inclusion of 2500 to 3000 ppm zinc in diets for post-weaning pig leads to tissue retention from 220 μg/g (Jensen-Waern et al., 1998; Carlson et al., 1999) to 445 μg/g (Zhang & Guo, 2007) in liver, and retentions up to 3020 μg/g have been reported (Case and Carlson, 2002). In a study carried out with metallic silver, no silver retention was detected in renal or muscular (semimembranous) tissue in weaned piglets given 20 or 40 ppm silver for 35 days (n=18), and only 0.435 and 0.837 μg per g were recorded in liver (Fondevila et al., 2009). Another experiment repeated in the same conditions (Gonzalo, Latorre & Fondevila, unpublished) showed minimal silver retention in muscles (0.036 and 0.033 μg/g with 20 and 40 ppm silver in diet) and kidney (0.034 and 0.039 μg/g, respectively) that was observed in 6 out of 8 animals, whereas silver was detected in liver of all animals at 0.400 and 0.557 µg/g for 20 and 40 ppm, respectively. It has to be considered that these concentrations are more than 3000-fold lower than in the case of zinc and the range is below the EPA recommendation, as it has been commented above. Further, pigs are not given silver additive during their growth and finishing phases (from 20 to 90-100 kg, commercial slaughter weight), and our group did not detect any traces of silver in muscles, kidneys or liver of 90 kg pigs receiving the additive up to 20 kg weight, thus showing the detoxifying capacity of liver to excrete silver (Lansdown, 2006).

In an experiment with broiler chicks as another animal productive species, dosage of metallic silver nanoparticles (ARGENTA) for 5 weeks was continued by 7 days of non-supplemented period (Prieto & Fondevila, unpublished). Silver retention was 0.035, 0.031 and 0.045 µg/g in muscular tissue and 0.113, 0.086 and 0.185 µg/g for the same treatments in liver tissue for 20, 30 and 40 silver ppm in diet, respectively (n=10). Only 5 out of 10 animals given 20 and 30 ppm silver showed detectable concentration in muscles, while 6 and 7 out of 10 animals with the same treatments showed silver concentration in the liver.

Experimental conditions	Ag dose (mg/kg)	Intake (g/d)	Growth (g/d)	F:G (kg/kg)	Reference
weaned pigs, n=5, 28 to 35 d age	0	162	107		Fondevila et al. (2009)
	20	143	122		
	40	177	157		
	s.e.m.	--	41.3		
weaned pigs, n=5, 35 to 42 d	0	253	314b		Fondevila et al. (2009)
	20	313	393ab		
	40	365	461 a		
	s.e.m.	--	36.4		
weaned pigs, n=6 pens of 4 pigs, 21 to 35 d	0	154b	66	2.13	Fondevila et al. (2009)
	20	189a	102	1.95	
	40	148 b	93	1.70	
	s.e.m.	8.5	11.0	0.196	
weaned pigs, n=6 pens of 4 pigs, 35 to 56 d	0	527b	337	1.56b	Fondevila et al. (2009)
	20	670a	375	1.80a	
	40	630a	347	1.82a	
	s.e.m.	32.3	21.2	0.050	
weaned pigs, n=6 pens of 2 pigs, 21 to 147 d; silver was dosed from 21 to 56 d of age	0	1737	684	2.53a	Gonzalo, Latorre & Fondevila (unpublished)
	20	1638	677	2.42b	
	40	1734	693	2.50ab	
	s.e.m.	46.9	16.8	0.029	
broilers, n=8 pens of 28 chicks, 1 to 42 d; silver was dosed from 1 to 35 d of age	0	99.7	54.6	1.83	Prieto & Fondevila (unpublished)
	20	97.3	55.3	1.76	
	30	96.6	53.9	1.79	
	40	99.0	54.1	1.83	
	s.e.m.	0.74	1.28	0.030	

Table 1. Effect of inclusion of metallic silver nanoparticles (ARGENTA) on productive performances of animals

F:G, feed to gain ratio, a,b, letters show differences among means ($P<0.05$)

6. Acknowledgements

Research included in this work has been funded by Laboratorios Argenol, S.L. (Zaragoza, Spain), through the projects OTRI 2006/0279, OTRI 2007/0640 and DEX-600100-2008-23. J. Ducha (University of Zaragoza), M.A. Latorre (Centro de Investigación y Tecnología de Aragón), J. Prieto and E. Gonzalo have been involved in the experimental research included in this paper.

7. References

Abe, Y.; Ueshige, M.; Takeuchi, M.; Ishii, M. & Akagawa, Y. (2003). Cytotoxicity of antimicrobial tissue conditioners containing silver-zeolite. *International Journal of Prosthodontology*, 16, 141-144.

Agency for Toxic Substances and Diseases Registry (1990). Toxicological profile for silver. U.S. Public Health Service, 145 pp.

Atiyeh, B.S.; Costagliola, M.; Hayek, S.N. & Dibo, S.A. (2007). Effect of silver on burn wound infection control and healing: review of the literature. *Burns*, 33, 139-148.

Bard, A.J. & Holt, K.B. (2005). Interaction of silver ions (I) with the respiratory chain of *Escherichia coli*: an electrochemical and scanning electrochemical study of the antimicrobial mechanism of micromolar Ag+. *Biochemistry*, 44, 13214-13223.

Blomberg, L.; Henrikson, A. & Conway, P. L. (1993). Inhibition of adhesion of *Escherichia coli* K88 to piglet ileal mucus by *Lactobacillus spp. Applied and Environmental Microbiology*, 59, 34–39.

Brandt, D.; Park, B.; Hoang, M. & Jacobe, H.T. (2005). Argyria secondary to ingestion of homemade silver solution. *Journal of American Academy of Dermatology*, 53, S105-S107.

Broom, L.J.; Miller, H.M.; Kerr, K.G. & Knapp, J.S. (2006). Effects of zinc oxide and *Enterococcus faecium* SF68 dietary supplementation on the performance, intestinal microbiota and immune status of weaned piglets. *Research in Veterinary Science*, 80, 45-54.

Burt, S. (2004). Essential oils: their antibacterial properties and potential applications in foods: a review. *International Journal of Food Microbiology*, 942, 223-253.

Carlson, M.S.; Hill, G.M. & Link, J.E. (1999). Early- and traditionally weaned nursery pigs benefit from phase-feeding pharmacological concentrations of zinc oxide: effect on methallothionein and mineral concentrations. *Journal of Animal Science*, 77, 1199-1207.

Case, C.L. & Carlson, M.S. (2002). Effect of feeding organic and inorganic sources of additional zinc on growth performance and zinc balance in nursery pigs. *Journal of Animal Science*, 80, 1917-1924.

Choi, O.; Deng, K.K.; Kim, N.J., Ross, L.Jr.; Surampalli, R.Y. & Hu, Z. (2008). The inhibitory effects of silver nanoparticles, silver ions and silver chloride colloids on microbial growth. *Water Research*, 42, 3066-3074.

Cowan, M.M. (1999). Plant products as antimicrobial agents. *Clinical Microbiology Reviews* 12, 564-582.

Cromwell, G.L. (1991). Antimicriobial agents. In: *Swine nutrition*, E.R. Miller, D.E. Ullrey, A.J. Lewis (Eds.), pp. 297-315, CRC Press, Boca Raton.

Fondevila, M.; Herrer, R.; Casallas, M.C.; Abecia, L. & Ducha, J.J. (2009). Silver nanoparticles as a potential antimicrobial additive for weaned pigs. *Animal Feed Science and Technology*, 150, 259-269.

Gardiner, G.E.; Casey, P.G.; Casey, G.; Lynch, P.B.; Lawlor, P.G.; Hill, C.; Fitzgerald, G.F.; Stanton, C. & Ross, R.P. (2004). Relative ability of orally administered *Lactobacillus murinus* to predominate and persist in the porcine gastrointestinal tract. *Applied and Environmental Microbiology*, 70, 1895–1906.

Gopinath, P.; Gogoi, S.K.; Chattopadhyay, A. & Gosh, S.S. (2008). Implications of silver nanoparticle induced cell apoptosis for *in vitro* gene therapy. *Journal of Nanobiotechnology*, 19, 075104.

Hahn, J.D. & Baker, D.H. (1993). Growth and plasma zinc responses of young pigs fed pharmacologic levels of zinc. *Journal of Animal Science*, 71, 3020-3024.

Hedemann, M.S.; Jensen, B.B. & Poulsen, H.D. (2006). Influence of dietary zinc and copper on digestive enzyme activity and intestinal morphology in weaned pigs. *Journal of Animal Science*, 84, 3310-3320.

Hill, G.M.; Cromwell, G.L.; Crenshaw, T.D.; Dove, C.R.; Ewan, R.C.; Knabe, D.A.; Lewis, A.J.; Libal, G.W.; Mahan, D.C.; Shurson, G.C.; Southern, L.L. & Veum, T.L. (2000). Growth promotion effects and plasma changes from feeding high dietary concentrations of zinc and copper to weanling pigs (regional study). *Journal of Animal Science*, 78, 1010-1016.

Hogberg, O.; Canibe, N.; Poulsen, H.D.; Hedemann, M.S. & Jensen, B.B. (2005). Influence of dietary zinc oxide and copper sulphate on the gastrointestinal ecosystem in newly weaned piglets. *Applied and Environmental Microbiology*, 71, 2267-2277.

Hwang, M.G. ; Katayama, H. & Ohgaki, S. (2007). Inactivation of *Legionella pneumophila* and *Pseudomonas aeruginosa*: evaluation of the bactericidal ability of silver cations. *Water Research*, 41, 4097-4104.

Jensen, LS; Peterson, RP & Falen, L. (1974). Inducement of enlarged hearts and muscular dystrophy in turkey poults with dietary silver. *Poultry Science*, 53, 57-64.

Jensen-Waern, M.; Melin, L.; Lindberg, R.; Johanisson, A.; Peterson, L. & Wallgren, P. (1998). Dietary zinc oxide in weaned pigs-effects on performance, tissue concentrations, morphology, meutrophil functions and faecal microflora. *Research in Veterinary Science*, 64, 225-231.

Jung, W.K.; Koo, H.K.; Kim, K.W.; Shin, S.; Kim, S.H. & Park, Y.H. (2008). Antibacterial activity and mechanism of action of the silver ion in *Staphylococcus aureus* and *Escherichia coli*. *Applied and Environmental Microbiology*, 74, 2171-2178.

Kim, J.S.; Kuk, E.; Yu, K.N.; Kim, J.H.; Park, S.J.; Lee, H.J.; Kim, S.H.; Park, Y.K.; Park, Y.H.; Hwang, C.Y.; Kim, Y.K.; Lee, Y.S.; Jeong, D.H. & Cho, M.H. (2007). Antimicrobial effect of silver nanoparticles. *Nanomedicine: Nanotecnology, Biology and Medicine*, 3, 95-101.

Klasen, H.J. (2000). Historical review of the use of silver in the treatment of burns. I. Early uses. *Burns*, 26, 117-130

Lam, P.K.; Chan, E.S.; Ho, W.S. & Liew, C.T. (2004). *In vitro* cytotoxicity testing of nanocrystalline silver dressing (Acticoat) on cultured keratinocytes. *British Journal of Biomedical Science*, 61, 125-127.

Lansdown, A.B. (2006). Silver in health care: antimicrobial effects and safety in use. *Current Problems in Dermatology*, 33, 17-34.

Li, B. T.; Van Kessel, A. G.; Caine, W. R.; Huang, S. X. & Kirkwood, R. N. (2001). Small intestinal morphology and bacterial populations in ileal digesta and feces of newly weaned pigs receiving a high dietary level of zinc oxide. *Canadian Journal of Animal Science*, 81, 511-516.

Li, Y.; Leung, P.; Yao, L.; Song, K.W. & Newton, E. (2006). Antimicrobial effect of surgical masks coated with nanoparticles. *Journal of Hospital Infection*, 62, 58-63

Lok, C.N.; Ho, C.M.; Chen, R.; He, Q.Y.; Yu, W.Y.; Sun, H.; Tam, P.K.H.; Chiu, J.F. & Che, C.M. (2006). Proteomic analysis of the mode of antibacterial action of silver nanoparticles. *Journal of Proteome Research*, 5, 916-924.

Mull, A.J. & Perry, F.G. (2001). The role of fructo-oligosaccharides in animal nutrition. In *Recent developments in pig nutrition 3*, J. Wiseman y P.C. Garnsworthy (eds.), pp. 79-105, Nottingham University Press, Nottingham.

Partanen, K.H. & Mroz, Z. (1999) Organic acids for performance enhancement in pig diets. *Nutrition Research Reviews, 12*, 117-145.

Percival, S.L.; Bowler, P.G. & Russell, D. (2005). Bacterial resistance to silver in wound care. *Journal of Hospital Infection*, 60, 1-7.

Ravindran, V., Kornegay, E.T.,1993. Acidification of weaner pig diets: a review. Journal of the Science of Food and Agriculture 62, 313-322.

Ricketts, C.R.; Lowbury, E.J.L.; Lawrence, J.C.; Hall, M. & Wilkins, M.D. (1970). Mechanism of prophylaxis by silver compounds against infection of burns. *British Medical Journal*, 2, 444-446.

Rungby, J. & Danscher, G. (1984). Hypoactivity in silver exposed mice. *Acta Pharmacologica Toxicologica*, 55, 398-401.

Sawosz, E.; Binek, M.; Grodzik, M.; Zielinska, M.; Sysa, P.; Szmidt, M.; Niemec, T. & Chwalibog, A. (2007). Influence of hydrocolloidal silver nanoparticles on gastrointestinal microflora and morphology of enterocytes of quails. *Archives of Animal Nutrition*, 61, 444-451.

Schreurs, W.J. & Rosenberg, H. (1982). Effect of silver ions on transport and retention of phosphate by *Escherichia coli*. *Journal of Bacteriology*, 152, 7-13.

Singh, M.; Singh, S., Prasad, S. & Gambhir, I.S. (2008). Nanotechnology in medicine and antibacterial effect of silver nanoparticles. *Digest Journal in Nanomaterials and Biostructures*, 3, 115-122.

Smith, J.W.; Tokach, M.D.; Goodband, R.D.; Nelssen, J.L. & Richert, B.T. (1997). Effects of the interrelationship between zinc oxide and copper sulfate on growth performance of early-weaned pigs. *Journal of Animal Science*, 75, 1861-1866.

Sondi, I. & Salopek-Sondi, B. (2004). Silver nanoparticles as antimicrobial agent : a case study on *E. coli* as a model for Gram-negative bacteria. *Journal of Colloid Interface Science*, 275, 177-182.

Spencer, R.C. (1999). Novel methods for the prevention of infection of intravascular devices. *Journal of Hospital Infection*, 43, S127-S135.

Wadhera, A. & Fung, M. (2005). Systemic argyria associated with ingestion of colloidal silver. Dermatology Online Journal 11, 12 (http://dermatology.cdlib.org/111).

Warriner, R. & Burrell, R.E. (2005). Infection and the chronic wound: a focus on silver. *Advances on Skin Wound Care*, 18, 2-12.

Wright, J.B.; Lam, K. & Burrell, R.E. (1994). Wound management in an era of increasing bacterial antibiotic resistance: a role for topical silver treatment. *American Journal of Infection Control*, 26, 572-577.

Wright, J.B.; Lam, K.; Buret, A.G.; Olson, M.E. & Burrell, R.E. (2002). Early healing events in a porcine model of contaminated wounds: effects of nanocristalline silver on matrix metalloproteinases, cell apoptosis and healing. *Wound Repair Regeneration*, 10, 141.

Wright, J.B.; Lam, K.; Hansen, D. & Burrell, R.E. (1999). Efficacy of topical silver against fungal burn wound pathogens. *American Journal of Infection Control*, 27, 344-350.

Yang, W.; Shen, C.; Ji, Q.; An, H.; Wang, J.; Liu, Q. & Zhang, Z. (2009). Food storage material silver nanoparticles interfere with DNA replication fidelity and bind with DNA. *Nanotechnology*, 20, 085102 (7 pp).

Yoon, K.Y.; Byeon, J.H.; Park, J.H. & Hwang, J. (2007). Susceptibility constrants of *Escherichia coli* and *Bacillus subtilis* to silver and copper nanoparticles. *Science of the Total Environment*, 373, 572-575.

Zhang, B. & Guo Y. (2007). Beneficial effect of tetrabasic zinc chloride for weanling piglets and the bioavailability of zinc in tetrabasic form relative to ZnO. *Animal Feed Science and Technology*, 135, 75-85.

Zhao, G. & Stevens, J.R. (1998). Multiple parameters for the comprehensive evaluation of the susceptibility of *Escherichia coli* to the silver ion. *Biometals*, 11, 27-32.

Zhou, W.; Kornegay, E.T.; Lindemann, M.D.; Swinkels, J.W.G.M.; Welton, M.K. & Wong, E.A. (1994). Stimulation of growth by intravenous injection of copper in weanling pigs. *Journal of Animal Science*, 72, 2395-2403.